Solid-State and Molecular Theory:

A Scientific Biography

Solid-State and Molecular Theory: A Scientific Biography

John C. Slater

*Institute Professor Emeritus,
Massachusetts Institute of Technology*

*Graduate Research Professor of Physics and Chemistry
University of Florida*

A WILEY-INTERSCIENCE PUBLICATION

John Wiley & Sons

New York · London · Sydney · Toronto

Copyright © 1975, by John Wiley & Sons, Inc.

All rights reserved. Published simultaneously in Canada.

No part of this book may be reproduced by any means, nor transmitted, nor translated into a machine language without the written permission of the publisher.

Library of Congress Cataloging in Publication Data:
Slater, John Clarke, 1900–
  Solid-state and molecular theory.

  "A Wiley-Interscience publication."
  Bibliography: p.
  Includes index.
  1. Energy-band theory of solids. 2. Molecular theory. 3. Wave mechanics—History. 4. Physics—History—United States. 5. Slater, John Clarke, 1900-   I. Title.
  QC176.8.E4S55   1975   530.1′24   74-22367
  ISBN 0-471-79681-6

Printed in the United State of America

10 9 8 7 6 5 4 3 2 1

PREFACE

The decades of prosperous industrial and technological growth which the United States enjoyed following World War II were based on scientific discoveries made before the war. Everyone has heard of the advances in nuclear physics and fission which led to the atomic bomb and the nuclear reactor. But on the whole the more important discoveries were those which led to solid-state electronics, and thus to television, microwave and satellite communications, the electronic systems which made space flight possible, the computer, and the laser. These were the foundations of the postwar industrial strength of the country. And they can all be traced back to the development of quantum theory and wave mechanics during the decade from 1923 to 1933. That was the decade too in which American science was really coming of age, advancing from a fairly timid start in 1923 to a point where the country was clearly a scientific leader by 1933.

I am fortunate enough to have been associated with this advance during the whole 50 years from 1923 to 1973. As a faculty member first at Harvard, then at MIT, I have had a chance to observe the development of the science and technology of the study of atoms, molecules, and solids during this whole period. This book is the story of that development, with particular attention to a very recent feature of the theory, the so-called $X\alpha$-SCF method. This is a method of applying to the problem of large molecules, containing heavy atoms, the same sort of theoretical techniques which had been at the basis of the development of the theory of solid-state electronics. The large molecules are involved in the type of problems which will be faced in the next great crisis of technology: the catalysts which are needed for antipollution and energy-conversion devices, the transformation of radiation energy into chemical or electrical energy, and endless other problems that will be met in connection with the energy crisis. Fortunately the theory has advanced in the last several years to a point where it is now developing with enormous rapidity, much as the solid-state theory which led to solid-state electronics developed in the 1930s.

My scientific career, and the scientific developments we are talking about, are so interwoven that is seemed worthwhile to make this book into something half-way between autobiography and history of science and technology on the one hand, and pure science on the other. It is the aim of this book to have the science comprehensible enough so that the reader with a general scientific education will be able to get the main points, and yet far-reaching enough so that the more technical reader will get a general view of the background of the new method. It touches so many aspects of the quantum theory of atoms, molecules, and solids that one can get a rather good view of that whole field by following the history of the past 50 years, as it is presented in this book.

The author would like to acknowledge several useful comments regarding early days of wave mechanics from Professor J. H. Van Vleck and Mrs. Katherine J. Sopka of Harvard University, who were good enough to go through the manuscript.

J. C. SLATER

Gainesville, Florida
August 1974

CONTENTS

BOOK I. WAVE MECHANICS IN THE CLASSICAL DECADE, 1923–1932

1.	Introduction	3
2.	1923–1924, Waves and Particles	8
3.	1925, Matrix Mechanics	15
4.	1925, The Spinning Electron and the Exclusion Principle	22
5.	1926, Schrödinger's Equation	31
6.	1926–1927, The Two-Electron Atom	43
7.	1928, Hartree and the Self-Consistent Field	52
8.	1929, The "Gruppenpest" and Determinantal Wave Functions	58
9.	1929, The Theory of Complex Spectra	66
10.	1929–1930, The Hartree-Fock Method	78
11.	1926, The Fermi-Dirac Statistics	81
12.	1927, The Thomas-Fermi Atom Model	86
13.	1927, Heitler and London and the Hydrogen Molecule	90
14.	1927–1932, Molecular Orbitals	95
15.	1928–1930, Bloch Sums and Brillouin Zones	112
16.	1927–1930, Ferromagnetism	120
17.	1926–1932, Quantum Electrodynamics	130
18.	1928–1932, Dirac and the Theory of the Electron	145
19.	1927–1932, The Helium Atom	151
20.	1929–1931, Bethe and Frenkel, Localized States in Crystals	156

BOOK II. TRANSITIONAL YEARS, 1933–1940

21.	MIT and Princeton, 1930–1940	163
22.	1933–1934, The Cellular Method and Energy Bands	173
23.	Further Energy Bands, 1934–1940	185
24.	General View of Energy Bands, 1933–1940	192
25.	Localized States and Magnetism, 1936–1937	200

BOOK III. WAR AND POSTWAR YEARS, 1941–1951

26. Radar, 1940–1945 — **209**
27. Postwar Development at MIT, 1945–1951 — **217**
28. Postwar Physics, 1945–1951 — **226**

BOOK IV. THE MIT SOLID-STATE AND MOLECULAR THEORY GROUP, 1951–1964

29. The SSMTG and Brookhaven, 1951–1952 — **237**
30. The Statistical Exchange, 1951–1955 — **243**
31. Computers, Energy Bands, and Molecules, 1951–1964 — **255**
32. The MIT Materials Center, 1956–1964 — **268**

BOOK V. THE Xα-SCF METHOD, 1964–1973

33. Energy Bands, Magnetism, and the Transition State, 1964–1970 — **275**
34. The Multiple-Scattering Method, 1965–1973 — **284**
35. Total Energy and Magnetic Problems, 1967–1973 — **293**
36. The Transition State and Localized Excitations — **304**
37. Improving the Muffin-Tin Potential — **312**

BOOK VI. LOOKING AHEAD

38. Big Molecules, Energy, and the Computer — **325**
39. Computers and the Next Millennium — **333**

SUGGESTED READING — **339**

INDEX — **343**

Solid-State and Molecular Theory:

A Scientific Biography

BOOK

I

Wave Mechanics in the Classical Decade, 1923–1932

1. Introduction

Seldom in the history of science has there been a more exciting decade than that from 1923 to 1932. It was then that the theory of wave mechanics took form. In 1923, many ideas about the quantum theory were in the air, waiting for a unifying idea which would bring them together into a full-fledged theory. Many features of atomic structure could be explained, but many more could not, and the detailed nature of molecules and crystals was still a mystery. By 1932, we had a theory, wave mechanics, which gave every indication of being the fundamental basis for understanding all types of matter, just as Newtonian mechanics was the fundamental basis of the study of the motion of the heavenly bodies. The story of the way in which all the many aspects of this theory were worked out, so that by 1932 they formed a consistent whole, is one of the most remarkable in the whole history of human thought.

In 1932, the more glib scientists were inclined to say that the whole of chemistry, metallurgy, crystallography, even biophysics and biochemistry and the study of life, was essentially solved. Schrödinger's equation, which had been postulated in 1926, would give all the answers. This is the fundamental equation of wave mechanics, just as Newton's laws of motion are the fundamentals of classical mechanics, or Maxwell's equations of electromagnetic theory.

It is a rather remarkable fact, perhaps an accident of history, that two quite separate events of 1933 had a profound effect in slowing down the development of wave mechanics and its applications to atomic, molecular, and solid-state physics. First, Hitler came to power in Germany. This led to the persecution of the Jews, and to the departure from continental Europe of a large fraction of the scientists who had been most active in the development of the quantum theory. Second, new discoveries in experimental nuclear physics had turned the attention of a large number of scientists from the study of atoms, molecules, and solids to the study of the nucleus. Many of the displaced scientists migrated to the United States or to western Europe, and decided that when they moved, they might as well take up what seemed the newer and more exciting field of nuclear science. It was not long before the discovery of nuclear fission led them to think of the possibility of the atomic bomb, and the tyranny of Hitler led them to think of using it. Everyone knows the result in its effect on history and politics. But the effect on science was that out of a great number of brilliant research workers who had collaborated on the development of wave mechanics and its application to the study of atoms, molecules, and solids during the classical decade from 1923 to 1932, relatively few were left to carry on the application of quantum theory to the problems of molecular or chemical physics which had been the main concern of the preceding decade.

It was certainly true, as had been glibly stated by 1932, that Schrödinger's equation would give all the answers to these questions. But wave mechanics proved to be the most intractable of all branches of mathematical physics, and when one combines this with the fact that a great many of the ablest workers left the field, it was only natural that progress in working out the detailed applications of wave mechanics to chemical physics was much slower than one would have anticipated in 1932. We are talking in the last part of this volume about a method of approximating to exact solutions of Schrödinger's equation, excellently adapted to solving the types of problems that one meets in the study of molecules and solids. It is only in the last two decades that this method was formulated, and only in the last three or four years that it has been giving results so promising that it could not be disregarded. Why has it taken so long?

We have already suggested two of the reasons, the inherent difficulty of the problem and the exodus of many able workers from the field. Next of course comes the almost total stoppage of work on pure science during World War II. In the days after that war, new students had very little knowledge of the prewar work, and they naturally went into new fields which had evolved from wartime research. The nuclear and high-energy fields attracted a great many workers, and in addition there were exciting new experimental fields such as magnetic resonance and neutron diffraction, which were capable of giving valuable information about molecular and solid-state problems. Comparatively few went into the type of theory that had been popular in the 1920s and 1930s, though the development of the transistor and of solid-state electronics showed that the application of the quantum theory to molecules and solids had great practical importance.

But the greatest obstacle to the development of the theory was the inherent complication of the calculations which had to be made. Methods were known that were capable of giving good descriptions of very simple molecules or simple crystals, but the calculations simply grew too difficult for all problems involving more than a very few electrons per atom or per unit cell. It was the development of the electronic digital computers during the 1950s which really made it possible to bring the theory to its present promising state. Even with the largest computers which are available to research scientists, the calculations are still so involved that it takes great ingenuity to devise approximations usable for problems as complicated as those of practical interest. The decade of the 1960s was the period in which methods of approximation were being tried out and programmed, and it is only in the 1970s that these methods have arrived at the point where they are really valuable. Here we have the answer to our question, why did it take so long?

I am one of the very small number of scientists who have been connected with the field through the whole 50 years from 1923 to 1973, and who have been constantly trying to bring the theory to the state it has reached at present. For that reason, more personal reminiscences will be appropriate than if I were simply recording events about which I had heard at second hand. In June 1923, I took my doctorate in physics at Harvard, working with Percy W. Bridgman, a pioneer in the experimental study of phenomena at high pressure, who later received a Nobel Prize for his studies. He was not only an experimenter of the greatest ingenuity, but also a very clear theoretical thinker, specializing particularly in thermodynamics and electromagnetic theory. My knowledge of those fields came largely from him. The training in quantum theory at Harvard was given by Edwin C. Kemble, a very able young man a few years older than I, one of the earliest Americans to establish really good instruction in the quantum theory of the day. I shall not go into the details of the theory as we knew it in 1923, but it was very definitely up to date as of the time, as I discovered later when I visited European laboratories and found I was as well trained as the students there.

At Harvard we not only had excellent teaching, but an inspiring research atmosphere as well. The head of the department at the time was Theodore Lyman, discoverer of the Lyman series in the far-ultraviolet spectrum of hydrogen, one of the most important experimental facts which had helped to show the correctness of the theory of the hydrogen atom proposed by Niels Bohr in 1913. Another department member was William Duane, an x-ray pioneer, who had made some of the most accurate measurements of Planck's constant h. Frederick A. Saunders, later to be well known for the discovery of the Russell-Saunders coupling in atomic theory, became department head after Lyman. Edwin H. Hall, who had discovered the Hall effect in the last century, was a retired but still active department member. Harvard was one of the best places to learn physics and the quantum theory in the 1920s.

My doctoral research was on the compressibility of the alkali halide crystals of the sodium chloride type. Bridgman was just perfecting his technique for measurements of compressibility, and I not only had to construct apparatus, but to make my own crystals, crystallizing from the melt, and to make measurements on them. But I was not satisfied with merely the experimental results, I wanted to try to explain them by use of such theory as was then available. Max Born and Alfred Landé had done theoretical work on the problem some five years earlier, explaining how the electrostatic attractions between the positive alkali ions and the negative halide ions held the crystals together, opposed by repulsive forces whose origin they were not able to explain. I tried very hard to see if anything that was known about quantum theory at the time would lead to these repulsions, and concluded that nothing in the theory could explain them. In fact, it was already well known that the existing form of quantum theory could not even explain the attractive and repulsive forces involved in the structure of the hydrogen molecule. I was convinced by these facts that the quantum theory of 1923 was not adequate to describe the nature of molecules and solids, and resolved to do my best to help in the development of more adequate theories.

Fortunately I was able to get a traveling fellowship for the year 1923–1924, and decided to go to Europe to be in the center of the quantum-mechanical developments which I was convinced would come very shortly. By talking things over with my friend John H. Van Vleck, who had been a graduate student at the same time I was, and then had become an instructor at Harvard, it appeared that Copenhagen with Niels Bohr would be the best place to go. Professor Lyman sought information from many sources which led him to agree that this was a wise choice. Since Bohr was away during the fall of 1923, I spent that time at the Cavendish Laboratory at Cambridge, then the abode of Rutherford, Aston, C. T. R. Wilson, and other famous pioneers in the field of nuclear physics, with occasional visits from J. J. Thomson, who had already retired. I worked on various ideas relating to the quantum theory, which I shall mention in the next section, getting valuable advice from Ralph H. Fowler. Then in December 1923 I went to Copenhagen, where I spent the spring with Bohr and Hans A. Kramers, Bohr's assistant at the time, and got to know as visitors to the laboratory Werner Heisenberg, Wolfgang Pauli Jr., and various others of the brilliant young men who were making a name for themselves in quantum theory. I also met the various other American visitors at Bohr's laboratory at the time, including Harold C. Urey, who later won the Nobel Prize for his discovery of deuterium.

After leaving Copenhagen in June 1924, I went back to Harvard as a staff member, and remained there until 1930, when Karl T. Compton, who had just been named president of the Massachusetts Institute of Technology, induced me to go to MIT as head of the physics department. This meant that during the decade 1923–1932, as well as for many years thereafter, I was in the environment of Cambridge, which then as now was one of the centers of science in this county. I had a second trip to Europe, in 1929, when I spent some months at Leipzig on a Guggenheim Fellowship. Heisenberg was there then, as was Friedrich Hund, who had shortly before spent some months at Harvard. Peter Debye, a genial and helpful member of a somewhat older generation, was there too, and in Leipzig and on various visits during that period I had a chance to meet Felix Bloch, Paul Scherrer, Douglas R. Hartree, Eugene Wigner, Albert Einstein, Erich Hückel, Edward Teller, Nevill F. Mott, John E. Lennard-Jones, as well as many others. Through these visits to Europe, summer school experiences at Stanford, Berkeley, and Chicago, and through visitors to Harvard and MIT, I got to know practically all the scientists who were active during the classical decade of 1923-1932. I formed particularly close friendships with some of the American scientists of the time, including Van Vleck, whom I had known since graduate school, Robert S. Mulliken, whom I knew well during two years while he was in Cambridge, Arthur Compton and Samuel K. Allison of Chicago, Edward Condon of Berkeley and Princeton, Linus Pauling of California Institute of Technology, as well as many others. The group of scientists working on the quantum theory in those days was a small one—maybe 50 to 100 well-known names—and practically all of us knew each other personally.

During all this time, almost countless momentous discoveries had been made in quantum theory and its application to molecular and solid-state problems, and we shall now go on in the following sections to describe the most important of these discoveries, so as to understand how they fitted together to form a complete theory.

2. 1923–1924, Waves and Particles

The scientific event of the year 1923 was the discovery of the Compton effect, by Arthur Compton who was then at Washington University, St. Louis, and later at the University of Chicago (the younger brother of Karl Compton, who later became president of MIT). Compton, an x-ray scientist, had been studying the scattering of x-rays by atoms. It was already well known that x-rays were electromagnetic waves, like light, but of much smaller wavelength. It was also known that Einstein's theory of the photoelectric effect, dating from 1905, suggested that the energy in x-rays, as well as in ordinary light, was not carried directly by the electromagnetic waves which led to interference and diffraction effects, but instead consisted of small particles of energy, which we then called quanta but now call photons, each having an energy $h\nu$, where h is Planck's constant and ν is the frequency of the radiation. Compton's discovery consisted first of the observation that the frequency of the scattered radiation (now called the Compton modified radiation) had a slightly smaller frequency than the incident radiation, the frequency difference depending on the angle of scattering. But it was not this discovery which excited physicists as much as Compton's very simple theory to explain it.

Compton's theory took the point of view of the extreme corpuscular theory of radiation. It was easy to show by relativistic arguments that if the energy of a photon was $h\nu$, its momentum must be $h\nu/c$, where c was the velocity of light. Compton considered a collision of a photon with an electron in an atom. In such a collision, there would be conservation of energy and conservation of momentum. If the photon was scattered in the collision, it would change its momentum, which of course is a vector quantity, so that there could be a large change if it were scattered through a large angle. There must have been an equal and opposite change in the momentum of the electron. The magnitudes were such that this momentum of the scattered electron was great enough to eject it from the atom, and it could be observed in cloud-chamber experiments as a recoil electron. Its considerable recoil velocity meant that it went off with considerable kinetic energy. By conservation of energy, this kinetic energy of the electron had to be compensated by an equal loss of energy of the scattered photon. Since the photon's energy was $h\nu$, this meant a corresponding decrease of frequency ν, and the computed value checked exactly the loss of frequency observed by Compton, verifying his theoretical picture. One even found a distribution of frequencies in the scattered beam, which could be explained in terms of the initial distribution of electronic momenta among the electrons of the atom. This effect furnishes a valuable way of observing the momentum distribution of the atomic electrons.

Here we had what looked like definite experimental evidence of the correctness of the corpuscular theory of light, a theory which had had some followers ever since the days of Newton. But we had equally definite experimental evidence of the correctness of the wave theory, which dated back to Huygens: The wave theory had to be used to explain the interference and diffraction effects, whose reality was observed in x-ray diffraction just as much as in ordinary optics. Which theory was right? Some scientists tried to believe one theory rather than the other, and to devise methods of explaining interference and diffraction in terms of a corpuscular theory. Newton himself had devised a rather fantastic theory, a theory of "fits" of easy reflection and easy refraction, to explain by means of the corpuscular theory how a beam of light incident on a glass surface was partly reflected, partly refracted, and how the Newton's rings he had discovered, a typical interference effect, could be explained. Other scientists tried to explain both types of phenomena in terms of a wave theory; if waves are reflected by a moving mirror or scattered by a moving object, their frequency is modified by the Doppler effect, and one can set up rules for the velocity of this imaginary moving object which would explain the Compton effect in terms of the wave theory. But all such attempts were rather artificial. The straightforward deduction from everything that was known was that both theories were simultaneously to be used, one for one type of phenomena, the other for others.

This was not as absurd a situation as it seemed at first. A number of scientists—W. F. G. Swann, among others—had suggested that the purpose of the electromagnetic field was not to carry a continuously distributed density of energy, but to guide the photons in some manner. This was the point of view which appealed to me, and during my period at the Cavendish Laboratory in the fall of 1923, I elaborated it. In electromagnetic theory one finds an energy density, the electrical energy density being proportional to the square of the electric field, the magnetic energy density to the square of the magnetic field. In an interference or diffraction pattern where there are bright and dark fringes, the energy density is high in the bright fringes, low in the dark fringes. It is perfectly possible to assume that this energy density serves to determine the probability of finding a photon at a given point of space: The probable number of photons per unit volume times the energy $h\nu$ of a photon must equal the continuous energy density as calculated by electromagnetic theory. I was willing, as were many other scientists, to accept this type of relation between the continuous electromagnetic wave and the point carriers of radiant energy, the photons.

One had, then, to assume the existence of these electromagnetic waves, but there was a great difficulty associated with them. Bohr's theory of the hydrogen atom assumed the existence of electronic stationary states, and assumed that radiation was emitted when the atom jumped from an upper state of energy E_2, to a lower state of energy E_1. The emitted energy $E_2 - E_1$ was assumed to be the energy of the photon $h\nu$, which therefore determined the frequencies of the emitted radiation in terms of the energy levels E_2 and E_1. The Bohr-Sommerfeld quantum conditions, which had been known in their simplest form since 1913, and in the later more sophisticated form suggested by Sommerfeld since 1916, predicted the energy levels, and the energy differences gave frequencies which agreed with the known spectrum of hydrogen.

All of this was simple enough, but the difficulty was with the assumption that the radiation was emitted in the form of a photon at the instant the atom jumped from the state E_2 to the state E_1. Any student of physics or of mathematics knows that a wave train of finite length has a frequency spectrum which is not strictly monochromatic, but which instead has a frequency breadth $\Delta\nu$ which is of the order of magnitude of $1/T$, where T is the length of time during which the train is emitted. I had first learned this years earlier, through reading A. A. Michelson's fascinating book *Light Waves and Their Uses*, which my father had in his library. If this time is much longer than the period of oscillation, which is $1/\nu$, the breadth $\Delta\nu$ will be very small compared to the frequency ν. The observed sharpness of spectral lines shows that this must be the case. The experiments are consistent with emitted wave trains which have perhaps the order of 10^5 or more waves in the train. How, I asked myself, could a physicist with the insight of Bohr have suggested that the radiation was emitted instantaneously? Surely it must have taken long enough for 10^5 waves to be emitted.

A good deal was known even then about the lifetimes of atoms in their stationary states. An atom is excited to a level above the ground state, stays in that state for a while, and then falls back to the ground state. The interesting fact is that the lifetime in the excited state is of the same order of magnitude as the time required to emit the 10^5 waves, more or less, which are needed to produce the observed sharpness of the spectral lines. The situation thus appeared to me to be perfectly clear. It must be that all the time an atom is in an excited state, it must be emitting electromagnetic waves of all the frequencies corresponding to transitions to lower states which would be allowed by Bohr's theory. The intensities of these waves would have to be determined by Bohr's correspondence principle, which had been worked out some years earlier, and which tied together the quantum probability of emission with corresponding amplitudes in the classical oscillation of the electron in the atom. One would not only have these spontaneous oscillators producing the waves, but also induced oscillations produced by external radiation falling on the atom, which by interference with the external radiation would result in absorption of radiation according to the electromagnetic theory.

These electromagnetic waves, in my view, would not directly carry energy, as they would in the classical electromagnetic theory, but would be connected with the probability of finding photons at the given point, as was described in an earlier paragraph. If one found the flux of electromagnetic energy outward from the atom, as determined by Poynting's vector, one could connect this with the probability of emitting a photon with consequent jumping of the atom to a lower state, while an inward flux in the presence of an external field would be connected with the probability of absorbing a photon. These probabilities of emission and absorption had been introduced by Einstein in his derivation of Planck's radiation law in 1917.

This represents substantially the view I had when I went to Copenhagen at Christmas time, 1923. While at the Cavendish Laboratory in the fall of 1923 I had talked over the ideas with Fowler, who liked them, and had put them on paper not only in memoranda of my own, but in a letter to Kemble. As soon as I discussed them with Bohr and Kramers, I found them enthusiastic about the idea of the electromagnetic waves emitted by oscillators during the stationary states—they at once coined the name "virtual oscillators" for them. But to my consternation I found that they completely refused to admit the real existence of the photons. It had never occurred to me that they would object to what seemed like so obvious a deduction from many types of experiments. The result was that they insisted on our writing a joint paper in which the electromagnetic field was described as having a continuously distributed energy density whose intensity determined the probability of transition of an atom from one stationary state to another. One had then to assume only a statistical conservation of energy between the continuously distributed energy density in the electromagnetic field and the quantized energy of the atoms. They grudgingly allowed me to send a note to *Nature* indicating that my original idea had included the real existence of the photons, but that I had given that up at their instigation.

This conflict, in which I acquiesced to their point of view but by no means was convinced by any arguments they tried to bring up, led to a great coolness between me and Bohr, which was never completely removed. During the rest of my time in Copenhagen, I worked on the detailed formulation of my ideas on optical phenomena, which I published in early 1925 without any approval from Bohr or Kramers. This paper gave a consistent picture of the virtual oscillators, of their function of determining the probabilities of emission and absorption of photons, and of the effect of the finite lifetime on breadth of spectral lines, which I still believe was essentially correct, and which I suspect will be more recognized as time goes on.

It was astonishing to see how rapidly my point of view was vindicated. Various experimentalists pointed out that if the intensity of the electromagnetic field determined only the probability of transition, there was no good reason to expect that the recoil electron, whose recoil was given by probability arguments, would come off at the same time as the scattered photon, which also came off by probability arguments. Bothe and Geiger, and slightly later Compton and Simon, in 1925 set up ingenious experiments with counters to measure simultaneously the recoil electron and the scattered photon, and they showed clearly that they actually appeared simultaneously. This settled the argument, and I believe that no one continued to doubt the real existence of the photons, conservation of energy between photons and atoms, and the probability relation between the motion of the photons and the intensity of the electromagnetic wave.

But the real vindication of my point of view came from a very different direction. None of us in Copenhagen in the spring of 1924 had known of the work of Louis de Broglie in Paris. A man of famous family, he had turned somewhat after the usual age to the study of physics, and was working for his Ph.D. in Paris. His brother Maurice de Broglie was an x-ray scientist, and through him Louis de Broglie was aware of the Compton effect and its relation to the existence of photons. He started publishing papers in late 1923, at the same time as my work in the Cavendish, but his main publications did not reach general attention until early 1925. His point of view about the relation of photons and the electromagnetic field was essentially the same one to which I had come practically simultaneously. But he did not have the antagonism of Bohr to contend with, and consequently he followed his ideas to their obvious conclusion. If there were an electromagnetic wave to guide the scattered photon in the Compton effect, why should there not also be a wave of some sort to guide the recoil electron? The two were inextricably tied together. Thus came the origin of wave mechanics.

But de Broglie's work went far beyond this simple parallelism between photons and their waves and mechanical particles and their waves. For he had the insight to realize that Sommerfeld's quantum conditions, which had formed a mysterious part of physics up to that point, followed straightforwardly from the material waves, which we now call de Broglie waves. He stated his arguments in a rather cumbersome form derived from relativistic mechanics, but this is not necessary. We have seen that the momentum of a photon was $h\nu/c$, which equals h/λ, where $\lambda = c/\nu$ is the wavelength of the electromagnetic wave. To find the wavelength of the material waves, de Broglie assumed that the momentum p was likewise given by $p = h/\lambda$. Suppose, he said, that not only can we use material waves to predict scattering processes, but also that we have some sort of standing waves in atoms. We should then be able to write the condition for the existence of standing waves in the form of a statement that the number of wavelengths around the complete circuit of the standing wave should be an integer. The number of waves in a distance dq is $dq/\lambda = p\,dq/h$. Thus the condition for standing waves is $\int p\,dq/h = n$, an integer, or $\int p\,dq = nh$. But this was just the famous quantum condition of Sommerfeld, which had formed one of the postulates of quantum theory from 1916 to 1923. Here was a consequence of the simultaneous existence of waves and particles which was bound to bring a new day in quantum physics. It is only proper that we reckon 1923 as the beginning of wave mechanics, when de Broglie published his first notes on these profoundly fascinating ideas.

There is an interesting and often-told story of how the ideas of de Broglie turned into Schrödinger's equation and its consequences in 1926. de Broglie's ideas became known rather slowly. I have stated that as far as I know, no one in Copenhagen was aware of them in 1924, when I was there. Even in 1925 they were not common knowledge; I noted them myself by reading one of de Broglie's articles in the Harvard physics library, rather than by having them pointed out to me by anyone else. The story is that Debye had heard of them, and became one of the first to realize the importance of the new ideas. Debye at the time was at the Federal Institute of Technology (ETH) in Zurich, and he was well acquainted with Erwin Schrödinger, a first-class theoretical physicist at the University of Zurich. Debye suggested to Schrödinger that he look into the new ideas, and report them in a colloquium. Schrödinger immediately saw that the idea was crying out for mathematical formulation in the form of a wave equation, much as Clerk Maxwell had realized that Faraday's ideas of fields of force had been crying out for mathematics in the middle 1800s. But instead of taking decades for the mathematical development to take place, as with electromagnetic theory, Schrödinger with a magnificent burst of speed produced the whole thing in a series of monumental papers all published in 1926. This was an even more important date than 1923 in the history of wave mechanics.

But so fast was quantum theory moving, that there are a number of very important developments that came between 1924 and 1926. First there were the outgrowths of the Copenhagen school, which were largely the work of Heisenberg, and which arose from the virtual oscillators of 1924. Second were the profoundly important discoveries of the spinning electron and the exclusion principle, which put the study of atomic multiplets on a sound basis. We shall take up these two developments in the next two sections, before we go on to Schrödinger and his work.

3. 1925, Matrix Mechanics

On or about January 17, 1924, while the Bohr-Kramers-Slater paper was still being finished, Kramers showed me a theorem which he had proved shortly before, and which later proved to be practically as important in the development of matrix mechanics as the Compton effect was in the development of wave mechanics. This was the theorem which later became known as the Kramers-Heisenberg dispersion formula. Heisenberg did not enter the picture until later, in June 1924, when he came to Copenhagen to spend a year. I believe I was one of the few people who were told this theorem at such an early date, and it is not always realized what an important role Kramers played in these early discoveries. Fortunately I still have the piece of paper on which Kramers wrote down the formula in substantially the form which he later published in two letters to *Nature* in the late spring of 1924. This formula, as I copy it from this memorandum, is

$$\frac{\partial}{\partial J_i}\left(\frac{C_i^2 \omega_i}{\omega^2 - \omega_i^2}\right) \rightarrow \frac{C_1^2 \omega_1}{\omega^2 - \omega_1^2} - \frac{C_2^2 \omega_2}{\omega^2 - \omega_2^2} \tag{3-1}$$

This will mean nothing to the reader until I give a certain amount of background.

First, this equation deals with what is called multiply periodic motion in classical mechanics. This is a type of motion in which several different frequencies can simultaneously exist, such as a vibrational and a rotational frequency. It had been known for a number of years that one had such motion for an electron moving in a central, or spherically symmetric, electrostatic field, with a superposed magnetic field, such as one had in the atomic models that were then current. The electron had one frequency for its radial, in-and-out motion, a second frequency for the rotation of the orbit, and a third for its precession in the magnetic field. It was only for such multiply periodic motion that the quantum theory could be applied, in the version we had in 1923.

For each of the frequencies, one could set up in classical mechanics a quantity called an action variable, which was $\int p\,dq$. This was the same quantity to which Sommerfeld applied a quantum condition, setting the quantity equal to nh in the quantum theory. The energy, generally called H, since it was identical with the Hamiltonian function met in classical mechanics, could be written in terms of the various action variables, denoted by J's. Thus, with the example I mentioned above, there was one J, J_r, for the radial motion; when the quantum conditions were applied, this quantum number was called the radial quantum number. Similarly, for the rotation of the orbit, there was a J_θ, whose quantized value was called the azimuthal quantum number, and for the precession there was J_ϕ, whose quantized value was the magnetic quantum number. The latter two quantum numbers measured the total angular momentum of the electron and its component along the direction of the magnetic field.

A very important theorem in classical mechanics formed the basis of Bohr's correspondence principle, which he had proposed several years earlier. This theorem stated that the vibration frequency ν_i, equal to $\omega_i/2\pi$, where ω_i is the corresponding angular frequency, was given by

$$\nu_i = \frac{\partial H}{\partial J_i} \quad (3\text{-}2)$$

Here ν_i is the vibrational or precessional frequency we have been mentioning. Bohr pointed out the very close relationship between this classical formula for frequency and the corresponding quantum-theoretical formula when the quantum number n_i changes from an initial value n_{i1} to a final value n_{i2}, equal to $n_{i1} - 1$. If E_1 is the value of the energy H when J_i is equal to $n_{i1}h$, and E_2 is the value when J_i is $n_{i2}h$, the change in quantum number being one unit, Bohr's frequency condition is $h\nu_i = E_1 - E_2$, or

$$\nu_i = \frac{E_1 - E_2}{J_{i1} - J_{i2}} \quad (3\text{-}3)$$

In other words, the quantum-mechanical frequency is the ratio of a finite difference of H to that of J_i, while the classical frequency is the partial derivative, or the ratio of the differences in the limiting case where h goes to zero.

Bohr had looked at the corresponding relation, not only for the case where just one quantum number changes by unity but for a case where J_1 changes by τ_1 units, J_2 by τ_2 units, and so on, and had shown that in this case the quantum frequency is related in a similar way to one of the overtone or harmonic frequencies in the classical motion. This was the overtone vibration of frequency $\tau_1 \nu_1 + \tau_2 \nu_2 + \cdots$. He had then examined the way in which the classical motion could be expressed as a linear combination of harmonic vibrations, one with each of these overtone frequencies, and with a specified amplitude.

If one found the radiation field emitted by such a classical oscillation, it would consist of radiation with the various overtone frequencies, and the rate of radiation of one of these overtone frequencies was proportional to the square of the amplitude of this frequency. Bohr's correspondence principle stated that the rate of radiation in the quantum theory would be proportional to the square of the classical amplitude of the corresponding overtone frequency, but now computed somehow for an intermediate energy between E_1 and E_2, or for corresponding intermediate values of the quantum numbers. This principle had proved to be widely obeyed. In cases where the amplitude of a classical overtone was zero, it was assumed that the corresponding quantum-mechanical radiation did not occur, the probability of having a quantum-mechanical transition was zero, and one had what was called a selection rule. When the amplitude was different from zero, the value deduced from the classical oscillation, computed for example for a state half-way between the initial and the final state, proved to give a reliable indication of the experimentally determined intensity. These probabilities of transition could be tied in with the well-known probabilities A and $B\rho$ of spontaneous emission and absorption of radiation which Einstein had introduced in 1917 to give a derivation of Planck's radiation law from the Bohr theory, where ρ is the density of electromagnetic radiation.

Bohr's argument had been applied only to the spontaneous radiation emitted by the free motion of the electron, which led to Einstein's A coefficient. However, there was a great interest in the dispersion of light around an absorption line, which demanded finding the forced oscillation of the electronic system in the presence of an external radiation field. This would lead both to Einstein's B coefficient and to the formula for anomalous dispersion, which was known to be very well represented by the classical electron theory of Drude and Lorentz, which treated the electrons as linear oscillators. Consequently Kramers had been working on the problem of the forced oscillation of a multiply periodic system in the presence of an external electromagnetic field with arbitrary frequency $\nu = \omega/2\pi$. This was an involved problem in the classical mechanics of multiply periodic systems, but Kramers had solved it before I arrived in Copenhagen, and I found it easy to reproduce his proof from the knowledge I already had of such systems. It turned out to be the case that a component of frequency ν would appear in the motion, with an amplitude proportional to the external field. The coefficient multiplying this external amplitude, to give the forced motion, had a form according to Kramers which was like the expression on the left of Eq. 3-1, summed over the various natural angular frequencies ω_i. The C_i's were quantities proportional to the amplitudes of the corresponding vibrations in the free motion of the electron.

The feature of this equation which attracted Kramers was the fact that, similarly to Eq. 3-2, it was written as a partial derivative with respect to J_i. It was natural to suppose that as in Eq. 3-3, the quantum-theoretical form of this result could be found by replacing the derivative by a finite difference, as indicated on the right side of Eq. 3-1. The two terms whose difference was indicated in this quantum-theoretical form had almost exactly the form of the forced amplitudes of linear oscillators in the Drude-Lorentz theory, though the partial derivative with respect to J_i in the classical expression was much more complicated. Thus Kramers had made it very plausible that he had found the true explanation of the reason why the Drude-Lorentz theory worked so well. There was just one difference between the resulting formula and the Drude-Lorentz formula. In the latter case, the forced amplitude looked like one of the terms on the right side of Eq. 3-1, but without the quantity similar to C_i^2 which was varying from transition to transition in the atom. Kramers' formula agreed with the experimental findings, which indicated that the Drude-Lorentz formula had to be multiplied by a dimensionless quantity, often much smaller than unity, to agree with experiment. These dimensionless factors had already been called oscillator strengths, and what Kramers had done was to find formulas for these oscillator strengths in terms of the C_i's, which in turn led to definite relations between the oscillator strengths and the A and B coefficients of the various transitions.

Naturally Kramers and I were very conscious of the relation between these results and the "virtual oscillators" we had been proposing in the Bohr-Kramers-Slater paper. The amplitudes of the spontaneous oscillations relating to a transition could be computed approximately from the correspondence principle. And in the presence of an external electromagnetic field, Kramers' formula of Eq. 3-1 would give the forced oscillation which would lead to dispersion and absorption of radiation. Kramers wished to have various features of the calculation carried out, and he asked me to work on them. Within several weeks I had produced the results he wanted.

Even in that short time, however, I realized that I was not temperamentally suited to be an assistant to the professor, in the European tradition of carrying out his calculations as he requested. I was very anxious, for instance, to start with the technique Kramers had used, but to use it to carry through the absorption probability, the relation to the lifetime in stationary states and the consequent breadth of the emission and absorption lines, and such points as I felt were necessary to convert the vague and general ideas of the Bohr-Kramers-Slater paper into a consistent and quantitative theory. The result of this was that, proceeding as amicably as possible, I let Bohr and Kramers know that I preferred to work on my own. I carried through these various calculations, in which Bohr and

Kramers indicated no interest, and incorporated them in a paper I finished shortly after returning to Harvard, and which was published in early 1925. I made use in this of Kramers' formula, and of similar work which Van Vleck was doing at the same time; Kramers published his two notes in *Nature* regarding his derivation in the spring of 1924, so that it would be available for me to use.

During the spring of 1924, Heisenberg paid a visit to Copenhagen, and I had a chance to get to know him, socially as well as scientifically. He was obviously brilliant, but at the same time unassuming and friendly, and I was much taken with him. Since I was returning to Harvard in June, Bohr and Kramers arranged to have him come to Copenhagen at the time I left, to assist Kramers with the work on dispersion. It was a result of this collaboration that the Kramers-Heisenberg paper was written, appearing at about the same time that my paper came out. Their results were essentially equivalent to mine, but they did not give any treatment of absorption, line broadening, and related topics, which I felt were a main feature of my work. It is interesting that this line-broadening feature was essentially an example of the principle of uncertainty, a consequence of any type of wave theory with finite wave trains, but I do not think that Bohr, Kramers, and Heisenberg were ready for the idea at the time. It was only some years later, in 1927, that Heisenberg published his paper on the uncertainty principle, and that he and Bohr made this such a great feature of their version of wave mechanics. And it was in 1930 that Weisskopf and Wigner derived the same results from wave mechanics and the Dirac radiation theory which I had discussed in my 1925 paper.

During 1925 W. Kuhn and W. Thomas independently arrived at an interesting result, called the oscillator sum rule, which stated that the sum of all the oscillator strengths for the various transitions of a single electron from a given stationary state had to equal unity. In this statement, it had to be noticed that Eq. 3-1 contains both positive and negative terms, the positive ones coming from transitions to higher energy levels, the negative from transitions to lower levels. The oscillator strengths had to be considered positive to the upper levels, negative to the lower ones, which were referred to as negative dispersion. Kuhn got at this sum rule from the limit of Kramers' formula for the case where all natural frequencies ω_i were small compared to the external frequency ω. In this limit, J. J. Thomson many years earlier had derived a scattering formula from very simple classical theory, which was known experimentally to be correct in the limit of x-ray scattering. This Thomson scattering formula led directly to the Kuhn-Thomas sum rule.

All of this work referred very simply and straightforwardly to optical dispersion and related topics. But at this point the great intuition and true genius of Heisenberg began to show themselves. He was trying to think of a mathematical framework into which stationary states and transitions of quantum theory would naturally fit. If we have stationary states of energy $E_1, E_2, \ldots$, carrying a single index, then we can have transitions from any state to any other, with probabilities determined by the amplitudes of the oscillators—"virtual oscillators," if one chose to denote them that way—given by the correspondence principle. Each one of these transitions must carry two indices, denoting the initial and final states, so that the oscillator amplitudes would be quantities with two indices, suggesting the matrices of algebra. Is it not possible, Heisenberg asked, to set up an algebra which would lead directly to these matrix components, such that each nondiagonal component would refer to a particular transition, each diagonal component to a stationary state?

Here it will not pay us to try to follow in detail the ideas of a genius. Heisenberg started with the matrix rule of multiplication, namely that the matrix product of matrices of two quantities p and q is

$$(pq)_{ij} = \sum (k) p_{ik} q_{kj} \qquad (3\text{-}4)$$

Furthermore, it is not necessarily true that two matrices commute with each other, or that $(pq)_{ij} = (qp)_{ij}$. He tried to set up matrix components for the displacement of the electron forming the virtual oscillator, which he called q_{ij}, and of its momentum, which he called p_{ij}. He then found that if he assumed that p and q did not commute with each other, but that instead

$$(pq)_{ij} - (qp)_{ij} = \frac{h}{2\pi i} \delta_{ij} \qquad (3\text{-}5)$$

where the i appearing in $h/2\pi i$ is $\sqrt{-1}$, and where δ_{ij} is defined to be unity if $i=j$, zero if $i \neq j$, and if he furthermore assumed that the matrix components p_{ij} were related to q_{ij} as they would be for a sinusoidal oscillation with frequency as given in Eq. 3-3, he could prove the Kuhn-Thomas sum rule.

The Kramers dispersion formula, in other words, and the Kuhn-Thomas sum rule, which had followed from it, were what led him to the so-called commutation rule of Eq. 3-5, which soon proved to play a central role in the matrix mechanics Heisenberg set up. By combining this commutation rule with rather obvious matrix expressions for the potential and kinetic energy of a linear oscillator, he showed that he could derive expressions for the energy of a linear oscillator which led to the correct quantized energy levels. He could get the matrix components leading to the oscillator strengths, and find these quantities for the oscillator. He went further, and applied similar methods to the components of an angular momentum. He was well on the way to a complete formulation of quantum mechanics in a form of mathematics which involved only the observable quantities, such as these amplitudes of the virtual oscillators responsible for the emission and absorption of radiation.

After leaving Copenhagen, he collaborated with Max Born and Pascual Jordan, of Göttingen, and the three of them made enormous progress in the study of the coupling of the angular momentum vectors in the atom. Landé had made a semiclassical theory of this vector coupling several years earlier, led on by empirical rules which he had arrived at through study of multiplet structure and Zeeman effect in atoms. But Landé had to modify his formulas slightly to agree with experiment. Classical mechanics suggested as the formula for the square of a vector the obvious quantity l^2, but Landé found that in place of this quantity he had to use $l(l+1)$. The methods of Born, Heisenberg, and Jordan using quantum mechanics automatically led to the modified Landé formulas which agreed with experiment.

I had a small opportunity to make contact with this work during the latter part of 1925. Born spent several months in Cambridge, Massachusetts, lecturing at MIT, on the behavior of crystals, expanding on the work I have referred to earlier in connection with my work on the alkali halides, and also on the new quantum mechanics of Heisenberg, Jordan, and himself. Naturally I became acquainted with him, and had a chance to work on some of the aspects of the quantum-mechanical theory. I made a little progress beyond what had already been published, but before I could get my work written up, a paper by a hitherto unknown genius appeared: P. A. M. Dirac's first paper on quantum mechanics. It included not only the small points I had worked out, but much more besides. But at least I got to know Born well, and kept scientific and personal relations with him during the rest of his life.

By now we had Born, Heisenberg, Jordan, and Kramers all working at quantum mechanics, together with Dirac, who plunged into the field with a succession of brilliant papers. They were immediately joined by Pauli, precocious son of a well-known Viennese medical man, who produced a result that was more of a tour de force than any of the others: He used the matrix mechanics to work out the structure of the hydrogen atom, and derived Bohr's formula for its spectrum from this entirely independent point of view. All this was going on quite independently of the de Broglie-Schrödinger line of research, which did not come to public knowledge until 1926. We shall come to it shortly, but we have not yet exhausted the wonders of the marvelous year 1925. The story of the spinning electron and the exclusion principle must still be told.

4. 1925, The Spinning Electron and the Exclusion Principle

I have mentioned that on the faculty at Harvard was Frederick A. Saunders, a spectroscopist who was concerned with the Russell-Saunders coupling in atomic spectra. He was an elderly and fatherly type, a spectroscopist of long experience. Henry Norris Russell was probably the real inspiration of the pair, a grandfatherly rather than a fatherly figure, a fragile-looking white-haired astrophysicist from Princeton, but with a spirit filled with fire and energy. We saw a good deal of him at Harvard during 1925, for that was the year when he and Saunders were collaborating on their study of the spectrum of calcium, important for its astrophysical aspects, but also the problem which provided the key to the understanding of the complex spectra of the atoms.

It was a topsy-turvy year. The paper of Russell and Saunders came out first, but one could not understand it without the exclusion principle and the spinning electron. Pauli's paper on the exclusion principle came out next, but one could not understand it without the spinning electron. Finally Uhlenbeck and Goudsmit's paper on the spinning electron appeared, and things began to fall into line. I shall not even try to describe the confused state of mind that all of us were in while these things were unfolding, when no one could understand what the various papers meant, but everyone felt that great things were in the air. I naturally spent most of my time trying to get things straight. Fortunately Kemble and I, who were running the Harvard theoretical seminar that year, had announced complex spectra as the topic for the year, and we had to understand these phenomena before we tried to explain them to the students.

The only way to present these things is backward, starting with the spinning electron, for otherwise they make no sense. Landé had been making his rules about the coupling of angular momentum vectors in the atom, and Sommerfeld had been systematizing them, inventing a quantum number which he called the inner quantum number, but no one understood what these vectors really were. It was interesting to read successive editions of Arnold Sommerfeld's invaluable book *Atombau und Spektrallinien*, which was our "bible" in the study of quantum mechanics at Harvard, and to see Sommerfeld's valiant efforts to understand and present the things that no one really understood. The point was that everyone had been thinking of these angular momenta, and their resulting magnetic moments, in terms of the orbits of electrons in the atom acting like little solenoidal currents. They were what Ampère had thought of in the 1820s to describe the magnetic behavior of the atoms. And yet the spectroscopists found angular momenta showing up in the atomic systems in addition to those of the various outer electrons of the atom. The theorists tried to ascribe these other angular momenta to what they called the "Atomrumpf," the inner core of the atom, and yet it was hard to see why these should have angular momenta and magnetic moments.

We must realize that by the years I am writing about, we had a rather good idea of the electronic shells in the atom and the formation of the periodic system of the elements. Bohr had written papers about them in 1922, describing work he had done on the subject during the preceding two years. I well remember the excitement with which I discovered Bohr's article newly arrived at the Harvard physics library, on a date which I can verify from my records as May 4, 1922. I read all 67 pages of it, and felt it was the most important thing that had come along for a number of years. This paper showed clearly that the "Atomrumpf" inside the sodium atom had the same electronic structure as the neon atom, which was believed to have no magnetic moment, and yet here were the spectroscopists trying to assign an angular momentum to it. This made no sense, as everyone was ready to admit.

I have said that we must present things backward in unraveling the complications of the atoms. Thus let us start with George Uhlenbeck and Samuel Goudsmit, who suggested at the end of 1925 that the electron had an intrinsic angular momentum and magnetic moment, as if it were rotating on its axis like the earth, besides rotating in an orbit around the nucleus as the earth moves around the sun. Here was an explanation for the extra angular momentum and magnetic moment that everyone was ready to accept, and it is what led to our whole understanding of the atomic problem. It is still rumored, however, that something approximating the idea was first suggested earlier, by Ralph de L. Kronig, and that he presented the idea to Pauli, who ridiculed it and kept it from being published. I can well believe this. Pauli was as set in his ways as Bohr had shown himself to be by his refusal to admit the existence of photons, and Pauli was a belligerent and dictatorial man, more than a match for the quiet and polite Kronig. Whatever may be the truth of this, the general scientific world learned of the spinning electron from two young Dutchmen, Uhlenbeck and Goudsmit. Uhlenbeck was a theorist, brought up in Java when that was a Dutch colony, and Goudsmit was a spectroscopist. I got to know them well soon after they moved to the University of Michigan in 1927, where they contributed greatly to building up the fine reputation of that university in quantum theory.

Uhlenbeck and Goudsmit not only suggested that the electron had an intrinsic angular momentum and magnetic moment, which at once became known as the electron spin, but they also were able to state the values of both the angular momentum and magnetic moment. The angular momentum of the electron orbit in the Bohr atom was assumed, according to the Sommerfeld quantum conditions, to be an integer times $h/2\pi$, where this quantity $h/2\pi$ was soon abbreviated by the symbol $\hbar$. To explain the experimental facts, Uhlenbeck and Goudsmit had to assume that the angular momentum of the spin was $\frac{1}{2}\hbar$. The magnetic moment of the Bohr atom was an integer times a quantity usually called μ_B, the Bohr magneton, proportional to the angular momentum. But Uhlenbeck and Goudsmit had to assume that the magnetic moment of the spin was μ_B rather than $\frac{1}{2}\mu_B$, which would have been the case if the ratio of magnetic moment to angular momentum for the spin had been the same as for the orbital motion. Instead, this ratio, sometimes called the gyromagnetic ratio, had to be assumed twice as big for the spin as for the orbital motion.

This assumption was devised to fit in with the theories Landé had postulated to explain the Zeeman effect, the splitting of spectral lines into several narrowly spaced components in a magnetic field. Landé had had to introduce a ratio g, different from unity, for the ratio of the actual gyromagnetic ratio to that predicted by the theory of orbital magnetic moments. He had worked out detailed formulas for the values of these g quantities, called Landé g factors, based on the assumption that one was coupling one vector for which g was equal to unity, with another for which g was equal to 2. It now became clear, through the work of Uhlenbeck and Goudsmit, that the vector with $g=1$ was composed of orbital angular momenta, and that with $g=2$ was composed of spin angular momenta.

With this understanding of the spin angular momentum and magnetic moment, we can begin to understand the assumptions Russell and Saunders had used to explain the experimental facts concerning the calcium spectrum. The calcium atom has two electrons outside the closed shell, which is constructed like the argon atom. If one assumed that this closed shell, the "Atomrumpf," had no net angular momentum or magnetic moment, then as far as the coupling of vectors was concerned, calcium was like a two-electron atom. The general theory of vector coupling, which had been worked out from his semiclassical point of view by Landé, and which was just being described on matrix mechanics by Born, Heisenberg, and Jordan, assumed that if one coupled two angular momentum vectors together, l_1 and l_2 (in units of $\hbar$), they could form a quantized resultant L (again in units of $\hbar$) which could take on integral values from $l_1 + l_2$ to $|l_1 - l_2|$. Russell and Saunders assumed that each of the two outer electrons of the calcium atom had a quantized orbital angular momentum, the l_1 and l_2 we have just mentioned. By then it was recognized that the values $0, 1, 2, 3, \ldots$, for l corresponded to the orbits the spectroscopists for some years had called $s, p, d, f, g, h, j, \cdots$ orbits. Russell and Saunders denoted the resultant quantum number by L, which could take on values $0, 1, 2, \ldots$, and assigned to these L values the corresponding large letters $S, P, D, F, \ldots$.

But in addition to these orbital angular momenta and magnetic moments, each electron would have its spin angular momentum and magnetic moment. The spin angular momentum, corresponding to l_1 and l_2, could be denoted as s_1 and s_2, but by Uhlenbeck and Goudsmit's hypothesis, each of these quantum numbers had to be $\frac{1}{2}$ unit. When they were coupled together, one could have the integral values of the resultant, which Russell and Saunders called S, of 1 or 0. Finally they assumed that the vectors L and S were coupled together to give a quantity J, representing the total angular momentum of the atom, measured by Sommerfeld's inner quantum number. This value of J could go from a maximum of $L + S$ to a minimum of $|L - S|$. Thus for $S = 1$, J could take on the values $L + 1, L, L - 1$ (provided L were at least as great as 1 unit), a total of three different levels. For $S = 0$, J had to have the single value equal to L. Experiment indicated that the energies associated with the various values of J for a given set of L and S were rather close together, while those with different L or different S were quite far apart. One would therefore expect to find in the calcium spectrum some groups of three closely spaced levels, and other levels standing by themselves, corresponding to $S = 1$ and 0, respectively. These were observed, and were called triplet and singlet levels, respectively.

These rather simple concepts proved to give a very good account of the calcium spectrum, and it was only a few years before the spectroscopists applied them not only to two-electron atoms but to atoms with many more electrons. The atoms of the iron group—Ti, V, Cr, Mn, Fe, Co, Ni—had been known for years to have very complicated spectra, which seemed at first too complicated ever to be understood. But by the use of the principles introduced by Russell and Saunders, the spectroscopists had explained all of them successfully by the end of the decade we are writing about, or shortly thereafter. However, there was one feature still to be described, which also was worked out in 1925. This was connected with the Pauli exclusion principle. Its result was that some of the levels predicted by the simple rules we have outlined failed to appear.

Bohr, in his work on the periodic system in 1922, had had no explanation of why various shells in an atom could contain only a finite number of electrons. Thus the shell which we call the K shell, containing $1s$ electrons, where 1 is the principal quantum number, has only two electrons, like the helium atom. The L shell, containing $2s$ and $2p$ electrons, has only eight electrons, so that the neon atom, with its K and L shells filled, has ten electrons. The M shell, with $3s$, $3p$, and $3d$ electrons, can hold 18 electrons, but the partially filled shell with eight $3s$ and $3p$ electrons is occupied at argon, with its 18 electrons. The $3d$ electrons only begin to appear in the iron group of atoms.

These facts had been pointed out by Bohr in 1922, but he had not assigned the numbers correctly to $2s$ and $2p$ separately, or to $3s$, $3p$ separately. Edmund C. Stoner, a young Englishman who later made a name for himself as an expert in the field of magnetism, published a paper in 1924 in which, on the basis of x-ray evidence, he showed that each s state, the $1s, 2s, 3s, \ldots$, contained only two electrons; each p state, $2p, 3p, \ldots$, contained only six electrons; each d state only ten; and so on. Sommerfeld at once noticed this paper, and mentioned it in the preface of his next edition of *Atombau und Spektrallinien*. It obviously gave an explanation of the ten electrons found in the neon atom, the 18 in argon, and so on. Stoner's correction to Bohr's theory was immediately accepted universally.

But this fact set Pauli to thinking. Though Pauli ridiculed the idea of an electron spin, nevertheless he had worked out the idea that for an electron with a given integral l value, there must really be two possible angular momentum values, $l+\frac{1}{2}$ and $l-\frac{1}{2}$. He wanted to ascribe some rather mystical significance to this doubling of l values, which was why he rejected the matter-of-fact Uhlenbeck-Goudsmit idea of the electron spin. But however that might be, he noticed that there was a relation between his hypothesis and the number of electrons in a given shell. We have hardly mentioned the idea of space quantization, according to which the component of the angular momentum of l units along an axis, the axis of an external magnetic field, has to be a quantum number m times the fundamental unit $\hbar$. This idea arose, as we mentioned earlier, from Sommerfeld's quantum conditions, and was at the foundation of the theory of the Zeeman effect. For an s electron, with $l=0$, there would be just one possible value for m. For a p electron, with $l=1$, there would be three m values, $1, 0, -1$. Similarly for a d electron, with $l=2$, there were five m values, and so on. Pauli at once thought of doubling these numbers, on account of his mystical doubling of the angular momentum l values, and came out with two states for an s electron, six for a p, ten for a d, and so on.

It was only a small step from this observation to the hypothesis that in nature only one electron could be found in a given state. This hypothesis, which came to be called the Pauli exclusion principle, was immediately accepted as an addition to the known facts about the quantum theory of electrons and as an explanation for the finite number of electrons in Bohr's theory of the periodic table. Of course, it fitted in perfectly with Uhlenbeck and Goudsmit's spinning electron, as we realize now. The orbital angular momentum l would have a component along the axis which began to be called m_l, while the spin would have a component m_s, which could take on either the value $\frac{1}{2}$ or $-\frac{1}{2}$. The product of the $2l+1$ possible values of m_l and the two possible values of m_s leads at once to $2(2l+1)$, the number of electrons allowed for states of given n and l. Looking back, we can hardly regard Pauli's refusal at first to accept the spinning electron in any more charitable way than as pure stubbornness.

A young German, Friedrich Hund, entered the field at this point, and showed that this way of regarding the exclusion principle led to definite spectroscopic results, aside from its role in limiting the number of electrons in a closed shell in an atom. He started his discussion of a given atom, for example of a two-electron atom like calcium, by assuming that the electrons moved independently of each other, so that we would find $2(2l+1)$ combinations of m_l and m_s for a single electron. These represented different electronic states, but with the same energy, or as we say degenerate with each other, since they differed only in the orientation of the angular momenta in space. This orientation would not affect the energy if the electrons moved in a spherically symmetrical field, as one was assuming for the atom; energy dependence on m_l would come in only if there were an external magnetic field. And energy dependence on the orientation of the l vector with respect to the spin presumably would vanish if there were no magnetic effects included in the theory.

But then he assumed the interactions between the electrons to be introduced as perturbations. On a classical model of the motion, these interactions between the electron orbits of electrons with different m_l would be of the nature of torques between the orbits, arising because the orbits were not spherically symmetric. They would result in the angular momentum varying with time, since the torque is the time rate of change of the angular momentum. Thus the angular momentum of an individual electron would no longer be quantized, or the l and m_l would no longer be what one called "good quantum numbers." However, the torques were internal to the system, so that the total angular momentum would still be conserved, and a "good quantum number" could still be assigned to the resultant angular momentum.

It was this resultant angular momentum which Russell and Saunders denoted by L. Similarly, the resultant of the spin angular momenta was assumed to be S, again a good quantum number. The angular momentum L would be quantized in space; its component along the axis, M_L, could take on any one of the $2L+1$ values from L to $-L$. And the spin angular momentum S would be space quantized with a component M_S along the axis. These "good quantum numbers" continued to be good as one went from the case of no interaction to that with interaction, so that one could establish a parallelism between two ways of writing the quantum numbers. Let us explain this by an example. We shall take that of the coupling of two p electrons, each having $l=1$, $s=\frac{1}{2}$.

Each electron will have six possible sets of quantum numbers: $m_l = 1, 0, -1$, coupled with $m_s = \frac{1}{2}$ or $-\frac{1}{2}$. Thus with the two p electrons, one has 36 ways of combining the quantum numbers of the first electron with those of the second. But now let us first couple the l's, and then separately the s's, after the fashion of Russell and Saunders. In Table 4-1 we indicate these ways of coupling the l's and s's separately. We see that there are nine possible arrangements of M_L as the sum of m_{l1} and m_{l2}, and four of $M_S = m_{s1} + m_{s2}$, which can again be combined to give 36 combinations. This table has exhibited the 36 states in a form appropriate to Russell-Saunders coupling. Thus that coupling would lead to $L=2$, 1, or 0 (that is, l_1+l_2 down to l_1-l_2). The $L=2$ would lead to $M_L=2, 1, 0, -1, -2$; $L=1$ to $M_L=1, 0, -1$; $L=0$ to $M_L=0$. We have just the right number of states of each M_L in the table. Similarly, the two s's would lead to $S=1$ or 0, having $M_S=1, 0, -1$ for $S=1$, $M_S=0$ for $S=0$, respectively. Again we have just the right values.

Hund first established this sort of parallelism between the two ways of describing the independent states with which we were dealing. But then he

Table 4-1. Coupling of Two p Electrons

m_{l1}	m_{l2}	M_L	m_{s1}	m_{s2}	M_S
1	1	2	1/2	1/2	1
1	0	1	1/2	$-1/2$	0
0	1	1	$-1/2$	1/2	0
1	-1	0	$-1/2$	$-1/2$	-1
0	0	0			
-1	1	0			
0	-1	-1			
-1	0	-1			
-1	-1	-2			

noticed that if we set up the 36 combinations, we have a good many of them which contradict the exclusion principle, provided both electrons have the same principal quantum number as well as the same azimuthal quantum number l. Such cases are known as equivalent electrons. For example, the first set of m_l's in the table, with the first set of m_s's, gives both electrons the same m_l and m_s, and if they are equivalent they have the same n (principal quantum number) and l. Thus according to Pauli they are excluded. This immediately excludes certain possible combinations, leading to the exclusion of certain sets of L and S. The simplest example comes, not for two equivalent p electrons, which our table illustrates, but for two equivalent s electrons. For that case all m_l's are zero, and we do not have the first half of Table 4-1, but the second half is unchanged. We should then have to exclude the first and last cases: $m_{s1} = m_{s2} = \frac{1}{2}$ and $m_{s1} = m_{s2} = -\frac{1}{2}$, leading to $M_S = 1$ and -1, respectively. Also certain other combinations are identical with each other. In the case of the two s electrons, the two combinations $\frac{1}{2}, -\frac{1}{2}$ and $-\frac{1}{2}, \frac{1}{2}$ simply refer to one electron with $m_s = \frac{1}{2}$, one with $m_s = -\frac{1}{2}$. The electrons have identical values of n, l, and m_l, and cannot be distinguished from each other, so that there is no distinction between these two cases. Hence the two equivalent s electrons lead only to the multiplet with $S = 0$, and that with $S = 1$ is excluded. In the spectroscopic notation devised by Russell and Saunders, in consultation with other spectroscopists, the first is called ^1S, which is allowed, and the second is ^3S, which is excluded; these are the singlet and triplet S states, respectively.

Similarly, for two equivalent p electrons it is not hard, when the whole table of 36 combinations is written down, to cross out those that are excluded by Pauli's principle. There are six of these: The quantum numbers of the first electron can take on any of the six possibilities, but those of the second must be equal to those of the first. Of the remaining 30 combinations, just half, or 15, are independent, the other 15 being duplicates of these on account of the identity of the electrons. It is easy to verify that the 15 surviving combinations lead to states which, in the Russell-Saunders notation, are ^3P (triplet P, with $S = 1, L = 1$), ^1D (singlet D, with $S = 0, L = 2$), and ^1S (singlet S, with $S = 0, L = 0$). It is more complicated, but not in principle more difficult, to work out similar results for equivalent d or equivalent f electrons. Hund did this, and the remarkable fact is that the experimental spectroscopists found that just those states allowed by the exclusion principle to exist were found in the spectra, while those predicted to be missing were not found. It was clear that this was a correct analysis of the situation, and as we have stated, it was needed to supplement the simple predictions of the Russell-Saunders coupling.

Thus we close the discussion of the doings of the momentous year 1925. The stage was set for Schrödinger's monumental papers which came out during 1926. Kemble and I and the students in our seminar had been barely able to keep up with the progress during the year. It is rarely that a subject that seemed almost completely mysterious at the beginning of a year is all cleared up by the end. Hund went on to write a book on line spectra of the elements, which did not appear until 1927, after Schrödinger's work, but which represented work done almost entirely before wave mechanics. I wrote a very prosaic paper on the dynamics of the coupling of the electrons in the atom, the sort of thing that I had been using in lecture notes for the students in the seminar. But I was very glad, later in 1929, that I had studied the complex spectra as soon and as thoroughly as I had, for it was in 1929 that I worked out the determinantal method for incorporating these results into wave mechanics, as will be described in Secs. 8, 9, and 10. I found much of my work already done for me, in my seminar notes of 1925.

5. 1926, Schrödinger's Equation

In Sec. 2, I described how de Broglie's work on wave mechanics led Schrödinger to undertake to put his ideas into more mathematical language. Of course, by the time he came to work through this problem in 1926, Schrödinger was well acquainted with Heisenberg's matrix mechanics, which had come out in the preceding year. Thus he was able in his work to incorporate both points of view. Schrödinger's work is so familiar that if I were to go into it in detail, I would be merely duplicating any existing text on wave mechanics. And here I do not have as much personal knowledge of the situation as I had with some other parts of the history I have been describing. Consequently I cannot outline the directions of Schrödinger's thought. In his papers, he indicates a line of argument closely allied to work of Sir William Rowan Hamilton, famous Irish astronomer and mathematician, in the 1830s. Hamilton was the first to point out the close analogy of the theories of optics and of mechanics. But Schrödinger then tied his work in with simpler and more familiar arguments. One can guess with fair plausibility how he must have come to the many aspects of wave mechanics which he covered in the great series of papers which came out in 1926.

His primary object was to incorporate de Broglie's ideas into the general framework of mathematical physics. Whenever a mathematician deals with any sort of waves, he meets a wave equation. A very simple example is that met with the vibration of a string. If u is the amplitude of the transverse displacement of the string, at a distance x along the string and at a time t, it was worked out in the 1700s that the differential equation determining u as a function of x and t is

$$\frac{\partial^2 u}{\partial x^2} - \frac{1}{v^2}\frac{\partial^2 u}{\partial t^2} = 0 \qquad (5\text{-}1)$$

Here v is the velocity of propagation of the wave along the string. Often one is interested in solutions that are sinusoidal functions of time, having a factor $\exp(2\pi i\nu t)$ or $\exp(i\omega t)$, where $\nu = \omega/2\pi$ is the frequency of oscillation. If one has a wave with this frequency, with a sinusoidal dependence on x, its wavelength λ will be related to the velocity v and frequency ν through the relation $\nu = v/\lambda$. If one used such a form and substitutes u as a function of x times $\exp(2\pi i\nu t)$ into Eq. 5-1, one gets an equation for the space dependence of u. Since in this case $\partial^2 u/\partial t^2 = -\omega^2 u = -4\pi^2\nu^2 u$, we have

$$\frac{d^2u}{dx^2} + \left(\frac{2\pi}{\lambda}\right)^2 u = 0 \qquad (5\text{-}2)$$

where now u represents merely the space-dependent part of u. This form of the equation is particularly valuable, for it can be shown that it holds even for a string whose density varies from one point to another. In this case, the wavelength λ, which is determined in terms of the tension in the string and its density, is a function of x, and yet, as we have just mentioned, Eq. 5-2 still proves to apply.

This might have been a natural point from which Schrödinger could have started. For he was interested in the de Broglie wave whose wavelength was given by h/p, where p was the momentum of the particle, and he needed the case where this varied from point to point. If one is considering a particle of mass m, its kinetic energy is $p^2/2m$. If it is located at a point where the local potential energy is V, and if its total energy is E, we have $E = p^2/2m + V$, so that $p = [2m(E-V)]^{1/2}$. The quantity $2\pi/\lambda$ appearing in Eq. 5-2 is $2\pi p/h = p/\hbar$, when we remember de Broglie's assumption that $\lambda = h/p$. Thus Eq. 5-2 is transformed into

$$\frac{d^2u}{dx^2} + \frac{2m}{\hbar^2}(E-V)u = 0 \qquad (5\text{-}3)$$

This is Schrödinger's equation for a particle moving along the x axis. There is nothing more to the derivation than this.

If Schrödinger reacted as most of us would at this point, he must have been very anxious to see if he could solve his equation for a case somewhat more complicated than a freely moving particle, for which a sinusoidal function was the obvious solution. The natural case would have been the linear oscillator, a particle for which the potential energy is proportional to x^2. Here he would at once meet a differential equation which was not treated in the elementary books on mathematics. Either he would have had an unusually thorough course in differential equations, in which case he would have found that this equation had been discussed many years ago and that its solution was known, or he would have gone to an erudite mathematical colleague who would have known of this solution.

In either case he would have found that if one sets up u as a product of a so-called Gaussian function $\exp(-ax^2)$, where a is a properly chosen constant, and a power series in x, one can find the solution. It is not very hard when one knows the trick, and in a few years every elementary text on wave mechanics would be using this problem as an example. It would then appear that for most values of the energy E, the function represented by the power series increased for large x faster than the decreasing function $\exp(-ax^2)$ could compensate for the increase, and u would go to infinity for infinite x. However, it would appear that for certain definite values of E, the function u went to zero at infinity. The mathematicians called these values "Eigenwerte" in German, and the English and Americans soon began calling them eigenvalues (though the mathematics professors in Britain and America for years had had perfectly good English names for them, such as proper values or characteristic values). And these eigenvalues or characteristic values proved to be $(n+\tfrac{1}{2})h\nu$, where n was an integer.

Presumably at this point Schrödinger would have given some thought to the meaning to be ascribed to u. This quantity, which was immediately christened the wave function, was the amplitude of the de Broglie wave, whose square, or the intensity of the wave, should measure the probability of finding the particle at a given point of space, if de Broglie's interpretation was to be believed. Hence one could not use the solutions which went to infinity at infinite values of x. They would indicate an infinite probability of finding the particle at infinite distance. Instead, one would use the eigenfunctions (again the German word for the wave function which goes with one of the eigenvalues), which gave zero probability of finding the particle at infinity, and hence demanded that it be found at a finite part of space. The energy, then, had to equal one of the eigenvalues. And these energies, $(n+\tfrac{1}{2})h\nu$, were precisely the values which had been found by Sommerfeld's quantum condition for the linear oscillator, except that Sommerfeld's formula was $nh\nu$ instead of $(n+\tfrac{1}{2})h\nu$. Since only energy differences appeared in any application which had been made of this formula, there was no reason to think that the half-quantum numbers which wave mechanics gave were not reasonable.

After succeeding with this example, one would suppose that Schrödinger would have been consumed with curiosity to find what the method would give for the hydrogen atom. Here, of course, one has a three-dimensional problem, and the obvious thing would be to replace the derivative d^2u/dx^2 in Eq. 5-3 by its three-dimensional form

$$\nabla^2 u = \frac{\partial^2 u}{\partial x^2} + \frac{\partial^2 u}{\partial y^2} + \frac{\partial^2 u}{\partial z^2}$$

This would convert Eq. 5-3 into a type of three-dimensional wave equation which the erudite mathematical colleague would again recognize as something that had been solved many years before. It was known that in any such wave equation with a spherically symmetrical potential, such as the coulomb potential proportional to $1/r$ which one finds in the hydrogen problem, the wave function can be written as the product of a function of

r, $R(r)$, times a spherical harmonic $Y_{lm}(\theta,\phi)$ of the angles θ,ϕ in spherical polar coordinates. The spherical harmonics had been studied in great detail, and the differential equation for $R(r)$ was no more complicated to solve than that for the linear oscillator.

In this case, the solution was known to be the product of an exponential $\exp(-ar)$, where a was properly chosen, and a power series in r. Here again, at least for negative values of the energy E, the solution remained finite for infinite r only for certain eigenvalues of E, and these proved to be precisely the values of energy which had been given by Bohr's theory. The integers l and m played precisely the same role as the corresponding azimuthal and magnetic quantum numbers in the older quantum theory. About at this point, one would suppose that Schrödinger, if he had been a demonstrative man instead of the rather quiet middle-aged professor that he was, would have flung his hat in the air, shouted Eureka!, and would have concluded that he had found the secret of wave mechanics.

Before we go on to the many amplifications of the theory which were to follow before all of the 1926 papers were written, it is worthwhile to report one quaint fact. In spite of the fact that Schrödinger had been led to his derivations through de Broglie's parallelism between waves and particles, Schrödinger never really believed in the coexistence of the two. His preconceptions about this question were as stubborn as Bohr's had been in 1924. He believed only in the waves, and felt that the square of the amplitude u directly measured the charge density of the electron, which he thought of as being continuously distributed in space. It was Max Born who, later in 1926, published two papers on collision problems by wave mechanics, and who insisted on the probability relation between the probability of finding the particle at a given point of space and the square of the amplitude of the wave function.

In either case, there was a simple consequence relating to the wave function which arose from its physical meaning. A linear differential equation like Schrödinger's has the property that any constant times a solution is itself a solution. Hence one must determine a multiplicative constant in some definite way. The obvious way to do this was to demand that the integral of the square of the wave function over all space, $\int u^2 dv$, should be unity. This is called the normalization condition. If one adopted the probability relation of de Broglie and of Born, this would mean that the total probability of finding the electron somewhere in space is unity. On Schrödinger's view that u^2 directly measured the charge density, he would use the normalized u times the value of the electronic charge to get the continuously distributed charge density of the electron.

Schrödinger by no means stopped when he had derived the equation which bears his name. There were many questions which arose from the postulate that the information about the electronic system could be found from a wave function which depended on the coordinates, but not directly on the velocities or momenta of the particles. How would one find the momentum from this wave function? And what was the relation between the methods Schrödinger was working out and the matrix mechanics which Heisenberg had been working out in the preceding year? Here again the author does not know how Schrödinger's mind was working, but some shrewd guesses can be made.

It does not take very much ingenuity to rewrite Eq. 5-3, or its three-dimensional form in which d^2u/dx^2 is replaced by $\nabla^2 u$, in the form

$$-\frac{\hbar^2}{2m}\nabla^2 u + Vu = Eu \qquad (5\text{-}4)$$

But as soon as we do this, we notice an analogy between it and the classical equation of conservation of energy. The kinetic energy is $p^2/2m$, or in three dimensions, $(p_x^2 + p_y^2 + p_z^2)/2m$. Equation 5-4 suggests the relation

$$\frac{p_x^2 + p_y^2 + p_z^2}{2m} u + Vu = Eu \qquad (5\text{-}5)$$

or the sum of the kinetic and potential energies, times u, equals the total energy, times u. If we try to get an interpretation of the interrelation between these two equations, we see a parallelism between p_x and the differential operator $\pm i\hbar \partial/\partial x$, where $i = \sqrt{-1}$; formally, the square of this operator is $-\hbar^2 \partial^2/\partial x^2$.

As soon as we notice this, we can try applying this differential operator itself to the wave function, rather than using the energy operator as in Eq. 5-4. The simplest case to examine is the freely moving particle, in which for a one-dimensional case the wave function u equals $\exp(2\pi ix/\lambda)$, if we choose to use the complex exponential form. If we apply the operator $-i\hbar \partial/\partial x$ to this function, we find

$$-i\hbar \frac{\partial}{\partial x}\exp\left(\frac{2\pi ix}{\lambda}\right) = \frac{h}{\lambda}\exp\left(\frac{2\pi ix}{\lambda}\right) \qquad (5\text{-}6)$$

Since according to de Broglie's assumption the quantity h/λ equals the momentum p of the electron, we see that the effect of applying this operator $-i\hbar\partial/\partial x$ to the wave function is equivalent to multiplying by the x component of the momentum. The three-dimensional case goes through equally simply. We note that if instead of the complex exponential $\exp(2\pi ix/\lambda)$ we had used $\exp(-2\pi ix/\lambda)$, we should have had to use the operator $i\hbar\partial/\partial x$, but the convention used in Eq. 5-6 is the common one. If we had used a real sinusoidal function, $\cos(2\pi x/\lambda)$ or $\sin(2\pi x/\lambda)$, the method would no longer have worked. The application of the differential operator would have changed the cosine into the sine, and vice versa. Here we begin to see that the use of complex wave functions and operators is essential in this sort of analysis.

We thus see how Schrödinger could well have been led to the concept that if one starts with a classical function of coordinates and momenta, the momentum components should be replaced in wave mechanics by differential operators. At this point, he could well have been reminded of the hitherto rather mysterious commutation rule which Heisenberg had introduced into matrix mechanics. For a momentum p and the corresponding coordinate x, we saw in Eq. 3-5 that Heisenberg had assumed the relation

$$(px)_{ij} - (xp)_{ij} = \frac{h}{2\pi i}\delta_{ij} = -i\hbar\delta_{ij} \tag{5-7}$$

It had not been obvious why the momentum and coordinate should not commute with each other. But if the momenta are to be replaced by differential operators, it automatically comes about that they will not commute with coordinates. Thus, let us write down the rather obvious equation

$$-i\hbar\frac{\partial}{\partial x}xu = -i\hbar x\frac{\partial}{\partial x}u - i\hbar u$$

$$[(p_x x)_{op} - (xp_x)_{op}]u = -i\hbar u \tag{5-8}$$

$$\text{where} \quad (p_x)_{op} = -i\hbar\frac{\partial}{\partial x} \tag{5-9}$$

Here we have started writing one of these operators in a form similar to the classical quantity, but with a subscript op (for *operator*). It begins to be really plausible that there is a relation between these differential operators and Heisenberg's matrix mechanics.

How would matrix elements, rather than operators, come into the theory? Before we can answer this, a matter of technique should be mentioned. We have stated that the wave functions u are assumed to be normalized, in the sense that $\int u^2 dv = 1$. But if we are dealing with complex wave functions, u^2 will not be real. The square of the absolute magnitude of u is u^*u, where u^* is the complex conjugate of u, the quantity that comes from u by changing i to $-i$ wherever it appears. Thus the complex conjugate of $\exp(2\pi ix/\lambda)$ is $\exp(-2\pi ix/\lambda)$, and u^*u is the product of the exponentials, or unity. This exponential function is normalized to unity over unit volume of space, if we adopt the definition of normalization given by

$$\int u^*u\, dv = 1 \qquad (5\text{-}10)$$

We shall refer to the various eigenfunctions as u_i, associated with eigenvalues E_i, and each one of these eigenfunctions is assumed to be normalized.

But in addition to this, there is a theorem which can be easily proved, and which holds for a great variety of wave equations similar to Schrödinger's equation. It is the property called orthogonality: If u_i and u_j are two eigenfunctions referring to different eigenvalues, one can prove that

$$\int u_i^* u_j\, dv = 0 \quad \text{if} \quad i \neq j \qquad (5\text{-}11)$$

We can combine Eqs. 5-10 and 5-11 in a single statement,

$$\int u_i^* u_j\, dv = \delta_{ij} \qquad (5\text{-}12)$$

where δ_{ij}, as in Eq. 5-7, is unity if $i=j$, zero if $i \neq j$.

Here we have a quantity with two indices, and it looks as if we were getting close to the concept of matrix components. If we take Eq. 5-8, and apply it to a wave function u_j, then multiply on the left by u_i^*, and integrate over the volume, we find that

$$\int u_i^* (p_x x - x p_x)_{op} u_j\, dv = -i\hbar \delta_{ij} \qquad (5\text{-}13)$$

This looks like the matrix equation 5-7 and it certainly suggests that the way to find the matrix component F_{ij} of an operator like $p_x x - x p_x$ must be according to the definition

$$\int u_i^* (F)_{op} u_j\, dv = F_{ij} \qquad (5\text{-}14)$$

This is Schrödinger's fundamental relation between wave mechanics and matrix mechanics.

The simplest case of a matrix is that of a quantity like the energy, or the Hamiltonian operator,

$$(H)_{op} = -\frac{\hbar^2}{2m}\nabla^2 + V \qquad (5\text{-}15)$$

For if Schrödinger's equation, Eq. 5-4, is satisfied, $(H)_{op} u_j = E_j u_j$. If we substitute this into Eq. 5-14, we find

$$H_{ij} = \int u_i^* (H)_{op} u_j \, dv = E_j \int u_i^* u_j \, dv = E_j \delta_{ij} \qquad (5\text{-}16)$$

Any matrix, like H_{ij} or δ_{ij}, whose matrix components for $i \neq j$ are all zero, is called a diagonal matrix. The only nonvanishing components are along the diagonal of the array, if matrix components are arranged in the usual way,

$$\begin{array}{cccc} H_{11} & H_{12} & H_{13} & \cdots \\ H_{21} & H_{22} & H_{23} & \cdots \\ H_{31} & H_{32} & H_{33} & \cdots \\ \vdots & \vdots & \vdots & \end{array}$$

On the other hand, the matrix components of a quantity like x, the displacement of a particle, involve nondiagonal matrix components. As in Heisenberg's original discussion of matrix mechanics, the diagonal matrix components, like the values E_i of the energy matrix, refer to stationary states, whereas the nondiagonal components refer to transitions. The matrix components of x are the quantities involved in finding the oscillator strengths of the virtual oscillators.

We shall not go further here into the manipulation of matrix components. It can be shown that in all respects they are identical with Heisenberg's matrix components appearing in his matrix mechanics. It is ordinarily simpler to derive them from Eq. 5-14 than by a direct manipulation, such as Born, Heisenberg, and Jordan used, though the matrix methods are rather simpler than the wave-mechanical derivations for the angular momentum matrices. We consequently ordinarily think of the matrices as something derived from Schrödinger's wave functions. But we should not forget that it was Heisenberg who first arrived at them and their significance, by methods a good deal more difficult than we now use. And one may doubt whether Schrödinger would have been able to derive the matrix theorems as promptly and completely as he did in his 1926 papers, if they had not already been worked out in 1925. Anyone who has tried to derive a complicated result or prove a complicated theorem knows that it helps immensely to know in advance what the answer is. Schrödinger had that advantage. His work did not stand by itself, but was part of the whole cooperative effort which had been going on since 1923, an effort which led by 1926 to a complete formulation of quantum and wave mechanics.

So far, we have based our discussion on a wave equation not involving the time, which we set up by analogy with the simple space-dependent differential equation for the vibrating string, Eq. 5-2. How about the time dependence? Naturally we shall hope that when this is brought in properly, it will be found that the oscillations of the wave function have the frequencies given by Bohr's frequency condition, $\nu_{ij} = (E_i - E_j)/h$. On Schrödinger's concept that u^*u represented the actual charge density, or density of the virtual oscillator's charge, we should expect to find real sinusoidal oscillations with these frequencies.

In going from Eq. 5-1 to Eq. 5-2, we assumed a disturbance whose time dependence was like $\exp(2\pi i\nu t)$. The differential equation 5-1 involving the time is a second-order differential equation, and the time dependence could equally well be $\exp(-2\pi i\nu t)$. Thus the wave function involving the time, for the ith eigenvalue, would be expected to be $u_i(x)\exp(2\pi i\nu t)$ or $u_i(x)\exp(-2\pi i\nu t)$, or a linear combination of the two. Here ν would be a frequency ν_i somehow associated with the ith state. The quantity $u_i^* u_j$ would then involve a time dependence involving a frequency $\nu_i - \nu_j$ or $\nu_i + \nu_j$. This suggests Bohr's frequency condition, but somehow the possibility of the sum must be eliminated, and only a difference of ν's, which presumably would equal E_i/h's, should be allowed. This cannot be achieved with a second-order differential equation involving $\partial^2 u/\partial t^2$. It requires instead a first derivative, $\partial u/\partial t$. Schrödinger might have justified this by the following sort of argument.

In Schrödinger's equation not involving the time, which we can write in the standard form

$$(H)_{op} u_i = E_i u_i \qquad (5\text{-}17)$$

where $(H)_{op}$ is the operator of Eq. 5-15, we are converting the three components p_x, p_y, p_z of the momentum into differential operators by the prescription of Eq. 5-9. But we must not forget the theory of relativity, according to which the three momentum components, and the negative of the energy, form the four components of a four-vector. If we are transforming p_x into $-i\hbar\partial/\partial x$, etc., why do we not transform E into $i\hbar\partial/\partial t$? ψ is the wave function involving the time, we might reasonably postulate that

$$(H)_{op}\psi = i\hbar \frac{\partial \psi}{\partial t} \qquad (5\text{-}18)$$

If then we assumed that ψ was equal to $u(x)\exp(-iEt/\hbar)$, we should find that $i\hbar\partial\psi/\partial t$ was equal to $E\psi$, and by canceling out the time dependence, we should have Eq. 5-17 for the differential equation not involving the time. Thus this postulate works out properly from that point of view. In Eq. 5-18 we have the general form of Schrödinger's equation involving the time.

This equation has an infinite number of possible solutions, one for each of the eigenvalues. We can make an arbitrary linear combination of all of these with constant coefficients, and still have a solution of Eq. 5-18. Thus a general solution of Schrödinger's equation involving the time is

$$\psi = \Sigma(i) C_i u_i(x) \exp\left(\frac{-iE_i t}{\hbar}\right) \qquad (5\text{-}19)$$

We should assume, following de Broglie, that the quantity $\psi^*\psi$, which is now a function of coordinates and time, represents the probability density, or the probability of finding the particle at point x at time t. If we followed Schrödinger, we should assume that it represented the real charge density, but it equals

$$\psi^*\psi = \Sigma(i,j) C_i^* C_j u_i^*(x) u_j(x) \exp\left[\frac{i(E_i - E_j)t}{\hbar}\right] \qquad (5\text{-}20)$$

This involves just the frequencies $(E_i - E_j)/h$, as in Bohr's frequency condition. We have obtained the correct time dependence in Eq. 5-18 by our simple argument based on relativity. It could well be that Schrödinger followed some such line of thinking in arriving at this general equation.

On either de Broglie's or Schrödinger's interpretation of the wave function, it is very interesting to find the average value of some quantum-mechanical operator. We start with the value of this quantity for a given value of x, and use $\psi^*\psi$ as a weighting function to get at the average. We know that the way to find the effect of an operator $(F)_{op}$ on ψ is simply to set up $(F)_{op}\psi$, and if we multiply this by ψ^* on the left [to indicate that $(F)_{op}$ acts on ψ, but not on ψ^*], we should be led to the result that

$$(F)_{av} = \int \psi^* (F)_{op} \psi \, dv = \Sigma(i,j) C_i^* C_j \int u_i^*(x)(F)_{op} u_j(x) \, dv \exp\left[\frac{i(E_i - E_j)t}{\hbar}\right]$$

$$= \Sigma(i,j) C_i^* C_j F_{ij} \exp\left[\frac{i(E_i - E_j)t}{\hbar}\right] \qquad (5\text{-}21)$$

where F_{ij} are the matrix components defined by Eq. 5-14.

We see that the results are quite different, depending on whether F is an operator leading to a diagonal matrix, like the energy, or is an operator having only nondiagonal matrix compoents. For a diagonal matrix, we have

$$(F)_{av} = \Sigma(i) C_i^* C_i F_{ii} \qquad (5\text{-}22)$$

in which the time dependence drops out. Hence the average of such a quantity as the energy is made up of the diagonal matrix components, in this case the energy values. The quantities $C_i^* C_i$ which appear in the average have the nature of weighting factors, giving the probability that a system is in the ith stationary state. For a nondiagonal matrix, Eq. 5-21 leads directly to oscillating probability density, or oscillating charge, with the correct quantum-mechanical frequency for a transition between the ith and jth states.

There is one property of the matrix components of ordinary mechanical operators which we shall merely mention, though we shall not give the proof. For such matrix components F_{ij}, we have what is called the Hermitean property: For a Hermitean operator,

$$F_{ij}^* = F_{ji} \qquad (5\text{-}23)$$

Under these circumstances, $(F)_{av}$ can be easily shown to be real, whereas otherwise it might have imaginary components.

We have now given enough of the general description of wave mechanics and its relation to matrix mechanics so that we can go on further. Textbooks on wave mechanics give much more detail in the matter of the theorems than we have given here, and Schrödinger gave many more details. We shall, however, meet further details during the course of the book. There is just one theorem, easily proved from what we have already found, which is of great practical value, the variation principle: If we compute the average value of the energy, $\int \psi^* (H)_{op} \psi \, dv$, for any arbitrary wave function ψ, the lowest possible value will come for the ground state of the system. The proof follows at once from Eq. 5-22. If the eigenvalues of the system are $E_1, E_2, \ldots$, arranged in order of ascending energy, the average energy will be $\Sigma(i) C_i^* C_i E_i$. Clearly the lowest possible value is found when $C_1 = 1$, all other C_i's are zero, so that the wave function is the lowest eigenfunction u_1.

Schrödinger made a number of applications to various problems in his 1926 series of papers. For instance, he discussed the interaction of an atomic system with radiation, the modifications made in wave mechanics by a relativistic treatment, and a number of other points. We shall postpone these until we take up these problems separately and discuss the work of other later workers. We shall go in the next sections to discuss a number of very important problems, starting with the one which for our purposes is the most important, the many-body problem, including the exclusion principle, which was largely neglected by Schrödinger in his papers. And instead of treating each year and its progress separately as we have been doing for the crowded years 1923 to 1926, we shall consider the evolution of each of these problems over a period of several years. Great progress was made during the years from 1927 to 1932, but the pace was not quite as hectic as it had been from 1923 to 1926.

Before we start these discussions, however, it is only proper to say something about the work of Dirac. During 1926 Dirac published papers that were as numerous and as important as Schrödinger's. He arrived at many of the same results, but by much more elusive reasoning, and he expressed them in much less comprehensible language. There seem to be among theoretical physicists two quite different types of thinkers. One type is the prosaic, pragmatic, matter-of-fact sort, who indicates the argument behind what he does, and tries to write or speak in the most comprehensible manner possible. The other is what we might call the magical, or hand-waving type, who like a magician, waves his hands as if he were drawing a rabbit out of a hat, and who is not satisfied unless he can mystify his readers or hearers.

The reader of this book will have realized already that the author is of the prosaic, matter-of-fact type. I believe that Heisenberg and Schrödinger were the same, and this is why I have been able to indicate, or guess at, the kinds of arguments they must have used in arriving at their results. Dirac is definitely of the other school. He tries to present beautiful results which stand on their own, without any suggestion of the scaffolding which had to be used in building them up. It is only natural that I have been able to derive much less benefit from his work than from the work of those authors whose arguments I can follow better. When I come to discoveries, and there are many of them, where Dirac has anticipated others, I shall not hesitate to describe what he did. But I do not feel that for the purposes of this volume a detailed description of his methods would be worthwhile.

6. 1926–1927, The Two-Electron Atom

Schrödinger, as I mentioned, paid very little attention to the many-body aspect of wave mechanics in his 1926 papers. Yet the most important result of his work, from the point of view of the properties of atoms, molecules, and solids, is that it has proved to give what is apparently the exactly correct method for handling their properties, at least to the approximation in which electrostatic forces only are involved. We shall take up that approximation now, postponing until Secs. 17 and 18 the discussion of magnetic forces, relativistic effects, and so on. In work during 1926 and 1927, followed by more elaborate calculations during the subsequent years, various workers studied the two-electron systems, the helium atom and the hydrogen molecule, in great detail. The striking success of this work convinced scientists that Schrödinger's methods were capable of handling these problems with great accuracy. It was this work, more than anything else, that convinced the scientific world that we now had the theoretical basis for a complete treatment of ordinary matter.

In the discussion of Schrödinger's equation in the preceding section, I described the method as if for a single particle. It is an almost trivial extension to write down the equation for a problem with any number of particles. Each of the particles has three coordinates, x, y, z, so that with N particles there are $3N$ space coordinates and the time. The wave function ψ must be a function of all these variables. Schrödinger's equation would still be written in the form of Eq. 5-18, but now the Hamiltonian operator $(H)_{op}$ would involve the sum of the kinetic energy operators $-\hbar^2\nabla^2/2m$ for all particles, and the potential energy would include not only the sum of the potential energies of the electrons in the field of all nuclei, but also the repulsive potential energies arising from electrostatic interactions between each pair of electrons. To write this down is trivial, but to interpret it involves new concepts, and to solve the resulting differential equation proves, as I intimated in earlier sections, to be an extremely intractable problem. However, reasonable approximations for the two-electron problem are not hard to carry out.

First, what is the interpretation of a wave function ψ which is a function of the coordinates of all N particles, as well as time? It is hard to see what the interpretation would be according to Schrödinger's view that $\psi^*\psi$ measured the charge density, and that is probably the reason why Schrödinger avoided the problem. However, with the probability interpretation most workers have preferred, the meaning is straightforward. One assumes that

$$\psi^*(x_1 y_1 z_1 x_2 \cdots z_N, t)\psi(x_1 y_1 \cdots z_N, t)\, dx_1 dy_1 \cdots dz_N \qquad (6\text{-}1)$$

gives the probability of finding the coordinate x_1 in the range dx_1, etc., up to finding the coordinate z_N in dz_N, at time t. This is the sort of probability that one is used to in the study of statistical mechanics, where one often uses a many-dimensional phase space, with three coordinates and three momenta for each particle.

It is well to stop and ask what we mean by probabilities, for this is central to the whole interpretation we are using for wave mechanics. Since the time of Willard Gibbs, in the latter part of the last century, we have been familiar with the idea of an ensemble. This represents a collection of many repetitions of the same experiment, agreeing as concerns the large-scale or macroscopic properties which we can control, but taking on all different values of those microscopic properties which are on such a small scale that we cannot experimentally determine or control them. It does not necessarily imply a system with a great many particles in it. The essence of the ensemble is the large number of repetitions of the experiment. One can then follow the way in which the ensemble develops as time goes on, and hence can find the way various average quantities change with time. The probability of finding certain coordinates in certain ranges, as we have introduced it in the preceding paragraph, means simply the fraction of all systems in the ensemble which lie within the prescribed limits.

In classical statistical mechanics, the only way to find out how the ensemble varies with time is to solve for the motion or orbit of many different systems in the ensemble, and to construct probability functions giving the fraction of systems in the ensemble within a certain range at any given time. There are certain powerful theorems, the most important one being Liouville's theorem, which help us with this calculation. In wave mechanics, on the contrary, the quantity $\psi^*\psi$ gives the probability distribution of a system with the wave function ψ, and Schrödinger's equation gives directly the time rate of change of ψ, and hence of $\psi^*\psi$. Thus in a sense wave mechanics leads more directly to a proper interpretation of the time evolution of the ensemble than we are able to give in classical mechanics.

Classical mechanics allows us to set up an ensemble whose systems are all arbitrarily close to a given point in phase space. Thus we can close in on an individual system of exactly determined properties, and can find how it moves. Wave mechanics cannot do this. We can set up a wave function ψ in the coordinate space $x_1 \cdots z_N$ which is as concentrated as possible, but this is necessarily a function which satisfies Schrödinger's equation, a form of wave equation. Waves are inherently limited by the diffraction phenomenon. If we try to set up a wave function ψ, at $t=0$, which is very sharply limited in space, equal to a large value when the x's are in the close neighborhood of a given point, and zero outside this neighborhood, it can be done. It is easy to show how to set up ψ as a function of time, if its value at $t=0$ is given. But we find that the wave function will very promptly spread out, and give a probability describing a situation where the various systems of the ensemble rapidly move to different points of space. This implies that they must have been going with a wide variety of velocities. This is an elementary example of the uncertainty principle, which as we have already mentioned Heisenberg discussed in 1927.

This principle, which can be demonstrated by very simple means from diffraction theory, states, that if one of the coordinates q_i is limited to a range Δq_i in the ensemble, all values of q_i outside this range being excluded from the ensemble, then the momenta p_i will prove to be filling up a range Δp_i, where

$$\Delta p_i \Delta q_i \geq h \qquad (6\text{-}2)$$

A solution of Schrödinger's equation, concentrated in this way at $t=0$ and allowed to develop according to wave mechanics as time goes on, is called a wave packet. As Heisenberg has shown, no possible physical method of preparation of an exsemble of systems can simultaneously localize the coordinates and momenta more sharply than is allowed by the uncertainty principle.

Subject to the uncertainty principle, however, is a very important theorem relating wave mechanics and classical mechanics. This theorem, whose truth was mentioned by Schrödinger, was proved by Paul Ehrenfest in 1927. Ehrenfest, a modest and diffident man of an older generation, felt that he could hardly keep up with wave mechanics enough to contribute anything that the young workers did not already know. Nevertheless he gave a simple proof that the center of gravity of a wave packet, developing according to Schrödinger's equation, moves precisely as a system would in classical mechanics. It is this theorem which allows us most directly to show how classical mechanics follows from wave mechanics, subject only to the restriction that the wave packet cannot be made smaller than is allowed by the uncertainty principle.

Now that we have some understanding of the physical meaning of the wave function ψ, let us ask what form it is going to have for a many-electron problem. The work during the years 1926 and 1927 was limited almost completely to the two-electron problem, and that means the helium atom or the hydrogen molecule. Late in 1926, Heisenberg published the first of a series of three papers, extending through 1927 into 1928, which laid the foundation for our whole understanding of the wave mechanics of the many-body problem. The first paper discussed the general problem, the second the application to atoms with two electrons, and the third ferromagnetism. It is the first two of these extremely important papers which concern us in this section. Many of the same ideas appear in Dirac's papers of the same period, but our discussion will follow more closely the methods of Heisenberg.

Suppose we have a two-electron problem like the helium atom, and have a wave function ψ. It will be a function of the coordinates $x_1 y_1 z_1$ of the first electron, and of $x_2 y_2 z_2$ of the second, as well as of time. For purposes of abbreviating the notation, let us write this as $\psi(x_1, x_2, t)$. If we have been able to solve Schrödinger's equation and find such a solution, we notice that there must be another function, $\psi(x_2, x_1, t)$, which would be equally valid, since the two electrons are identical with each other. The second differs from the first only in that the electrons have changed places. If we wish to symbolize the situation in terms of a single variable x_1 instead of the three quantities $x_1 y_1 z_1$, we can show x_1 as the axis of abscissas, and x_2 as the axis of ordinates. Then the operation of interchanging x_1 and x_2 is equivalent to reflection in the 45° diagonal line, whose equation is $x_1 = x_2$. This is called a symmetry operation, an operation which leaves the energy, or Hamiltonian function, unchanged, since as far as the energy is concerned, it is immaterial which of the two electrons is at x_1, and which at x_2.

Now there is a simple theorem which proves that if we have such a simple symmetry operation, in the nature of a reflection, all solutions of the differential equation must be either symmetric or antisymmetric with respect to the reflection. If we consider two points which are mirror images of each other in the 45° diagonal, a symmetric function will have identical values at these two points, whereas an antisymmetric function has equal numerical values, but opposite signs. Since this theorem is quite general, we can be sure from the beginning that the solutions of Schrödinger's equation for the helium atom must have this property of being either symmetric or antisymmetric. Heisenberg proceeded to examine the consequences of this prediction.

He started by disregarding the coulomb repulsion, proportional to $1/r_{12}$, where r_{12} is the distance between the electrons, which will act between the electrons. The potential energy of the two-electron problem is the sum of this coulomb repulsion and the attractions of each electron for the nucleus. If we generalize slightly, and think of the K shell of an atom with atomic number Z, rather than the helium problem with $Z=2$, these two attractive terms in the potential energy are proportional to $-Z/r_1$ and $-Z/r_2$, respectively, where r_1 and r_2 are the distances of the two electrons from the nucleus. It is obvious that though the coulomb repulsion $1/r_{12}$ is of the same order of magnitude as these attractive terms for helium, it becomes less important proportionally as we go to a high atomic number. Heisenberg's approximation of disregarding the repulsion thus becomes very reasonable in the K shell of a heavy atom, though it is a poor approximation for helium.

If the potential energy were just $-Z/r_1 - Z/r_2$, in the proper units (ordinarily called atomic units), we could solve Schrödinger's equation for the two-electron atom at once. For then the Hamiltonian for the two-electron problem would be just the sum of Hamiltonians for the one-electron problems of an electron moving in the field of a nucleus of charge Z units. Schrödinger's equation would become

$$(H_1 + H_2)_{op}\psi = i\hbar \frac{\partial \psi}{\partial t} \qquad (6\text{-}3)$$

where H_1 would depend on the coordinates of electron 1, H_2 on the coordinates of electron 2. We can then carry out what is called a separation of variables: The wave function ψ of the two-electron problem can be written as the product of a function $u_a(x_1)$ (where we shall use x_1 to stand for x_1, y_1, z_1) and a function $u_b(x_2)$ and a time-dependent factor $\exp(-iEt/\hbar)$.

The operator H_1 in Eq. 6-3 will operate only on the factor $u_a(x_1)$, the operator H_2 only on $u_b(x_2)$, and the operator $i\hbar \partial/\partial t$ only on the time-dependent factor. Thus when we substitute our product function into Eq. 6-3, we find

$$[(H_1)_{op} u_a(x_1)]u_b(x_2)\exp\left(\frac{-iEt}{\hbar}\right) + [(H_2)_{op} u_b(x_2)]u_a(x_1)\exp\left(\frac{-iEt}{\hbar}\right)$$

$$= E u_a(x_1) u_b(x_2) \exp\left(\frac{-iEt}{\hbar}\right) \qquad (6\text{-}4)$$

The next step in this process of separation of variables is to divide both sides of Eq. 6-4 by the function $u_a(x_1)u_b(x_2)\exp(-iEt/\hbar)$. We then have

$$\frac{1}{u_a(x_1)}(H_1)_{op} u_a(x_1) + \frac{1}{u_b(x_2)}(H_2)_{op} u_b(x_2) = E \qquad (6\text{-}5)$$

Here we have the statement that the sum of two functions, one depending only on the coordinates x_1 of the first electron and the second only on the coordinates x_2 of the second electron, must add to a constant E. This cannot be satisfied for all values of x_1 and x_2 unless each of these quantities is a constant. Let these constants be E_a and E_b, respectively. Then we have

$$\frac{1}{u_a(x_1)}(H_1)_{op}u_a(x_1) = E_a, \quad \text{or} \quad (H_1)_{op}u_a(x_1) = E_a u_a(x_1)$$

$$\frac{1}{u_b(x_2)}(H_2)_{op}u_b(x_2) = E_b, \quad \text{or} \quad (H_2)_{op}u_b(x_2) = E_b u_b(x_2)$$

(6-6)

$$\text{with} \quad E_a + E_b = E \qquad (6\text{-}7)$$

The two equations 6-6 are simply those for the hydrogen-like problem of an electron in the field of a nucleus of Z units of charge, so that the solution is known. The energies E_a and E_b are hydrogenic energies, equal to $-Z^2/n^2$ rydbergs, if n is the principal quantum number. The subscripts a and b denote the quantum numbers n_a, l_a, m_{la} corresponding to one stationary state of the hydrogenic problem, and those of another state b. The wave function $u_a(x_1)u_b(x_2)$ is the wave function not involving the time which we have found for our approximate solution of Eq. 6-3. But as Heisenberg pointed out, a solution $u_b(x_1)u_a(x_2)$ is equally legitimate, and corresponds to the same energy $E_a + E_b$. These two states thus are degenerate with each other.

In case we have two degenerate solutions of Schrödinger's equation, it is simple to show that any linear combination of these solutions is itself a solution with the same eigenvalue. Thus we would be allowed to make the symmetric linear combination $u_a(x_1)u_b(x_2) + u_b(x_1)u_a(x_2)$, and the antisymmetric linear combination $u_a(x_1)u_b(x_2) - u_b(x_1)u_a(x_2)$, as Heisenberg suggested. In this limiting case where we are entirely disregarding the term $1/r_{12}$ in the Hamiltonian, we would not be required to use these particular linear combinations. However, Heisenberg went on to show that if the interaction energy between the electrons is included, we must use the symmetric or antisymmetric combinations.

We can carry out such a proof by setting up an arbitrary linear combination

$$u = A u_a(x_1)u_b(x_2) + B u_b(x_1)u_a(x_2) \qquad (6\text{-}8)$$

where A and B are coefficients to be determined. We then apply the variation principle, varying A and B to get the lowest possible value of the energy. We can derive a general expression for the energy, which is

$$\int u^*(H)_{op} u \, dv = E_a + E_b + (ab/g/ab) + 2AB(ab/g/ba) \tag{6-9}$$

where we use the abbreviations

$$(ab/g/ab) = \int u_a^*(x_1) u_b^*(x_2) \frac{1}{r_{12}} u_a(x_1) u_b(x_2) dv_1 dv_2$$
$$(ab/g/ba) = \int u_a^*(x_1) u_b^*(x_2) \frac{1}{r_{12}} u_b(x_1) u_a(x_2) dv_1 dv_2 \tag{6-10}$$

These abbreviations are not denoted by the same symbols used by Heisenberg, but fit in with notations which we shall use later for more complicated cases. Heisenberg called the integral $(ab/g/ab)$ a coulomb integral, and the integral $(ab/g/ba)$ an exchange integral. His paper was the first one in which the very important exchange integrals appeared in wave mechanics.

It is now easy to solve our problem. We wish to find the coefficients A and B which will give the lowest value of the energy of Eq. 6-9. These coefficients must be subjected to the condition $A^2 + B^2 = 1$ in order to keep the wave function of Eq. 6-8 normalized. It is not hard to show that the integral $(ab/g/ba)$ of Eq. 6-10 must be positive, and then to show that the lowest energy comes for $A = -B = 1/\sqrt{2}$. When we look at Eq. 6-8, we see that this gives the antisymmetric solution of Heisenberg. The symmetric solution, $A = B = 1/\sqrt{2}$, gives a second wave function which is orthogonal to the lowest solution, and which represents the other approximate wave function which we get by proceeding in this way.

We thus have two possible approximate wave functions

$$u = \frac{1}{\sqrt{2}} [u_a(x_1) u_b(x_2) \pm u_b(x_1) u_a(x_2)] \tag{6-11}$$

of which the one with the lower sign, the antisymmetric function, has a lower energy than the symmetric function, with the upper sign. But at this point in the argument, Heisenberg made a very significant observation: If both electrons are in the same orbital state, so that $u_a = u_b$, the antisymmetric function necessarily vanishes. Here was the key to the problem of the exclusion principle: If somehow the functions which necessarily vanished could be identified with the states forbidden by the exclusion principle, that principle would be transformed into a statement concerning the symmetry of the wave functions. Let us see how Heisenberg handled this situation.

We have seen in Sec. 4 that Pauli's principle states that no wave functions exist in which more than one electron has the same values of all four quantum numbers, n, l, m_l, and m_s. In the treatment we are giving here, the wave functions u_a, u_b involve the three quantum numbers n, l, and m_l. We have found that if these quantum numbers are the same for both one-electron wave functions, no antisymmetric function exists. The symmetric function, the only one we can have in this case, must then be the one arising from the case where one electron has $m_s = \frac{1}{2}$, the other has $m_s = -\frac{1}{2}$, which is not forbidden by the exclusion principle according to Table 4-1. In the ground state of the two-electron atom, we saw in that table that the allowed wave function was a singlet, ^1S, whereas the ^3S, which would have been allowed by the Russell-Saunders coupling scheme, was not allowed. We now see that the wave function of the ^1S state must be assumed to be symmetric in interchange of the electron's coordinates. The forbidden state, the triplet, would have the antisymmetric combination.

Heisenberg considered other states of the two-electron atom, such as the $1s2s$. Here the rules of Sec. 4 would lead to both a ^1S and a ^3S state. Heisenberg naturally assumed that since the symmetric wave function was the one for the ^1S state in the ground state of helium, we should expect to find the other singlet states with symmetric wave functions in Eq. 6-11, and the triplets with antisymmetric wave functions. The energies, as we find from Eq. 6-9, should be

$$E = E_a + E_b + (ab/g/ab) \pm (ab/g/ba) \qquad (6\text{-}12)$$

where the plus sign goes with the singlets, the minus sign with the triplets. Here was a formula which could be checked against experiment. Hund, in his study of the complex spectra, had made a generalization from experiment, which is known as Hund's rule: Of the various multiplets found from a given electronic configuration, the one with lowest energy will be the one with the largest value of both the quantum numbers S and L. Very few observed cases contradicted this rule. In this case of two s electrons, there is no quantum number L, but since as we have said the exchange integral $(ab/g/ba)$ can always be proved to be positive, Eq. 6-12 definitely predicts that the triplet states of the two-electron atoms should lie below the singlets, thus verifying Hund's rule.

Furthermore, it was not hard to compute the integrals $(ab/g/ab)$ and $(ab/g/ba)$, using the hydrogenic wave functions for u_a and u_b, and hence to get some ideas as to the total energy to be expected for the helium atom and for the energy differences between the singlets and triplets in the helium spectrum. Heisenberg carried out such calculations, and obtained reasonable agreement with experiment. We shall not go further into these points here, because later much more elaborate calculations were made, showing that the agreement with experiment gets better, the more accurately the calculations are carried out.

But there was one qualitative point which Heisenberg made, and which proved to be of great importance. In the vector coupling language of Sec. 4, the energy difference between the singlet and triplet, in such a case as $1s2s$, would come from interactions of the spin angular momentum vectors of the two electrons. No mechanism could be imagined at the time for this coupling energy to arise, except from magnetic interactions between the magnetic moments of the two spins. But these could be computed, and were thousands of times too small to explain the observations. However, the exchange integral $(ab/g/ba)$, which is responsible for the energy separation between singlet and triplet according to Heisenberg's theory, is a quantity arising from electrostatic interactions, many orders of magnitude greater than the magnetic effects, and of just the right order of magnitude to explain the observations. This was the first great achievement of the theory relating the symmetry of the atomic wave function to the spin orientation. It was several years later, however, before the way to generalize this to arbitrary atoms was worked out, and we shall come back to it in Secs. 8 and 9.

7. 1928, Hartree and the Self-Consistent Field

We have already mentioned that in the days before 1923, it had been found that quite a good account of atomic spectra could be given by assuming one or more electrons to move in a central field, their angular momentum vectors being coupled together according to the rules first worked out by Landé. Bohr's theory of the periodic system of the elements was based on this general concept. A good account had been given of the so-called Rydberg states in atomic spectra. Many years earlier, Rydberg had found that simple atomic spectra, in particular those of the alkali metals, had sequences of energy levels which could be expressed approximately by the simple rule

$$\text{energy} = -\frac{1}{(n-d)^2} \text{ rydbergs} \quad (7\text{-}1)$$

where n is an integer and d is a constant called the quantum defect for a series of levels, such as the s, p, or d series. Of course, Rydberg, who worked many years before the days of energy levels, did not express his rule in terms of energy levels, but his formula came down to the equivalent of the statement of Eq. 7-1. It was only natural to think of the central field in which the electrons moved as arising from the nuclear charge plus the field from the swarm of electrons surrounding the nucleus. To get the central field, one would have to assume a charge density of this swarm of electrons which was spherically symmetrical. Bohr and others made a distinction between several different sorts of electron orbits in such a field. First, there were the x-ray electrons, with orbits wholly inside the swarm of electrons. Second, there were the outer electrons, particularly the excited electrons in cases of optical excitation, whose orbits were partly inside, partly outside the swarm of electrons. It was these outer, so-called penetrating, orbits which led to the Rydberg terms. Bohr in 1920 had examined the energy levels of such orbits, on the basis of the old quantum mechanics, and had given a simple and straightforward proof of the approximate correctness of Eq. 7-1 in such a case. The quantum defect came from the penetration of the electron orbit into the swarm of electrons: the greater the penetration, the larger was d. A third class of orbits that were so large that they never penetrated had a quantum defect which was practically zero.

Naturally a number of scientists in the early 1920's tried to make these arguments more quantitative, investigating whether it was really possible to set up a central field in which the quantized energy levels reproduced fairly accurately the observed levels found in the alkali spectra. Two of the leading men among these scientists were Schrödinger and Douglas Hartree, a young Englishman two or three years older than I. They, and several other workers, found that in fact they could work back rather unambiguously from the observed spectra to find what the potential must be like, and hence what must be the spherical charge distribution of all the electrons inside the valence electron, or in other words of the "Atomrumpf" which the vector-coupling people were talking about. It was a rather natural question to ask: How did this distribution of charge compare with the actual charge density of the electrons in the inner orbits of the atom?

As far as I am aware, the first person who actually tried to answer this question quantitatively was R. B. Lindsay, in a paper in 1924, based on his Ph.D. thesis at MIT. Lindsay, a man who has made a distinguished career since then in other branches of mathematical physics, took the orbits found from the type of analysis I have been describing, assumed the occupancy of these orbits by electrons to be given by our knowledge of the periodic table, and then computed the amount of charge found between spheres of radius r and $r+dr$. When he compared this with the charge density required to produce the proper orbits, he found a rough agreement for the alkali metals. But of course the charge density which he found had infinitely high peaks at those particular radii characteristic of circular Bohr orbits, and other singularities at the radii giving the turning points, the perihelion and aphelion, of the other orbits. There was no indication that such singularities of charge density were really present in the potential field which would lead to the proper quantized orbits. This was the state of the problem until wave mechanics came along.

Wave mechanics of course made this picture much more plausible, because the wave functions even for a state which would have been a circular orbit on Bohr's theory were spread out in space, and gave continuous charge distributions. It was rather obvious that if one were to take the charge densities of the occupied levels, find the charge densities by assuming they were equal to the appropriate values of $u_i^* u_i$, and added to get the total charge density, the result would be much like that which was already known to be needed to lead to the proper central field. Hartree, who had never lost interest, at once started work on this problem, and his first papers involving wave mechanics, published in 1928, showed the power of his method. He defined what he called a self-consistent field: a potential field such that, if one solved Schrödinger's equation in this field for one-electron wave functions which we now refer to as orbitals, summed the densities $u_i^* u_i$ for the occupied orbitals, and computed by electrostatics the potential arising from the charge density, it would be identical with the original potential. He proceeded to calculate these self-consistent fields for a number of atoms, both light and heavy, found that the results showed a remarkable agreement with experiment, and was so impressed with the importance of the method that he made it his life work.

These papers impressed me so much that I immediately started correspondence with Hartree, beginning a friendship which lasted until his death in 1958. I first met him in 1929, on my trip to Europe in that year, and became well enough acquainted with him and his wife so that I frequently visited him at his home. He often visited Harvard and MIT, and we constantly kept up a correspondence, even during wartime, when by extraordinary coincidence we were working on the same problem in microwave radar. He came of an interesting family. His father was a retired businessman, with enough liking for mathematics so that he helped his son Douglas with his calculations on the self-consistent field, and made mathematical computation a hobby during his retirement. His mother had been at one time Lady Mayoress of Cambridge, England, in a day when it was distinctly out of the ordinary for a lady to take part in politics. Douglas Hartree was very distinctly of the matter-of-fact habit of thought that I found most congenial. The hand-waving magical type of scientist regarded him as a "mere computer." Yet he made a much greater contribution to our knowledge of the behavior of real atoms than most of them did. And while he limited himself to atoms, his demonstration of the power of the self-consistent field for atoms is what has led to the development of that method for molecules and solids as well.

There are several points of detail which should be discussed about his method. In the first place, he did not use identical spherical potentials for each orbital in the atom. He realized that in an N-electron atom, each electron is acted on only by the $N-1$ other electrons. Hence he added the charge densities of these $N-1$ other electrons to get the charge producing the potential acting on the remaining electron. Since he used different potentials for each orbital, the ordinary proof that different eigenfunctions of the same potential are orthogonal to each other did not apply in his case, and in fact there were slight nonvanishing integrals $\int u_i^* u_j dv$ between different orbitals, an inconvenient feature in making further calculations with the orbitals. This is one of the points where the Hartree-Fock method and the more recent $X\alpha$-SCF (self-consistent-field) method are superior to his original method.

Second, he assumed a spherical potential, and made a spherical average of the potential in cases where the charge density $\Sigma(i) u_i^* u_i$ was not spherically symmetrical. In this connection, we should mention an important feature of wave mechanics, called Unsöld's theorem, proved by

Albrecht Unsöld in 1927. This theorem states that if one adds the charge densities $\Sigma(i)u_i^*u_i$ for all three orbitals in a p state, corresponding to $m_l = 1$, 0, -1, or for the five in a d state, corresponding to $m_l = 2, 1, 0, -1, -2$, or more generally for all orbitals with a given n and l, the result is strictly spherically symmetric. The theorem can be proved from properties of spherical harmonics, or from general group theory. This means that for an atom all of whose electrons are in closed shells, the wave-mechanical charge density is strictly spherical. Hartree's spherical averaging affected only those few electrons which were not in closed shells. Of course, here we are using charge density in the sense simply of $\Sigma(i)u_i^*u_i$. It is this sort of interpretation of charge density which made Schrödinger's identification of $\psi^*\psi$ as a real charge density, rather than a probability, not completely without foundation.

Third, we must ask how Hartree achieved self-consistency. He did this by a method called iteration. He started with a spherical potential which was believed to be approximately correct. He solved Schrödinger's equation in that potential. Since it was a central field, the solution was known to be a spherical harmonic of angles times a radial function which could be found by numerical integration of a radial wave equation. Hartree was a master of methods of numerical calculation, and was able to devise very accurate techniques for this process. He computed the charge density $\Sigma(i)u_i^*u_i$ from the wave functions determined by numerical integration, and from its spherical average he computed a potential by electrostatic methods, again involving a numerical integration. This final potential would not agree with the initially assumed value, but proved to be closer to the desired self-consistent value than the initial assumption. Consequently it could be used to give an improved starting point for a next repetition of the same process. After enough of these iterations, the potential converged to a value which was self-consistent to any desired accuracy. Clearly this was a tedious process to carry out with a desk calculator, and it was fortunate that Hartree's father regarded such work as a diversion. But naturally Hartree kept his eyes open for all methods of mechanical computation. We shall come back to his interest in different types of computers.

When Hartree had achieved self-consistency, we should next ask what was the nature of his results, and how could they be checked against experiment. The two things he found directly from his calculations were the eigenvalues and eigenfunctions of the occupied orbitals in the atom. From the line of argument that had led to his method, it was naturally assumed that the energy required to remove an electron from the self-consistent field representing an atom should be a good approximation to the actual ionization energy of the corresponding electron in an atom.

These energies were well known experimentally. For the inner electrons they were the x-ray term values, and for the outer electrons the optical term values, and they had been studied by the experimenters for many years. The agreement between Hartree's calculated values and the experiments was uniformly remarkably good. At once it was possible to get good theoretical values not merely for the few lightest atoms, like helium which was as far as Heisenberg had gone, but for the heavy atoms as well. Hartree used rubidium as one of his examples in his first paper.

As for the eigenfunctions, they allowed one to compute the total charge density of the atom as a function of radius. This could also be found experimentally. The treatment of x-ray scattering allowed one, from the intensity of scattered radiation in different directions, to make a Fourier transform which gave a direct calculation of the charge density. It should be understood that this scattered radiation was what is called by the x-ray workers the coherent radiation, not the Compton-modified radiation which was concerned in the Compton effect. This distinction between the two types of scattering and other features of x-ray scattering were worked out in detail in a joint paper by Hartree and Ivar Waller, a Swedish theorist, in 1929.

In the early 1920s, the x-ray experimenters had been getting their techniques in shape to make accurate measurements of the required intensities of scattered radiation, and to carry out the Fourier transforms to get the radial charge densities in the atoms. Pioneer work had been done by W. H. and W. L. Bragg and their associates in England. Other work along the same line was done by Duane and his associate Havighurst at Harvard. Hartree worked closely with the experimenters in the comparison of theory and experiment. For example, there had been an early paper by W. L. Bragg, James, and Bosanquet in 1921, on the experimental determination of the intensities in the NaCl crystal, which was used to get the charge distribution around the sodium and chlorine atoms in that crystal. Comparisons with the calculations of Hartree showed very accurate agreement. Similar close agreement between theory and experiment was found in all cases. There was every evidence that the self-consistent-field method could give a remarkably accurate picture of the constitution of real atoms.

Nevertheless, there was something not completely satisfying about Hartree's procedure. He had arrived at it completely intuitively, using essentially the simple arguments we have been describing. It was a far cry from these methods to the sort of analysis which Heisenberg had been using for the helium atom. Could one derive Hartree's method from Schrödinger's equation for the many-body problem? I asked myself this question as soon as Hartree's first paper came out. I had been working, since 1927, on the helium atom, with results which I shall describe in Sec. 19, but my methods were quite different from Hartree's. However, it seemed to me that his method must treat the various electrons in the atom as being largely independent of each other, subject only to an overall potential field determined by electrostatics. If we really had independent electrons, we should be able to set up an approximate wave function simply as a product of the various orbitals, as Heisenberg had approximated by the function $u_a(x_1)u_b(x_2)$, only now Hartree's orbitals u_a and u_b would be more accurate than hydrogenic functions. If this were the case, we should be able to compute the average of the many-electron energy for this product function, and get an approximation to the many-electron energy of the atom. Then we should be able to do the same thing for the ion, and subtract to get an ionization ion energy.

I did this, and it proved to be the case that one could get from my analysis a quite respectable demonstration that Hartree's one-electron eigenvalues should be expected to lead to ionization energies, when used in the way I have sketched. There were small correction terms coming into the theory, depending on the approximation of spherical symmetry which Hartree had used, but in general the demonstration was satisfactory. However, I realized that one should properly use not the product function, but something more like the symmetric or antisymmetric function which Heisenberg had used for his singlet and triplet states. I published my results in the fall of 1928, but started work at once trying to look into the proper way of handling the symmetry question. This had advanced very little during the two years since Heisenberg's papers in 1926. It was in 1929 that I solved the problem, and published the paper on complex spectra which suggested the determinantal method for handling the many-electron problem. This is the topic of the next two sections.

8. 1929, The "Gruppenpest" and Determinantal Wave Functions

Following Heisenberg's papers of 1926 and 1927, several other workers—Eugene Wigner, Hund, Walter Heitler, Hermann Weyl—attempted to extend them to the case of more than two electrons. This was not as simple as one had hoped. It was easy enough if all electrons had the same value of m_s. In this case the wave function had to be one in which no two electrons could have the same values of all three quantum numbers n, l, and m_l. This case had been considered by Dirac in a paper in 1926, and had also been thought of by Heisenberg. One would wish to have a wave function with the same properties as the antisymmetric function $u_a(x_1)u_b(x_2) - u_b(x_1) \times u_a(x_2)$ of Eq. 6-11. Dirac and Heisenberg both noticed that this function could be written in the form of a determinant,

$$\begin{vmatrix} u_a(x_1) & u_a(x_2) \\ u_b(x_1) & u_b(x_2) \end{vmatrix} = u_a(x_1)u_b(x_2) - u_b(x_1)u_a(x_2) \qquad (8\text{-}1)$$

A determinant with N rows and columns (N being 2 in the case above) is by definition the following quantity. One first forms all $N!$ products of an element from the first column, one from the second, and so on, with no two elements from the same row. One then assigns to each a plus or minus sign, depending on whether it requires an even or an odd number of interchanges of columns to bring the product in question to the principal diagonal of the array. In the case above, $u_a(x_1)u_b(x_2)$ is already in the principal diagonal, and thus has a plus sign. The other product, $u_b(x_1) \times u_a(x_2)$, would be in the principal diagonal, if the determinant were written with the two columns interchanged, as

$$\begin{vmatrix} u_a(x_2) & u_a(x_1) \\ u_b(x_2) & u_b(x_1) \end{vmatrix}$$

so that it requires one interchange to bring it to the principal diagonal, and the sign is minus. The determinant by definition is the linear combination of all $N!$ products, with coefficients ± 1 determined by the rules we have stated.

The determinant has several fundamental properties. First, an interchange of two columns changes the sign of the determinant. The reason is that this involves just one more or one less interchange of columns to bring any particular product to the principal diagonal, so that the coefficient of each product changes sign. Second, an interchange of two rows changes the sign of the determinant, for there is nothing in the definition of the determinant which distinguishes the rows from the columns. Third, a determinant with two equal rows, or two equal columns, is necessarily zero. For if two rows or two columns are equal, interchanging them cannot change the value of the determinant, and yet by the theorem just stated it must change its sign. The only number whose value is unchanged when it changes its sign is zero.

This last property is required to lead to the exclusion principle. Thus, let us have N different orbitals, $u_a, u_b, \ldots, u_N$, and N electrons. Set up the determinant

$$\begin{vmatrix} u_a(x_1) & u_a(x_2) & \cdots & u_a(x_N) \\ u_b(x_1) & u_b(x_2) & \cdots & u_b(x_N) \\ \cdot & \cdot & & \cdot \\ \cdot & \cdot & & \cdot \\ \cdot & \cdot & & \cdot \\ u_N(x_1) & u_N(x_2) & \cdots & u_N(x_N) \end{vmatrix} \quad (8\text{-}2)$$

Then if the orbitals u_a and u_b, for example, are identical, the first two rows of the determinant will be equal, and the determinant vanishes. The same is true if any two of the orbitals are identical. No such determinantal function, in other words, can be set up if two identical orbitals are included in the set $u_a \cdots u_N$. This means that the exclusion principle will automatically hold for such a determinantal function, provided we are considering only electrons of one spin.

Heisenberg and Dirac recognized this property of the determinantal functions. They pointed out that it implied another property: The wave function must be antisymmetric, or change its sign, if the coordinates of any two electrons are interchanged. This is the property discussed for the two-electron case in Sec. 6. The reason is that interchange of the coordinates of two electrons will interchange the corresponding two columns of the determinant, and hence will change its sign. We see therefore that the determinantal function forms the proper generalization of Heisenberg's antisymmetric function in the two-electron case.

There is still another property of such a function, which we have not pointed out before, and which is what leads to Hund's rule. If the coordinates of two electrons approach each other, the determinant approaches zero. The reason is that when the coordinates are exactly the same, then as we have just pointed out the corresponding two columns of the determinant will be identical, and the determinant must vanish. But this means that two electrons with the same m_s have a small probability of approaching each other in space, for the square of the wave function gives the probability of finding the coordinates with particular values. This is then a straightforward result of the exclusion principle. But it leads directly to Hund's rule. In Sec. 6, the energy separation between singlet and triplet states came from the average value of the $1/r_{12}$ coulomb repulsive term in the energy of the atom. Obviously if two electrons can approach each other closely when their spins are not parallel, which is the case with the symmetric function $u_a(x_1)u_b(x_2) + u_b(x_1)u_a(x_2)$, this repulsive coulomb energy will be greater than if they are kept apart by the exclusion principle, as they are with parallel spins. Thus the parallel spin case, which will correspond to the higher value of S, will have a lower energy than the antiparallel spin case.

Heisenberg and Dirac thus had replaced the exclusion principle by the statement that the wave function for the electrons for $m_s = \frac{1}{2}$, or with spin up, as we shall express it, had to be an antisymmetric function of the coordinates, and thus a determinantal function, and that for electrons for $m_s = -\frac{1}{2}$, or with spin down, had to be another determinantal function. But this formulation left the spin dependence of the wave function in an awkward state. How are we to explain the situation that if two electrons have the same orbitals, they must have opposite spins? Somehow, we should work into the theory an antisymmetric function of the spins, as well as the antisymmetric function of the orbitals we have expressed by use of the determinantal method.

It was at this point that Wigner, Hund, Heitler, and Weyl entered the picture, with their "Gruppenpest": the pest of the group theory, as certain disgruntled individuals who had never studied group theory in school described it. The mathematicians have made a great study of what is called the symmetric group, or the permutation group. This is a study of the effect of the $N!$ possible permutations of a set of N objects. The authors of the "Gruppenpest" wrote papers which were incomprehensible to those like me who had not studied group theory, in which they applied these theoretical results to the study of the many-electron problem. The practical consequences appeared to be negligible, but everyone felt that to be in the mainstream of quantum mechanics, one had to learn about it. Yet there were no good texts from which one could learn group theory. It was a frustrating experience, worthy of the name of a pest.

I had what I can only describe as a feeling of outrage at the turn which the subject had taken. The simple process by which Heisenberg and Dirac had replaced the exclusion principle by the statement that the wave function had to be antisymmetric in the coordinates of the electrons was an obvious step in the right direction, and Heisenberg had shown that it led to an understanding of Hund's rule. Its only lack was that it dealt with only three of the four quantum numbers of an electron, namely n, l, and m_l, but disregarded the fourth. Was it not possible, I asked, to generalize this antisymmetry to a case in which the fourth quantum number m_s was included as well, and so to get the complete statement of the exclusion principle in this simple form? When I started work in December 1928, to try to answer this question, the first thing I did was to find how to incorporate the spin and its quantum number m_s into the wave function.

In 1927, Pauli had shown how this could be done. We use m_s as a coordinate as well as a quantum number. When we regard it as a coordinate, we say that the square of the wave function for a single electron should give the probability of finding an electron at a given point of space, with spin up or with spin down. The spin, which can take on only two possible values, is thus like a coordinate. But clearly m_s is also a quantum number. If we have a wave function for which m_s is $\frac{1}{2}$, regarded as a quantum number, then we shall find a wave function $u(x,y,z)$ such that u^*u represents the probability of finding the electron at x, y, z, provided its spin is $\frac{1}{2}$, but the probability of finding it anywhere with spin $-\frac{1}{2}$ is zero. Thus if we set up a function $u_{n,l,m_l}(x,y,z)u_{m_s}(m_s)$, where the subscripts are quantum numbers and the quantities in parentheses are the coordinates, we see that the function $u_{m_s}(m_s)$ must equal unity when m_s (the coordinate) equals m_s (the quantum number), but it must equal zero when m_s (the coordinate) is different from m_s (the quantum number). Pauli had defined

$$u_{1/2}(m_s) \equiv \alpha(m_s) \qquad u_{-1/2}(m_s) \equiv \beta(m_s) \qquad (8\text{-}3)$$

I did not use this notation in my paper in 1929, noting instead that the function has the properties of a δ-function. We use Pauli's notation, however, since it is familiar. We now call a quantity like $u_{n,l,m_l}(x,y,z)\alpha(m_s)$ a spin orbital.

It was then an obvious step to replace Dirac's determinantal function of Eq. 8-2 by an identical expression in which the orbitals of Eq. 8-2 are replaced by spin orbitals. This is antisymmetric in interchange of the four coordinates, x, y, z, and spin, of any two electrons. And it obeys the exclusion principle, in that it automatically vanishes if any two spin orbitals are identical in all four quantum numbers. This method, though not familiar, was not really new; it had been used by Pauling in 1928, in discussing the repulsion of two helium atoms. It is the only step which had to be taken to incorporate the spin properly into the problem; no further use of the group theory was needed.

Naturally I had to work out enough examples to see how the method would be applied, and to give generalized formulas for the coulomb and exchange integrals of the type defined in Eq. 6-10. In the paper I published on the theory of complex spectra in 1929, I did all these things, and found that the theory could then be applied far beyond anything which had been done previously. One could give a complete parallel to the theory of complex spectra which Hund had worked out in 1925, and which was described in Sec. 4. But at the same time one could give detailed formulas for the separations between the energies of the various multiplet states, in terms of integrals which could be calculated from the same sort of orbitals which Hartree had been computing. Enough examples were worked out in the paper to show that one obtained reasonable results. We shall come back in the next section to some of the details of the calculation, for it is a good way to introduce the reader to some of the mathematical techniques which we need for the many-electron problems.

As soon as this paper became known, it was obvious that a great many other physicists were as disgusted as I had been with the group-theoretical approach to the problem. As I heard later, there were remarks made such as "Slater had slain the 'Gruppenpest.'" I believe that no other piece of work I have done was so universally popular. It was fortunate that the paper had been sent in to the *Physical Review* shortly before I left for my trip to Europe in the summer of 1929, first to attend a conference in Zurich, and then to go to Leipzig with Heisenberg and Hund for a half-year with a Guggenheim Fellowship. Everyone knew of the work. Hund had spent the preceding months in the spring of 1929 at Harvard, and of course had learned of my work there, and was much pleased with it. I first met Hartree at the Zurich meeting, though I had been corresponding with him, and found that he and Waller had independently arrived at much the same approach to the problem. Felix Bloch whom I also first met there, was greatly taken with the method, and I shall speak in Sec. 16 of the paper on ferromagnetism which he almost immediately worked out, using the determinantal techniques.

When I arrived in Leipzig at the end of September 1929, Heisenberg as well as Hund proved to be much interested in the new methods, which in a very real sense were extensions of the work they had done several years earlier. I worked on applications of the method there, as I describe later in Sec. 14. Late in November many members of the Leipzig group visited Berlin, and there I had my first chance to meet Wigner, a young Hungarian physicist about two years younger than I, originally trained as a chemical engineer, and already a great expert on group theory and the nature of symmetry. I found that he had been studying my paper with much interest, that he was entirely in sympathy with it, and that in fact he had thought of doing the same thing himself, but had not got around to it. Shortly before I left Leipzig, in December, I gave a seminar on the work I had been doing there, the only lecture I ever ventured to give in German. As it happened, not only Heisenberg, Hund, and Debye, as well as Edward Teller, a young Hungarian graduate student were present, but also Wigner, Pauli, Rudolph Peierls, and Fritz London, all on their way to Christmas holidays. Pauli of course lived up to his usual reputation, and broke in with remarks to the effect that "he couldn't understand a word of it," but Heisenberg came to my aid and helped explain things, and before we were through everyone was satisfied with the state the theory was in. I had a chance to meet several other scientists there, among them Erich Hückel, the theoretical chemist, who was at the point of making the calculations on the molecules of ethylene and benzene for which he became famous. There could not have been a more opportune time for me to visit Europe and spread the news of the new method.

This may be as good a time as any for me to describe various events in my personal history which had happened about this time, and which had considerable effect not only on my future plans, but perhaps in a wider area. Karl Compton, who in 1930 was chosen president of MIT, had been for a number of years the head of the physics department at Princeton. He had tried in 1927 to get me to leave Harvard and join the Princeton department. I had turned down this offer. But in the spring of 1929, he had made a second and much more serious effort to get me to come. They had had Edward Condon at Princeton during the year 1928–1929. He was a brilliant young American theorist whom I had first met in 1926, when I was teaching summer school at Stanford and he was getting his degree at Berkeley. He went to Europe on a traveling fellowship, spending time both at Göttingen and at Munich, working on the famous Franck-Condon principle in molecular theory. Then he had gone to Princeton in 1928, presumably to fill the position I had not accepted in 1927. But then, after a year, he had agreed to go to the University of Minnesota. Van Vleck had gone to Minnesota in 1923, but was called in 1928 to his alma mater, the University of Wisconsin, where he stayed until he went back to Harvard in 1934. This left a vacancy at Minnesota in theoretical physics, and Condon was going there. This moved the Princeton people to try again to induce me to go there.

I visited Princeton, and they worked very hard to get me to come. But I was not sure that I liked the small-town atmosphere of Princeton, and in addition Harvard made strong efforts to keep me, so that eventually I decided to stay. Some of these Harvard efforts are of more than personal interest. Theodore Lyman, who had been department head when I first went there, was a wealthy bachelor from a famous Brookline family, who stood very high in the Harvard ranks. He was very anxious to see the department develop, and was trying to get the university to help with the financing of a new research laboratory. When he heard of my Princeton offer, he was able to make a good enough case with President Lowell to induce the university to promise more help toward the new laboratory than they had otherwise intended. Lyman went on to raise a large sum of money among his wealthy friends; he converted the stocks and bonds they had contributed into cash shortly before the stock market crash of October 1929, and was able to go ahead with the construction of the laboratory in the early days of the depression, when money would go further. That is now the Lyman Laboratory of Physics.

Moves were also under way in the chemistry department. By then I had been long enough at Harvard so that I knew the faculty pretty well, and James B. Conant, who was still a chemistry professor, was trying very hard to get Linus Pauling to join the department. I had known Pauling only by reputation, but I helped Conant with the entertainment of Pauling when he came to visit, and tried my best to induce him to come to Harvard. He didn't, but I got to know him, and was much impressed. And all of these activities on the part of Harvard convinced me that I would do as well at Harvard as at Princeton. All this was coming to a head just before I left for Europe in June 1929.

This stabilized my plans for the duration of the European trip, but it was not long after my return, in the spring of 1930, that the news went around that Compton was coming to MIT as the new president. This of course pleased me, and made me very glad I had not left to go to Princeton. But it came to me as a complete surprise when on May 20, 1930, Compton called on me at Harvard, and asked me to come to MIT as head of the physics department. This struck me as a very different situation from the Princeton offer, and I accepted with only a short period for consideration. In Sec. 21, I describe the steps I took with Compton's help to build up the MIT department; I felt from the beginning that I had in that way a chance to do much more for the development of physics in this country than I would have been likely to have at Harvard.

It probably was a fortunate thing that this move to a position which could have absorbed all my efforts in administrative work came at a time when I had just developed the new determinantal method. For that method opened possibilities in practically every branch of atomic, molecular, and solid-state theory. These had all been held up waiting for a manageable treatment of the many-body theory, and this was just what the method gave. I had so many ideas regarding applications of the method, some of which I worked out in Leipzig, that many of these were still in the process of being worked out when I went to MIT. Thus the reader who looks at my output of scientific papers will notice no falling off of papers at the time, as might have been anticipated. I had more students and postdoctoral workers at MIT than I had had at Harvard (I had not directed any doctoral students there), continued to teach courses, started to write books, and in general was able to make a greater scientific contribution there than I could have if I had stayed where I was.

One of the developments of the determinantal method came so promptly that I may as well mention it here. Condon had been working temporarily at the office of the *Physical Review* when the manuscript of my paper on complex spectra came to the office. He had refereed it, and had become so interested that he decided to work in that field. This was the beginning of his long interest, which culminated in the book *Theory of Atomic Spectra* by Condon and Shortley in 1935. Condon after going to Minnesota in 1929 returned to Princeton in 1930, where he stayed until he went to Westinghouse in 1937. At the same time, in 1930, Wigner moved from Berlin to Princeton. Thus those two were the mainstays of the Princeton quantum theory group, and Princeton did not lose by my having stayed at Harvard and MIT.

9.   1929, The Theory of Complex Spectra

Let us interject here a somewhat more mathematical section, giving some of the points in the theory of complex spectra which I had worked out in the 1929 paper. One starts with a set of atomic orbitals, solutions of a central-field problem, so that they can be written as products of radial functions $R_{nl}(r)$ and spherical harmonics $Y_{lm_l}(\theta,\phi)$. This atomic orbital is assumed to be normalized, and in the theory the orbitals are assumed to be orthogonal to each other. This would not be exactly true with Hartree's method, but it is with the later Hartree-Fock method, which we go into in Sec. 10, and with the $X\alpha$-SCF method, which is the subject of Book V. Then we multiply one of the atomic orbitals $u_{nlm_l}(r,\theta,\phi)$ by one of the spin functions $\alpha(m_s)$ or $\beta(m_s)$, to give a spin orbital. In the ordinary treatment of the problem we assume spin orbitals with both α spin and β spin formed from each of the atomic orbitals. With an N-electron problem, we then pick out N of these spin orbitals, and form a normalized determinantal function from them. It can easily be shown that the normalization constant must be $1/\sqrt{N!}$, so that the determinantal function is

$$\psi = \frac{1}{\sqrt{N!}} \begin{vmatrix} u_1(1) & u_1(2) & \cdots & u_1(N) \\ u_2(1) & u_2(2) & \cdots & u_2(N) \\ \cdot & \cdot & & \cdot \\ \cdot & \cdot & & \cdot \\ \cdot & \cdot & & \cdot \\ u_N(1) & u_N(2) & \cdots & u_N(N) \end{vmatrix} \qquad (9\text{-}1)$$

where $u_1,\ldots,u_N$ represent the N spin orbitals we have picked out, and $1,\ldots,N$ represent the coordinates and spins of the N electrons. It is such a determinantal function that we use in solving the problem.

As a rule, we use a linear combination of quite a number of these determinantal functions ψ to form our solution. For example, if we are dealing with a single partly filled shell of electrons, so that N is less than the number $2(2l+1)$ spin orbitals connected with a filled shell, there are many ways of picking N spin orbitals out of the $2(2l+1)$ possible ones. In fact, this number is

$$\frac{[2(2l+1)]!}{N![2(2l+1)-N]!} \qquad (9\text{-}2)$$

For s electrons, $l=0$, this gives two values for $N=1$, corresponding to a choice of α or β spin. For p electrons, $l=1$, we have six determinantal functions for $N=1$ or 5, 15 for $N=2$ or 4, and 20 for $N=3$. We must use all these determinantal functions in order to get an acceptable solution for an atom with a single partially filled shell. Our object is to make a linear combination of these functions which best represents a solution of Schrödinger's equation.

To do this, let us set up a linear combination of the determinantal functions ψ_k, in the form

$$\psi = \Sigma(k)C_k\psi_k \qquad (9\text{-}3)$$

We use the variation principle, in the form that if we compute the energy and vary it with respect to one of the C's, we must have a minimum for the correct ground state. In fact, it is easy to extend the variation principle to the form that if the energy is stationary with respect to variation of the wave functions, we shall find not only one ground state of minimum energy, but also other functions that will be good approximations to excited states.

If $(H)_{op}$ is the correct many-electron Hamiltonian operator of the system, the average energy is

$$\text{average energy} = \int \psi^* \psi \, dv = \Sigma(nk)C_n^* C_k \int \psi_n^* (H)_{op} \psi_k \, dv$$

$$= \Sigma(nk)C_n^* C_k H_{nk} \qquad (9\text{-}4)$$

where the integrations over dv indicate integrations over the x, y, z of each electron, and summation over the spins of each. The matrix components H_{nk} are the matrix components with respect to our determinantal functions ψ_k. We must recall, however, that for Eq. 9-4 to be correct, ψ must be normalized. That is, we must have

$$1 = \int \psi^* \psi \, dv = \Sigma(nk)C_n^* C_k \int \psi_n^* \psi_k \, dv = \Sigma(nk)C_n^* C_k \delta_{nk}$$

$$= \Sigma(n)C_n^* C_n \qquad (9\text{-}5)$$

on account of the assumed orthonormality (that is, orthogonality and normalization) of the ψ_k's.

Now we must vary one of the coefficients, for example C_n^*, and make the average energy stationary, but subject to the proviso that Eq. 9-5 continues to be satisfied, or that the function remains normalized. The method of doing this is called the method of Lagrange multipliers. We take the average energy plus an arbitrary constant times the integral of Eq. 9-5, and set an arbitrary variation of this quantity with respect to change of C_n^* equal to zero. That is, we have

$$\frac{\partial}{\partial C_n^*} \Sigma(nk) C_n^* C_k (H_{nk} - E\delta_{nk}) = 0 \qquad (9\text{-}6)$$

where we have called the arbitrary constant $-E$. This equation guarantees that if the partial derivative of the quantity of Eq. 9-5 is zero, which is required if this quantity is to remain equal to unity, then the partial derivative of the quantity of Eq. 9-4 must also be zero, which is just what we wish to achieve.

When we carry out the differentiation in Eq. 9-6, we find

$$\Sigma(k) C_k (H_{nk} - E\delta_{nk}) = 0 \qquad (9\text{-}7)$$

This is Schrödinger's equation in a matrix form. It can be easily proved that E is the energy. It is assumed that we are dealing with a certain definite number, say P, of determinantal functions, so that there are P coefficients in the summation of Eq. 9-3. That is, the summation in Eq. 9-7 goes from 1 to P, and there are P different equations, since n can take on any one of the P values. Here we then have P simultaneous equations for the P constants C_k. These are simultaneous linear homogeneous equations, and in general they have no nonvanishing solutions. The condition that they have solutions is that the determinant of the coefficients of the C_k's vanish. That is,

$$\begin{vmatrix} H_{11} - E & H_{12} & H_{13} & \cdots & H_{1P} \\ H_{21} & H_{22} - E & H_{23} & \cdots & H_{2P} \\ \cdot & \cdot & \cdot & & \cdot \\ \cdot & \cdot & \cdot & & \cdot \\ \cdot & \cdot & \cdot & & \cdot \\ H_{P1} & H_{P2} & H_{P3} & \cdots & H_{PP} - E \end{vmatrix} = 0 \qquad (9\text{-}8)$$

This equation is called a secular equation. Since E appears to the Pth and lower powers, it is an algebraic equation of the Pth degree for E, and has P roots. These are the eigenvalues of the problem; only for these values of E can we find solutions for the C_k's. Having obtained these energies, we solve Eq. 9-7 for the C's, and get the eigenfunctions by use of Eq. 9-3. We have enough flexibility in the C's so that they can be adjusted to give normalized eigenfunctions, which prove to be orthogonal.

This is the standard method of making a linear combination of determinantal functions to get an approximate solution of Schrödinger's equation. The ordinary complex spectrum theory assumes that we take P to be as small as possible to give a reasonable description of the multiplet levels. This value of P would be the product of numbers such as given in Eq. 9-2 for the various shells of the atom. It is noted that a closed shell contributes a factor of unity, so that it is only the partially filled shells that contribute to P. When we use the theory in this form, which is all that was considered in the 1929 paper, the resulting energy levels are qualitatively correct, but not very good quantitatively. It can be proved, however, that if we let P increase, taking more and more highly excited orbitals u_{nlm_l} into our determinantal functions until we approach what is called a complete set of orbitals, we can in principle approach the exact solution of Schrödinger's equation as closely as desired. This method, sometimes called configuration interaction, is too cumbersome to be very useful for anything except the lighter atoms. When applied to helium, however, it has now been carried to a sufficiently high accuracy that it agrees with experiment with a very small error.

The practical problem of solving a large secular equation was a very severe one in 1929, when the complex spectrum theory was first worked out. Now, however, electronic digital computers can be programmed to solve secular equations with P up to several hundred, without trouble. A large part of the earlier spectrum theory was devoted to various devices for simplifying these solutions, which are no longer as necessary as they seemed at that time. However, there is one form of simplification which is of great value. Each one of the solutions of Schrödinger's equation, in a problem such as the atomic one, has certain symmetry properties. It can then be shown that by choosing determinantal functions so that they all possess the same symmetry properties, the size of the determinant can be greatly decreased. The reason is that we need only solve a secular problem between those particular symmetry types of determinants. In our atomic case, these symmetry properties prove to be very simple: All states with the same $M_L = \Sigma m_l$ and the same $M_S = \Sigma m_s$ belong to the same symmetry type. Thus we have only secular equations between functions of the same M_L and M_S.

Let us now proceed to the practical methods we must use to find the eigenvalues. From Eq. 9-8, we need the matrix components H_{pq} of the Hamiltonian between two determinantal functions ψ_p and ψ_q. The Hamiltonian for the N-electron problem, following methods we sketched in Sec. 6, is

$$(H)_{\text{op}} = \sum_{i=1}^{N} f_i + \Sigma(\text{pairs } i,j) g_{ij} \qquad (9\text{-}9)$$

Here

$$f_i = -\frac{1}{2}\nabla_i^2 - \frac{Z}{r_i} \qquad g_{ij} = \frac{1}{r_{ij}} \qquad (9\text{-}10)$$

in the hartree atomic units we are using, in which the distances are measured in terms of the bohr unit, a_0, the radius of the first Bohr orbit in hydrogen, and energies in terms of the hartree unit, equal to 2 rydbergs. The quantity $-\frac{1}{2}\nabla^2$ is the kinetic energy in these units, $-Z/r_i$ is the potential energy of interaction between the nucleus of charge Z units and an electron at distance r_i from the nucleus, and r_{ij} is the distance between the ith and jth electrons, so that g_{ij} is the coulomb repulsion between the ith and jth electrons.

In setting up the matrix components of the Hamiltonian, we run into one-electron and two-electron integrals similar to those met in Sec. 6 in the two-electron case. We may write them in the form

$$(a/f/b) = \int u_a^*(1) f_1 u_b(1) dv_1 \qquad (9\text{-}11)$$

$$(ab/g/cd) = \int u_a^*(1) u_b^*(2) g_{12} u_c(1) u_d(2) dv_1 dv_2 \qquad (9\text{-}12)$$

where, since we are now dealing with spin orbitals, the integrations over dv_1 include a summation over the spin of electron 1 as well as an integration over its space coordinates. In terms of these integrals, we then have the following rules for the matrix components of the Hamiltonian between the two determinantal functions.

First we have the diagonal matrix component for a single determinant. This equals

$$\text{diagonal component} = \Sigma(i)(i/f/i) + \Sigma(\text{pairs } i,j)[(ij/g/ij) - (ij/g/ji)]$$

$$(9\text{-}13)$$

Next we have a nondiagonal matrix component between two determinants in which u_i appears in the determinant ψ_p in the matrix component

$\int \psi_p^*(H)_{op}\psi_q dv$, and is replaced by u_i' in ψ_q, all other u's remaining unchanged and in the same order. This equals

$$(i/f/i') + \Sigma(j \neq i)[(ij/g/i'j) - (ij/g/ji')] \qquad (9\text{-}14)$$

Finally we have a nondiagonal matrix component between two determinants in which u_i and u_j appearing in the determinant ψ_p are replaced by u_i' and u_j' in ψ_q, all other u's remaining unchanged and in the same order. This equals

$$(ij/g/i'j') - (ij/g/j'i') \qquad (9\text{-}15)$$

The reason we have to specify the order of the u's in the determinants is that the sign changes when the order of two u's is interchanged, as we noted earlier. There are no nondiagonal matrix components between two determinantal functions differing in more than two spin orbitals. It should be emphasized that it is necessary that the spin orbitals be orthonormal for these simple rules to hold; for nonorthogonal orbitals the rules are very much more complicated.

Next we must have rules for finding the integrals. The integrals $(a/f/b)$ are simple enough so that they do not need to be discussed, except that they are zero if u_a and u_b have different spins m_s. However, the two-electron integrals, of which the general form is given in Eq. 9-12, involve the product of four wave functions of the form of $R_{nl}(r)Y_{lm_l}(\theta\phi)$, where $R_{nl}(r)$ is a radial wave function and the Y's are spherical harmonics. This is to be multiplied by $1/r_{12}$ and integrated over the coordinates of electrons 1 and 2, as well as being summed over spins. Naturally it is a considerable feat to take care of this integration, but it can be handled by means of well-known theorems involving spherical harmonics. The result can be stated in a simple form, involving only a two-dimensional integration over the radial coordinates of the two electrons:

$$(ab/g/cd) = \sum_{k=0}^{\infty} c^k(l_a m_{la}; l_c m_{lc}) c^k(l_d m_{ld}; l_b m_{lb}) R^k(ab;cd) \qquad (9\text{-}16)$$

Here the c's are quantities sometimes called Gaunt coefficients, in a form defined by Condon and Shortley, and are given in Table 9-1 for s and p orbitals, $l=0$ and 1. They can be given by a formula, but when calculations are made with the digital computer, they are usually generated as part of the program. However, a table like 9-1 allows us to make simple calculations without use of the digital computer.

Table 9-1. Quantities $c^k(l_i m_{li}; l_j m_{lj})$ for s, p Electrons

	m_{li}	m_{lj}	$k=0$	$k=1$	$k=2$
ss	0	0	1	0	0
sp	0	±1	0	$-\sqrt{\frac{1}{3}}$	0
	0	0	0	$\sqrt{\frac{1}{3}}$	0
pp	±1	±1	1	0	$-\sqrt{\frac{1}{25}}$
	±1	0	0	0	$\sqrt{\frac{3}{25}}$
	±1	∓1	0	0	$-\sqrt{\frac{6}{25}}$
	0	0	1	0	$\sqrt{\frac{4}{5}}$

Note: In cases where there are two ± signs, the two upper or the two lower signs must be taken together.

The matrix components of Eq. 9-16 are different from zero only if u_a and u_c have the same spin quantum number m_s, if u_b and u_d have the same spin quantum number $m_{s'}$, and if $m_{la} + m_{lb} = m_{lc} + m_{ld}$. From this rule we can easily verify the statement we made earlier, that the Hamiltonian has a nondiagonal matrix component only between two determinantal functions which have the same value of $M_L = \Sigma m_l$, and the same $M_S = \Sigma m_s$.

The quantities $R^k(ab; cd)$ are the ones that involve the radial part of the spin orbitals, which are assumed to satisfy a normalization condition $\int_0^\infty R^2 r^2 dr = 1$. They can be found from the equation

$$R^k(ab;cd) = \int_0^\infty \int_0^\infty R^*_{n_a l_a}(r_1) R^*_{n_b l_b}(r_2) (r_<^k / r_>^{k+1}) R_{n_c l_c}(r_1) R_{n_d l_d}(r_2)$$

$$\times r_1^2 r_2^2 \, dr_1 \, dr_2 \qquad (9\text{-}17)$$

Here $r<$ is the smaller, $r>$ the larger of the two quantities r_1 and r_2. The special cases of $R^k(ab;cd)$ occurring in the diagonal matrix component are ordinarily given special names,

$$R^k(ab;ab) = F^k(ab) \qquad R^k(ab;ba) = G^k(ab) \qquad (9\text{-}18)$$

For R_{nl}'s given analytically as functions of r, it is easy to compute these integrals. For R's given as numerical tables of values, there are programs for computing them with the digital computer.

To give the reader some feeling for these methods, let us ask how the two-electron problem is handled by means of them. We shall take the simple case of s electrons, and start with the case of two nonequivalent s electrons, for instance $1s$ and $2s$. We have four spin orbitals, which we can symbolize as $1s\uparrow$, $1s\downarrow$, $2s\uparrow$, $2s\downarrow$, where $1s\uparrow$ is a product of the $1s$ orbital of the coordinates and Pauli's function α of the spin, and so on. We must choose two of these four spin orbitals to be occupied, and this can be done in $4!/2!2! = 6$ different ways. In Table 9-2, we list these six choices, and give the values of M_S for each. Only four of these choices correspond to the configuration $1s2s$; the first one has two equivalent $1s$ electrons, the last has two equivalent $2s$, so for the moment we disregard them. The four remaining choices are the same as those given in the second half of Table 4-1.

Table 9-2

		M_S
$1s\uparrow$	$1s\downarrow$	0
$1s\uparrow$	$2s\uparrow$	1
$1s\uparrow$	$2s\downarrow$	0
$1s\downarrow$	$2s\uparrow$	0
$1s\downarrow$	$2s\downarrow$	-1
$2s\uparrow$	$2s\downarrow$	0

We now know that we can set up four determinantal functions corresponding to these four choices, but that the Hamiltonian has no nondiagonal matrix components between the functions of different M_S. Since the determinant formed from $1s\uparrow 2s\uparrow$ is the only one with $M_S = 1$, it must represent the correct function for the ^3S state with $M_S = 1$. Let us write its wave function. The symbol $1s\uparrow$ stands for $u_1(1)\alpha(1)$, where u_1 is the $1s$ orbital, and $2s\uparrow$ stands for $u_2(1)\alpha(1)$.

The determinantal function then becomes

$$\frac{1}{\sqrt{2}}\begin{vmatrix} u_1(1)\alpha(1) & u_1(2)\alpha(2) \\ u_2(1)\alpha(1) & u_2(2)\alpha(2) \end{vmatrix} = \frac{1}{\sqrt{2}}\begin{vmatrix} u_1(1) & u_1(2) \\ u_2(1) & u_2(2) \end{vmatrix}\alpha(1)\alpha(2) \quad (9\text{-}19)$$

This is the product of the antisymmetric function of coordinates, as given in Eq. 6-11, and the function $\alpha(1)\alpha(2)$ of spins. The spin function is symmetric in the spins; interchange of electron coordinates and spins does not change it. This is a characteristic of the two-electron problem: The wave functions prove in each case to be products of a function of the coordinates and a function of the spins, and the first function is antisymmetric, the second symmetric, for a triplet state. When the electron coordinates and spins are interchanged for the two electrons, the first factor changes sign, the second is unchanged, and thus the whole wave function is changed in sign, or is antisymmetric.

Let us next find the energy for the triplet state, using the rules of the present section. From Eq. 9-13, the first term is $\Sigma(i/f/i)$, which is the same quantity $E_{1s} + E_{2s}$ we indicated by $E_a + E_b$ in Eq. 6-7. The remaining term is just the same as that given in Eq. 6-12 for the triplet case with the minus sign. Now, however, we have in Eqs. 9-16, 9-17, and 9-18, together with Table 9-1, the detailed directions for computing the integrals involved in this energy. We find

$$\text{triplet energy} = E_{1s} + E_{2s} + F^0(1s, 2s) - G^0(1s, 2s) \qquad (9\text{-}20)$$

where F^0 and G^0 can be readily found from Eq. 9-17. If we carry through the same calculation for the determinant formed from spin orbitals $1s\downarrow$ and $2s\downarrow$, we get the same energy, in this case for the $M_S = -1$ component of the triplet.

We are left with two determinantal functions for $M_S = 0$, namely those formed from $1s\uparrow 2s\downarrow$ and $1s\downarrow 2s\uparrow$. Thus we shall have a secular equation with two rows and columns for this case. We must then find the diagonal matrix components of energy for each of the determinants, and the nondiagonal component between them. If we look at Eq. 9-13, we find that the terms $\Sigma(i/f/i)$ and $(ij/g/ij)$ are identical for each of the two determinants, and identical with what we found for the triplet state. There is, however, a difference in the exchange integral $-(ij/g/ji)$. By Eq. 9-12, writing in more detail as far as the spins are concerned, this is

$$(1s\uparrow 2s\downarrow/g/2s\downarrow 1s\uparrow) = \Sigma(m_{s1}m_{s2}) \int u_{1s}(r_1)u_{2s}(r_2)g_{12}u_{2s}(r_1)$$
$$\times u_{1s}(r_2) dv_1 dv_2 \alpha(1)\beta(2)\beta(1)\alpha(2) \qquad (9\text{-}21)$$

Here we have the summation over m_{s1} of the quantity $\alpha(1)\beta(1)$, with a similar quantity for electron 2. But such a summation is necessarily zero. If $m_{s1} = \frac{1}{2}$, $\alpha(1)$ is unity, but $\beta(1)$ is zero. Similarly, if $m_{s1} = -\frac{1}{2}$, $\alpha(1)$ is zero, though $\beta(1)$ is unity. Hence we get nothing for this exchange integral.

This is a general situation: The exchange integral is always zero between two spin orbitals corresponding to different spins. In computing the integral according to Eq. 9-16, this leads to the statement that the integral is different from zero only if u_a and u_c (in this case $1s$ and $2s$) have the same quantum number m_x, and if u_b and u_d (in this case $2s$ and $1s$) have the same m'_s. We see finally, then, that the diagonal matrix component of the Hamiltonian for either of the determinantal functions with $M_S = 0$ is

$$\text{diagonal matrix component} = E_{1s} + E_{2s} + F^0(1s, 2s) \quad (9\text{-}22)$$

Next we need the nondiagonal matrix component of energy between the two functions with $M_S = 0$. This is a case of Eq. 9-15, since the two determinants differ in two spin orbitals. From Eq. 9-15, the nondiagonal matrix component is

$$(1s\uparrow 2s\downarrow / g / 1s\downarrow 2s\uparrow) - (1s\uparrow 2s\downarrow / g / 2s\uparrow 1s\downarrow) \quad (9\text{-}23)$$

From the discussion given in the preceding paragraph, we see that the first term of Eq. 9-23 is zero, but the second is not, and is in fact equal to $-G^0(1s, 2s)$.

We can then set up the secular equation, which is

$$\begin{vmatrix} E_{1s} + E_{2s} + F^0(1s, 2s) - E & -G^0(1s, 2s) \\ -G^0(1s, 2s) & E_{1s} + E_{2s} + F^0(1s, 2s) - E \end{vmatrix} = 0 \quad (9\text{-}24)$$

or

$$[E - E_{1s} - E_{2s} - F^0(1s, 2s)]^2 - [G^0(1s, 2s)]^2 = 0$$

$$E = E_{1s} + E_{2s} + F^0(1s, 2s) \pm G^0(1s, 2s) \quad (9\text{-}25)$$

This equation, which corresponds to Eq. 6-12, thus leads to two energies, of which the one with the minus sign equals the triplet energy given in Eq. 9-20, and the one with the plus sign corresponds to the singlet state.

Next let us find the wave functions corresponding to these two states, with $M_s = 0$. We must solve Eq. 9-7, which in this simple case consists of two equations,

$$C_1(H_{11} - E) + C_2 H_{12} = 0$$
$$C_1 H_{12} + C_2(H_{22} - E) = 0 \quad (9\text{-}26)$$

Let us first consider the triplet case, which by Eq. 9-25 has the energy $E = H_{11} - G^0(1s, 2s)$. We remember that $H_{12} = -G^0(1s, 2s)$. Thus each of the two equations of 9-26 becomes $C_1 - C_2 = 0$, or $C_1 = C_2$. It is because we have solved the secular equation that the two equations are consistent with each other. The triplet wave function is then the sum of the two determinantal functions times a normalizing factor $1/\sqrt{2}$. The two determinants are

$$\psi_1 = \frac{1}{\sqrt{2}}[u_1(1)\alpha(1)u_2(2)\beta(2) - u_2(1)\beta(1)u_1(2)\alpha(2)]$$

$$\psi_2 = \frac{1}{\sqrt{2}}[u_1(1)\beta(1)u_2(2)\alpha(2) - u_2(1)\alpha(1)u_1(2)\beta(2)]$$

(9-27)

so that the triplet wave function is

$$\frac{1}{\sqrt{2}}(\psi_1 + \psi_2) = \frac{1}{\sqrt{2}}[u_1(1)u_2(2) - u_2(1)u_1(2)]\frac{1}{\sqrt{2}}[\alpha(1)\beta(2) + \beta(1)\alpha(2)]$$

(9-28)

The function of coordinates which we find is the same antisymmetric function we found for the triplet state in Eq. 9-19, multiplied by another symmetric function of the spins, in this case corresponding to $M_S = 0$. The energy depends only on the orbital function, not on the spin function, which is why this function has the same triplet energy we found for the case $M_S = 1$.

For the singlet state, we proceed as before, taking the energy $H_{11} + G^0$, and find that $C_2 = -C_1$. Thus the wave function is

$$\frac{1}{\sqrt{2}}(\psi_1 - \psi_2) = \frac{1}{\sqrt{2}}[u_1(1)u_2(2) + u_2(1)u_1(2)]\frac{1}{\sqrt{2}}[\alpha(1)\beta(2) - \beta(1)\alpha(2)]$$

(9-29)

Here we have the symmetric function of coordinates which we met in Eq. 6-11, but multiplied by an antisymmetric function of the spin. Thus in this case the antisymmetry of the wave function is secured by the antisymmetry of the spin part, rather than of the orbital function. This same situation is found in the case of the singlet function arising from the configuration $1s^2$, the first determinantal function in Table 9-2. Its wave function is

$$u_1(1)u_1(2)\frac{1}{\sqrt{2}}[\alpha(1)\beta(2) - \beta(1)\alpha(2)]$$

(9-30)

The energy of the configuration $1s^2$, as we see from Eq. 9-13, is

$$\text{energy of } 1s^{2\,1}S = 2E_{1s} + F^0(1s, 1s)$$

(9-31)

There is no exchange integral in this case, since the two orbitals correspond to different spin in the determinantal function. As we have already pointed out, these two-electron problems are unusually simple, in that the whole wave function is the product of a function of coordinates times a function of spins. Unfortunately this simple situation is not met with in cases of more than two electrons.

10. 1929–1930, The Hartree-Fock Method

At the end of Sec. 7, where we discussed Hartree's self-consistent field, I suggested that it was really necessary to work out the symmetry situation for an arbitrary atom before Hartree's method could be put on a more fundamental basis. One of the things I did in Leipzig in 1929 was to consider this problem. It was natural to start with an expression for the energy of a whole atom or other system, in terms of the occupied spin orbitals, and then to vary these spin orbitals to minimize the total energy. This would give the best orbitals which could be found, subject to the assumption that the many-electron wave function was built up out of spin orbitals. The first thing I did was to start with a simple product function, and to ask what this would give.

The energy of a product function can be easily expressed in terms of the notation we have used in the preceding section. It is

energy of product function

$$= \Sigma(i/f/i) + \Sigma(\text{pairs } i,j)(ij/g/ij)$$

$$= \Sigma(i) \int u_i^*(1) f_1 u_i(1) dv_1$$

$$+ \Sigma(\text{pairs } i,j) \int u_i^*(1) u_j^*(2) g_{12} u_i(1) i_j(2) dv_1 dv_2 \qquad (10\text{-}1)$$

Suppose we demand that this energy be minimized, subject to any arbitrary modification of the orbitals, so long as they remain normalized, which is necessary for Eq. 10-1 to hold. Let us make an arbitrary variation in a particular u_i^*, and let us handle the normalization by Lagrange multipliers. We then have

$$0 = \int \delta u_i^*(1) f_1 u_i(1) dv_1 + \Sigma(j \neq i) \int \delta u_i^*(1) u_j^*(2) g_{12} u_i(1) u_j(2) dv_1 dv_2$$

$$- \epsilon_i \int \delta u_i^*(1) u_i(1) dv_1 + \text{complex conjugate} \qquad (10\text{-}2)$$

Here $\delta u_i^*(1)$ is an arbitrary small variation function to be added to the original u_i^*, $-\epsilon_i$ is a Lagrange multiplier, and the complex conjugate takes care of the fact that we must vary u_i as well as u_i^*, since one is the complex conjugate of the other.

It is impossible to satisfy a condition like Eq. 10-2, holding for all possible variation functions δu_i^*, unless the integrand by which it is multiplied is everywhere zero. Hence we arrive at the equation

$$f_1 u_i(1) + \Sigma(j \neq i) \int u_j^*(2) u_j(2) g_{12} dv_2 u_i(1) = \epsilon_i u_i(1) \qquad (10\text{-}3)$$

But this is just Hartree's equation. The first term, $f_1 u_i(1)$, is the kinetic energy operator plus the potential energy in the field of the nucleus, operating on $u_i(1)$. The second is the potential of all other electrons, in all orbitals except the ith, acting on $u_i(1)$. The Lagrange multiplier ϵ_i plays the part of an eigenvalue. Thus we have a straightforward justification of Hartree's method from the variation principle.

While I was in Leipzig I sent in a short note pointing out this justification of Hartree's method, and I included in it the remark that if one used the energy expression of Eq. 9-13, for the case of a determinantal function, one would get additional terms from the exchange integrals, which should improve the method. I had planned to work out these additional terms, but did not have the opportunity on account of other things I was working on, and in the meantime V. Fock, in the Soviet Union, independently suggested the method and worked out the details. Fock did not use exactly the same notation we are using, but I shall give his equation, called the Hartree-Fock equation, in the form suggested by Eq. 9-13:

$$f_1 u_i(1) + \Sigma(j \neq i) \int u_j^*(2) u_j(2) g_{12} dv_2 u_i(1)$$

$$- \Sigma(j \neq i; \text{spin } j = \text{spin } i) \int u_j^*(2) u_i(2) g_{12} dv_2 u_j(1) = \epsilon_i u_i(1) \qquad (10\text{-}4)$$

Hartree immediately started using this equation, and obtained results that agreed perhaps slightly better with experiment than with the original Hartree equation, 10-3. Hartree also sometimes set up energy expressions for specific multiplets, and varied the orbitals in these energy expressions so as to minimize the energy of a particular multiplet. This is a useful procedure, but in this book, to avoid ambiguity, I restrict the term Hartree-Fock equation to the case of a single determinant, leading to Eq. 10-4.

We shall come later to the interpretation of the exchange terms in Eq. 10-4, which are missing in Hartree's original method, Eq. 10-3. There are, however, two properties of the Hartree-Fock equations which are very useful, and which should be pointed out. First, it can easily be proved that all solutions of the Hartree-Fock equations, corresponding to different eigenvalues and eigenfunctions, are exactly orthogonal to each other. This means that they are adapted to use in the multiplet theory we have been describing, which demands orthogonal orbitals.

Second, in 1933 Koopmans, a student of Kramers, proved a theorem which was the analog of the work I had done in 1928, connecting the eigenvalue ϵ_i with the ionization energy of an atom. If one multiplies Eq. 10-4 on the left by $u_i^*(1)$, and integrates over the coordinates of the first electron, one finds

$$\epsilon_{iHF} = (i/f/i) + \Sigma(j \neq i)[(ij/g/ij) - (ij/g/ji)] \qquad (10\text{-}5)$$

which consists of just those terms of the diagonal energy, Eq. 9-13, which are present in the atom, but would be missing in the ion in which the electron in the ith spin orbital is removed, on the assumption that the remaining spin orbitals were not changed. This is Koopmans' theorem. It means that the eigenvalues of the Hartree-Fock method form approximations to the negatives of the ionization potentials of the electrons in the atom. As we shall point out in Sec. 33, the major error in this calculation is the fact that the orbitals of the ion will really be appreciably different from those of the atom, so that properly we should make a separate self-consistent calculation of the total energy of the ion, and subtract the two. It is in fact found that this procedure of subtracting one energy from the other leads to results in better agreement with experiment than does the use of Koopmans' theorem.

# 11. 1926, The Fermi-Dirac Statistics

By going as far as we have with our discussion of the Hartree and Hartree-Fock methods, we have been getting a little ahead of a number of other important developments. Let us go back to 1926, and to a study of the behavior of a perfect gas composed of particles obeying the exclusion principle. This was made independently by Dirac, and by a hitherto rather unknown but very brilliant young Italian, Enrico Fermi. Fermi had been considering the problem for a couple of years, but his earlier papers had been published in Italian, and it was only in 1926 that he started publishing in German, and his work became generally known.

Both Fermi's and Dirac's treatments were based on the quantization of the perfect gas problem, as indicated by Schrödinger's equation. Suppose we have independent particles moving in a cubical cavity bounded by perfectly impenetrable walls at $x=0$, $x=L$, $y=0$, $y=L$, $z=0$, $z=L$. It is easily shown that Schrödinger's equations demand that the wave function go to zero at an impenetrable wall. Hence the wave function of an electron or other particle moving freely in such a box must be

$$u = \sin\frac{n_x \pi x}{L} \sin\frac{n_y \pi y}{L} \sin\frac{n_z \pi z}{L} \qquad (11\text{-}1)$$

where n_x, n_y, n_z are positive integers, since these sine functions go to zero on the proper planes. We have not bothered to include a normalization factor, since we shall not be using it anyway. We can now find the kinetic energy by operating on this wave function with the kinetic energy operator $-\hbar^2 \nabla^2/2m$, where m is the mass of the particle. We find

$$-\frac{\hbar^2}{2m}\nabla^2 u = \frac{h^2}{8mL^2}(n_x^2 + n_y^2 + n_z^2)u = Eu \qquad (11\text{-}2)$$

where we have equated the kinetic energy $p^2/2m = (h^2/8mL^2) \times (n_x^2 + n_y^2 + n_z^2)$ to the total energy E, since there is no potential energy in this problem. We have, in other words, a quantized problem, with energy levels corresponding to all positive integral values of n_x, n_y, n_z.

These energy levels are extremely close together, if the box is of ordinary size. To find the spacing, let us set up a space in which n_x, n_y, n_z are rectangular coordinates. Let us set up a sphere in this space, whose equation is

$$n_x^2 + n_y^2 + n_z^2 = \frac{8mL^2 E}{h^2} \qquad (11\text{-}3)$$

The radius of this sphere is the square root of the quantity on the right side of Eq. 11-3, or equals $(L/h)(8mE)^{1/2}$. In the space, there is one set of integral n's for each unit volume, and all volume in the first octant, corresponding to positive n's, is occupied by points which represent quantized states. The volume of the sphere is $\frac{4}{3}\pi r^3$, where r is the radius, and the octant has a volume one-eighth of this, or $\frac{1}{6}\pi r^3 = \frac{1}{6}\pi[(L/h)(8mE)^{1/2}]^3$. Since this volume is equal to the number of stationary states whose energy is less than E, the number of stationary states with energy less than E is given by

$$\frac{4\pi L^3}{3h^3}(2mE)^{3/2} \qquad (11\text{-}4)$$

The number of states per unit volume per unit energy range is found by differentiating with respect to E, and by dividing by the volume L^3 of the cavity:

$$\frac{dN}{dE} = \frac{2\pi}{h^3}(2m)^{3/2}E^{1/2} \qquad (11\text{-}5)$$

If one puts in numerical values, and assumes a volume of ordinary macroscopic dimensions, one finds, as was stated above, that the number of energy levels per unit energy range is enormous, or the levels are very close together. We note that if we are dealing with electrons, we have found the number of states per unit volume per unit energy range of each spin; the total number of levels is then twice the value of Eq 11-5, half of spin up, half of spin down.

The problem which Fermi and Dirac considered was the behavior of such a gas of electrons or other particles obeying the exclusion principle, at a given temperature. We here meet a problem of statistical mechanics. The stable state of a system at a given temperature is that which minimizes the Helmholtz free energy $A = U - TS$, where U is the internal energy, T is the temperature, and S is the entropy. We must assume that some states are completely occupied at this temperature, up to the limit imposed by the exclusion principle. Others will be completely empty, and still others will be partly occupied. For the ith state (specified by the ith set of values n_x, n_y, n_z), let us assume that the occupancy will be f_i, a quantity which on account of the exclusion principle must be between the limits 1 for complete occupancy, and 0 for a completely empty state. We shall now explain just what we mean by the partial occupancy with a fractional value of f_i.

Let us subdivide our very large number of one-electron energy levels into individual groups of levels, one group having G levels in it, a number large compared to unity, but still finite. We assume that all G of these energy levels have energies which are substantially equal. There will then be assumed to be $f_i G$ electrons in these G levels, some levels being occupied, some empty. There are many ways of choosing $f_i G$ levels to be occupied, the remaining $(1-f_i)G$ empty. This number of ways is

$$\frac{G!}{(f_i G)!(G-f_i G)!} \qquad (11\text{-}6)$$

which is computed in the same way we used in Eq. 9-2. However, here, in contrast to the case of multiplet theory we were considering in Sec. 9, we are dealing with a number G which is large compared to unity. In this case, there is a well-known formula known as Stirling's approximation to the factorial, which to a first approximation, holding for sufficiently large G, is

$$G! = \left(\frac{G}{e}\right)^G \qquad (11\text{-}7)$$

If we use this approximation, the number of ways of assigning $f_i G$ electrons to G levels, from Eq. 11-6, becomes

$$\left[\frac{1}{f_i^{f_i}(1-f_i)^{1-f_i}}\right]^G \qquad (11\text{-}8)$$

If we carry through a similar calculation for each of the groups of levels into which we have subdivided all of our energy levels, we shall have a product of quantities like that of Eq. 11-8, for each group. Since Eq. 11-8 is a certain quantity raised to the G power, this is just as if we multiplied functions $[f_i^{f_i}(1-f_i)^{1-f_i}]^{-1}$ for each one of the energy levels. Then the complete number of ways of choosing the assignments of electrons to orbitals, consistent with the assumed f_i's, is a product over all one-electron states,

$$W = \Pi(i)\left[f_i^{f_i}(1-f_i)^{1-f_i}\right]^{-1} \qquad (11\text{-}9)$$

This quantity W is called the thermodynamic probability. According to Boltzmann, one can calculate the entropy in statistical mechanics by using the equation

$$S = k \ln W \qquad (11\text{-}10)$$

where k is the Boltzmann constant, which appears in the gas law. That is, a state with a high thermodynamic probability also has a large entropy. If we combine Eqs. 11-9 and 11-10, we then find for the gas of particles obeying the exclusion principle the value

$$S = -k\Sigma(i)[f_i \ln f_i + (1-f_i)\ln(1-f_i)] \tag{11-11}$$

This formula, with substantially the same derivation, was given by both Fermi and Dirac.

We can now apply the condition for thermal equilibrium, $dA = d(U - TS) = 0$. This will minimize the free energy A, if the f_i's are properly chosen. We then find

$$dA = 0 = \Sigma(i)\left[\frac{\partial U}{\partial f_i} + kT\ln\frac{f_i}{1-f_i}\right]df_i \tag{11-12}$$

This must be satisfied for any choice of df_i's, subject only to the condition that the total number of particles should remain constant. Since the total number of particles is $\Sigma(i)f_i$, this means that we must have

$$\Sigma(i)df_i = 0 \tag{11-13}$$

The only way to satisfy Eq. 11-12, subject to Eq. 11-13, is to have

$$\frac{\partial U}{\partial f_i} + kT\ln\frac{f_i}{1-f_i} = \epsilon_F = \text{constant} \tag{11-14}$$

independent of i. For then the bracket in Eq. 11-12 can be taken outside the summation, and Eq. 11-13 immediately makes the whole expression vanish. We then solve Eq. 11-14 for f_i, and obtain the Fermi distribution law

$$f_i = \frac{1}{\exp(\epsilon_i - \epsilon_F)/kT + 1} \tag{11-15}$$

where

$$\epsilon_i = \frac{\partial U}{\partial f_i} \tag{11-16}$$

Substantially this derivation was given by Fermi and Dirac, except that they worked by maximizing the entropy subject to the condition of constant total energy, rather than by minimizing the free energy at constant temperature. When the variation is handled in this way, the result is the same, but the temperature is introduced into the calculation through a Lagrange multiplier.

In the Fermi theory of the perfect gas, the quantity ϵ_i is simply the one-electron energy of an electron in the state i, as given by Eq. 11-2. The reason is that the total energy U is the sum of the energies of all the particles, or

$$U = \Sigma(i)\epsilon_i f_i \qquad (11\text{-}17)$$

The quantity ϵ_F is known as the Fermi energy. The behavior of f_i as a function of energy $\epsilon_i - \epsilon_F$ is shown in Fig. 11-1 for a number of temperatures. As we go down toward the absolute zero, the function f_i equals unity for all energies less than ϵ_F, and it equals zero for all energies greater than ϵ_F. Fermi, Dirac, and Sommerfeld investigated the thermal properties of a Fermi gas in detail, and in 1928 Sommerfeld and his students explained a number of the thermal properties of a metal by assuming that the electrons of the metal acted like a Fermi gas. In particular, the specific heat of a Fermi gas at low temperatures is very much smaller than that of a gas of ordinary particles not subject to the exclusion principle. The reason is that as the temperature is raised above the absolute zero, it is only the electrons in the immediate neighborhood of the Fermi energy which change their energies at all, unlike a classical gas in which each electron or particle has a kinetic energy proportional to the absolute temperature. This behavior of the specific heat agrees with experiment, and removes a difficulty with the older Lorentz-Drude theory.

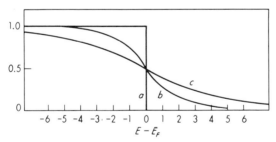

Fig. 11-1. Fermi function f_i from Eq. 11-15. (*a*) Absolute zero of temperature; (*b,c*) higher temperatures. From *Quantum Theory of Matter*, 2nd ed., by J. C. Slater. Copyright 1968 by McGraw-Hill, Inc. Used with Permission of McGraw-Hill Book Company.

12. 1927, The Thomas-Fermi Atom Model

Shortly after Fermi set up his theory of the perfect gas, it occurred to him, and independently to a young Englishman named L. H. Thomas, that one could apply the method to a gas in an external potential field, and if the field were that of a nucleus, one could set up an atomic model. This, as we note, was before the publication of Hartree's self-consistent field, and the Thomas-Fermi model was able to give quite good information even about heavy atoms, though it was never supposed to be as good as the Hartree calculations. It was sufficiently easier to use, however, so that it has had wide applications.

There is nothing in our derivation of the Fermi function f_i, as given in Eqs. 11-6 to 11-16, which demands that we be dealing with a perfect gas in the absence of an external force field. Suppose we had an electron gas whose electrons were in the field of many nuclei, as in a crystal, and in some sort of self-consistent field arising from the other electrons. We should still have a very closely spaced set of energy levels, and could still carry out the calculation of entropy which we have made in Eqs. 11-6 to 11-11. The condition for thermal equilibrium, $d(U-TS)=0$, is based on very fundamental thermodynamics and statistical mechanics, and would still be valid. Thus Eqs. 11-15 and 11-16 would still be valid. Only Eq. 11-17, $U=\Sigma(i)\epsilon_i f_i$, would fail to be correct if the potential in which the electrons moved were determined by the electrons, and hence by the f_i's, as would be the case in a self-consistent field. For in this case U would not be a linear function of the f_i's. These points are very important for our later discussion of the $X\alpha$-SCF method, for we shall find that parts of that method are based on the applicability of the Fermi-Dirac statistics. Equally important applications of the Fermi-Dirac statistics to electrons in a self-consistent field are those in solid-state electronics.

The approximations that Thomas and Fermi made in developing their atomic model were not in the Fermi statistics itself, as we can see from the preceding paragraph. Rather, they were in the calculation of the density of states and the kinetic energy. They considered the case of the absolute zero of temperature, in which the Fermi function of Eq. 11-15 and Fig. 11-1 indicates that all one-electron energy levels with energy below ϵ_F will be occupied with $f_i=1$, while all levels above ϵ_F will be empty, with $f_i=0$. To get the charge density at an arbitrary point in space, the straightforward way to proceed would have been as Hartree did, to find the self-consistent orbitals, and add the charge densities of all those that were occupied (or whose energy was less than ϵ_F). One would then find the potential from this charge density by electrostatics, and iterate to get self-consistent potential and orbitals. But we must not forget the work of Thomas and Fermi came before that of Hartree, and it probably seemed to them like an almost impossible task to find these self-consistent orbitals. They preferred instead to find the charge density at an arbitrary point in space by assuming that the electron distribution in the neighborhood of that point could be treated as part of a homogeneous electron gas.

In Eq. 11-4, we have seen that the theory of the homogeneous electron gas tells us that if all states with energy less than E are occupied, the number of states per unit volume will be $4\pi/3h^3(2mE)^{3/2}$. If we take account of the two orientations of spin and the charge $-e$ of an electron (where e is the magnitude of the electronic charge), the charge density would be $-8\pi e/3h^3(2mE)^{3/2}$. Here is a very straightforward way of estimating the charge density at a given point of space, provided the maximum kinetic energy of the electrons is known. Now suppose that the potential at a given point of space is $V(r)$, so that the potential energy of an electron at this point is $-eV(r)$. Let us choose the zero of potential energy to be the Fermi energy. Homogeneous electron gas theory would indicate that at the point where the potential is $V(r)$, we have electrons with all kinetic energies from zero up to $eV(r)$, so that the total energy of the electron with maximum kinetic energy is zero. Thus the assumption of Thomas and Fermi was that the charge density of electrons at this point was

$$\rho = -e\frac{8\pi}{3h^3}[2meV(r)]^{3/2} \qquad (12\text{-}1)$$

They then proceeded to apply a condition of self-consistency. They assumed that the potential $V(r)$ was that of the nucleus (they were working only with a single atom) plus that arising by electrostatics from the electronic charge density of Eq. 12-1. This led to a differential equation for the potential $V(r)$, which in an atom was a function of the radius only, which could be solved numerically, and which gave a universal potential function for use inside an atom.

As an example of the accuracy of this method, we show in Fig. 12-1 a function $U(r)$, defined by

$$V(r) = U(r) \times \text{potential of unscreened nucleus} \qquad (12\text{-}2)$$

as determined by the Thomas-Fermi method, compared with the corresponding function as determined by the Hartree-Fock method, for the zinc atom. We see that there is quite good agreement between the two methods, and can note the way in which the nuclear field is screened or shielded more and more by the electrons, as we go from the nucleus toward the outer part of the atom. In the Thomas-Fermi method, it proves to be the case that the function $U(r)$ is a universal function, holding for any atom, of a quantity

$$x = \frac{4r}{a_0}\left(\frac{2Z}{9\pi^2}\right)^{1/3} \tag{12-3}$$

where Z is the atomic number and a_0 is the bohr unit of length. The function U in Fig. 12-1 is plotted as a function of x rather than of r, so that it furnishes a universal screening function according to the Thomas-Fermi method.

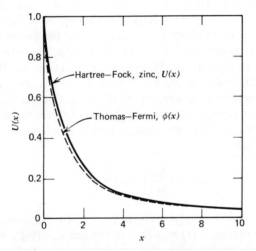

Fig. 12-1. Thomas-Fermi function $\phi(X)$, as function of x, defined in Eq. 12-3, compared with corresponding self-consistent-field function $U(x)$ of Eq. 12-2, for zinc. From *Quantum Theory of Matter*, 2nd ed., by J. C. Slater. Copyright 1968 by McGraw-Hill, Inc. Used with permission of McGraw-Hill Book Company.

A number of calculations were soon made using this approximate method, and it gives results of quite satisfactory accuracy for many properties. However, as we see from the discussion we have given, it is definitely not as good an approximation as the self-consistent field, and where that method can be used, it is to be preferred. But as we mention in Sec. 24, considerable development of the Thomas-Fermi method was carried out, particularly during the 1930s. One improvement in the method was to correct an obvious shortcoming. In the method as we described it, an electron in an N-electron atom would move in the field of the nucleus and all N electrons, rather than in the field of only $N-1$ electrons. Dirac in 1930 proposed a method of overcoming this difficulty, which is very important for the $X\alpha$-SCF method, for it led the way to the corresponding correction made in that method. On account of its importance in our future work, we postpone the discussion of this correction, leading to what is usually called the Thomas-Fermi-Dirac method, until Sec. 16 and 30.

13. 1927, Heitler and London and the Hydrogen Molecule

We have been grouping together in the preceding sections the work which was specifically directed toward atoms and their structure. However, work on molecules and solids was going on simultaneously. Shortly after Heisenberg's work on the helium atom, which we discussed in Sec. 6, two other young Germans, Walter Heitler and Fritz London, applied the same method to the other two-electron problem, the hydrogen molecule. They thought of two hydrogen atoms, each with its one electron, at a distance R apart. If they were far apart, the interaction between the two electrons was unimportant, and they could write the electronic wave function of the two electrons as a product of two one-electron functions, as Heisenberg had done. Then when the atoms approached, there would be integrals like the $(ab/g/ab)$ and $(ab/g/ba)$ we met in the helium problem, which would give an explanation of the molecular formation. As I mentioned much earlier, no theory of the hydrogen molecule based on the older quantum mechanics had achieved any success.

Heitler and London followed rather closely the pattern of Heisenberg's calculations, and we shall do the same. We still have the same situation of a wave function depending on the coordinates of the two electrons, and we still must assume that the wave function will be either a symmetric or an antisymmetric function of the coordinates of the two electrons. As with the helium atom, the symmetric function will represent a singlet state, the antisymmetric function a triplet. We must furthermore expect the ground state to be the symmetric one. The reason is that as the two nuclei are pushed together, R approaching zero, the problem will approach that of the helium atom, so that the lowest energy level of the hydrogen molecule will approach the lowest level of the helium atom in the limit. We have already seen that this ground state of helium is a ^1S state.

Heitler and London were interested particularly in the ground state, which obviously should approach the ground state of two hydrogen atoms in the limit as R became infinite. Thus it was natural to take for the function u_a a hydrogen $1s$ wave function on one atom, which we may call the atom a, while u_b is a hydrogen $1s$ wave function on the other atom, atom b. We shall refer to these wave functions as atomic orbitals.

The two-electron wave function will then be very much like the two functions of Eq. 6-11, with the symmetric combination, or singlet, referring to the ground state. There is one difference, however, which arises from the normalization. It is no longer true that the orbitals u_a and u_b will be orthogonal to each other. Heitler and London referred to the integral $\int u_a u_b \, dv$ (we do not bother with a complex conjugate, since both orbitals are real) as an overlap integral. If the atoms are very far apart, the orbitals both fall off exponentially as we go away from their respective nuclei, and their product is everywhere negligible. However, as the nuclei approach each other, u_a and u_b approach hydrogen nuclei on the same nucleus, and the integral approaches unity. It can be shown that this overlap integral, which is ordinarily called S, for two hydrogen $1s$ orbitals, is

$$S = \exp(-w)\left(1 + w + \frac{w^2}{3}\right) = 1 - \frac{1}{6}w^2 + \frac{1}{24}w^4 - \cdots \qquad (13\text{-}1)$$

Here w equals R/a_0. It is clear from Eq. 13-1 that S vanishes exponentially at infinite value of R or w, but approaches unity for small w.

One then finds that the normalized wave functions, instead of having the simple form of Eq. 6-11, are

$$u = \frac{1}{[2(1 \pm S^2)]^{1/2}} [u_a(x_1) u_b(x_2) \pm u_b(x_1) u_a(x_2)] \qquad (13\text{-}2)$$

The energy of the two states can be calculated much as in the helium case, but certain terms come in on account of the lack of orthogonality of u_a and u_b. It proves to be the following:

$$E = E_a + E_b + \frac{H_0 \pm H_1}{1 \pm S^2} \qquad (13\text{-}3)$$

where

$$H_0 = \int u_a^*(x_1) u_b^*(x_2) \left(\frac{1}{r_{12}} + \frac{1}{R} - \frac{1}{r_{1b}} - \frac{1}{r_{2a}}\right) u_a(x_1) u_b(x_2) \, dv_1 \, dv_2$$

$$H_1 = \int u_a^*(x_1) u_b^*(x_2) \left(\frac{1}{r_{12}} + \frac{1}{R} - \frac{1}{r_{1b}} - \frac{1}{r_{2a}}\right) u_b(x_1) u_a(x_2) \, dv_1 \, dv_2$$

$$(13\text{-}4)$$

Here $E_a = E_b$ is the energy of a hydrogen atom in its ground state $1s$. The resemblance to Eq. 6-10 is obvious, and the terms in $1/R$, $1/r_{1b}$, $1/r_{2a}$ are the ones that arise on account of the lack of orthogonality. The quantity r_{1b} is the distance from nucleus b to electron 1, r_{2a} is the distance from nucleus a to electron 2. The units, as before, are the atomic units.

The integrals of Eq. 13-4 are very much more difficult to calculate than the integrals involved with the problem of the helium atom. The so-called coulomb integral, H_0, is fairly simple, but the exchange integral H_1 is difficult enough so that Heitler and London had to approximate it, and it was somewhat later that Y. Sugiura in Japan showed how to compute it analytically. Unlike the case with the helium atom, this exchange integral proves to be negative rather than positive. The negative terms $-1/r_{1b}$ and $-1/r_{2a}$ in the integrand outweigh the positive terms $1/r_{12}$ and $1/R$. Both the integrals H_0 and H_1 are finite for small R, but go to zero as R becomes infinite, so that the energy approaches that of the two isolated atoms at infinite separation. At finite R, both singlet and triplet solutions are allowed, since the orbitals u_a and u_b in the wave function, Eq. 13-2, are different from each other at all finite values of R. Since H_1 is negative, the lowest energy comes for the singlet state, as we expected, not for the triplet as for an atom. The Heitler-London curves for the energies of these two states as functions of R are given in Fig. 13-1.

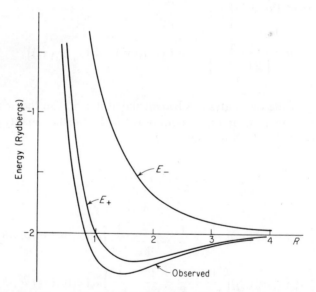

Fig. 13-1. Energies E_+ and E_- for attractive and repulsive states of hydrogen molecule, as determined by Heitler-London method, compared with observed value of energy in ground state. From *Quantum Theory of Molecules and Solids*, Volume 1, by J. C. Slater. Copyright 1963 by McGraw-Hill Book Company, Inc. Used with permission of McGraw-Hill Book Company.

The lower, singlet state has an energy qualitatively similar to the actual ground state of the hydrogen molecule. That is, the energy decreases as R

decreases, reaches a minimum at an equilibrium distance, then increases again at smaller R. The value of R where the energy has its minimum is of the right order of magnitude to agree with experiment, and the dissociation energy, the difference between the calculated energy and the energy of the separated atoms, again is of the right order of magnitude, but only something like half the experimental value. Since the calculated energy of the isolated atoms is in exact agreement with experiment, this means that the calculated energy of the molecule is not low enough. This is the right type of error, according to the variation principle. Any incorrect wave function, as we have seen, must lead to a higher calculated energy than the true wave function. The triplet state, with an energy that rises as R decreases, does not lead to a stable vibrational state of the molecule, but there is experimental evidence that it exists, and that again this simple calculation has led to the right order of magnitude to explain it.

The success of the Heitler-London theory produced a great sensation among physicists. Here was the first suggestion of a theory that could explain the covalent bond, that between two neutral atoms which did not form an ionic molecule. For some time the binding was regarded as a mysterious thing, arising from a nonclassical quantity, the exchange integral, which could not be given any ordinary physical meaning. As we shall see in Sec. 30, this point of view, which was held by most of the leading physicists, was entirely erroneous and misleading. We postpone discussion of this important question until Sec. 30.

Naturally there was an immediate desire to improve the accuracy of the calculations. One very simple method, which was used in 1927 and 1928, was to take advantage of the variation principle, and to try to vary the orbitals u_a and u_b as functions of R, to see if one could not get improvement in this way. The orbital of a $1s$ level in the field of a nucleus of charge Z units is proportional to $\exp(-Zr)$, if r is measured in bohrs. When the two hydrogen nuclei are pushed together to form the helium atom, Z increases to 2 units. Could one not take Z, then, to be a variable parameter, which might go from unity at $R = \infty$ to a larger value as R goes to zero, and thereby improve the calculations?

In 1927 G. Kellner tried this scheme for the limiting case of the helium atom. Heisenberg's scheme would have used hydrogenic orbitals with $Z=2$ for this atom, entirely disregarding the fact that the electron in this atom does not feel the complete attraction to the nucleus of charge 2 units, but is shielded or screened to some extent by the other electron. This is the main error Heisenberg would have removed by his integral $(ab/g/ab)$ in Eq. 6-10. Kellner tried the effect of using a Z intermediate between 1 and 2 for the orbital in the helium atom, and found that the optimum value was 1.6875, which gives quite a good energy for the ground state of the atom.

In 1928 C. S. Wang tried varying this parameter for the hydrogen molecule, and in 1931 N. Rosen extended this calculation to get a curve of energy versus R which reduced to the value 1.6875 for the limit $R = 0$. We show the original Heitler-London curve, Wang and Rosen's curve, and the experimental curve in Fig. 13-2. It is clear that this has made a very considerable improvement in the agreement with experiment. To go much further than this is a major job, and we shall return to the methods which have been used for it in Sec. 19. As we have stated several times, when the best methods of calculation which are now available are used, the calculated results for both the helium atom and the hydrogen molecule agree with experiment to a very high accuracy.

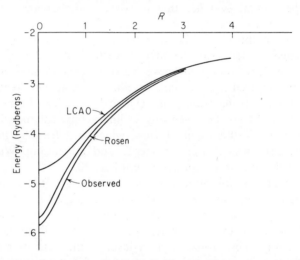

Fig. 13-2. Rosen's variational calculation of H_2 molecule (nuclear repulsion omitted), compared with Heitler-London LCAO calculation and observed curve. From *Quantum Theory of Molecules and Solids*, Volume 1, by J. C. Slater. Copyright 1963 by McGraw-Hill Book Company, Inc. Used with permission of McGraw-Hill Book Company.

14. 1927–1932, Molecular Orbitals

Heitler and London were not the only ones working on molecular theory in the early days of wave mechanics. Hund started in 1927, applying quite a different method which came in 1932 to be called the method of molecular orbitals. Robert Mulliken, a young American whom I had known well in the period of 1924–1926 when he was a postdoctoral worker at Harvard, followed almost immediately. And J. E. Lennard-Jones, a somewhat older Englishman who had been interested in interatomic attractions and repulsions for some years, wrote his first paper on the subject in 1929. I visited him in Bristol, where he then was, at the end of my European visit in 1929–1930.

Their ideas were very much alike, and quite along the lines of Hartree's self-consistent field. Hund's first papers came before Hartree's; he did not use the term self-consistent field, but his idea was the same: An electron in a molecule, just as in an atom, should move in the field of the nuclei and of the other electrons. The work in the 1920s, and in fact for many years thereafter, was almost entirely qualitative and descriptive, since no one in those days had any clear idea as to how to solve a self-consistent-field problem in any case which did not have the spherical symmetry of an atom. Nevertheless, there was much information which they derived from quite simple arguments.

The prewar work was largely on diatomic molecules, whose spectra had been studied experimentally for a long time. One must realize that molecular spectra are much more complicated than atomic spectra, in that they result from an interplay of electronic energy levels and energy levels arising from the rotation and vibration of the nuclear framework of the molecules. A first and important step toward the separation of these two parts of the problem from each other was taken in 1927 by Born and Oppenheimer. Robert Oppenheimer was a very brilliant physics undergraduate at Harvard during the 1920s, the period when I was there on the faculty, and we all recognized that he was a person of very unusual attainments. Rather than going on for his graduate work at Harvard, he went to Germany, and worked with Born, developing what has been known as the Born-Oppenheimer approximation. We can easily describe in words the idea behind this approximation, which Born and Oppenheimer justified by straightforward application of Schrödinger's equation to the problem.

One starts by assuming the nuclei, two in the case of the diatomic molecule, to be held rigidly in position, at a distance R apart. One then solves Schrödinger's equation for the motion of the electrons in the field of these fixed nuclei. It is this part of the problem which Heitler and London had attacked. The result, for each electronic state, is a wave function of the electronic coordinates, whose energy is a function of R, the internuclear distance, such as we have shown in Fig. 13-1. Next, one acts as if this energy as a function of R was a potential energy for the problem of the nuclear motion. This latter problem, leading for each electronic state to vibrational levels around the equilibrium value of R and rotational levels, had been studied very effectively by the older quantum mechanics, and it was easy to modify the older results to take account of the relatively small differences produced by wave mechanics. Hence it came about that practically the whole effort of the theorists in the period following 1926 was to solve for the electronic motion. Born and Oppenheimer investigated the errors arising from their approximate procedure, and showed that they were generally small enough so that they could be neglected in a first approach to the problem. The reason why they were so small was that the terms disregarded in the Born-Oppenheimer approximation had factors proportional to the ratio of the mass of the electron to the mass of the nucleus, and this was small on account of the enormously large mass of the nucleus in comparison with the electronic mass.

As a result of this theorem of Born and Oppenheimer, practically all the work we discuss in this volume assumes that the nuclei are held fixed, but we are interested in the behavior not only at the equilibrium internuclear distance, but also at larger and smaller distances. It is the curvature, or second derivative of the curve of energy versus R, in the neighborhood of the minimum, which leads to the vibrational frequency of the atoms in the molecule with respect to each other, which can be measured by many types of experiments. The equilibrium distance, at which the energy is a minimum, is found experimentally by the moment of inertia of the molecule, which in turn is known from rotational levels. We shall not go further into these questions, but should point out that to the practical molecular spectroscopist, they are of great importance. There was much activity in this field at Harvard in the 1920s. Kemble's research interest was in this field, and several other experts in molecular spectroscopy spent time at the laboratory.

The problem to which Hund, Mulliken, and Lennard-Jones addressed themselves was then that of an electron moving in a self-consistent field arising from two nuclei in fixed positions, at a distance R apart, and from the electrons of the two atoms. This problem has rotational symmetry about the axis. In a problem of rotational symmetry, one can prove very easily that a one-electron wave function must behave like $\sin m\phi$ or $\cos m\phi$ or $\exp(im\phi)$, where ϕ is the angle of rotation about the axis and m is an integer. Since the wave function must depend on three coordinates, of which φ is one, it must depend on two other coordinates, which could be chosen in many possible ways, but which fundamentally represented coordinates in a plane passing through the axis of rotation of the molecule. For example, one could use the distances from each of the two nuclei as coordinates.

The integer m (the molecular spectroscopists actually did not use this letter to describe it, but we do so on account of the similarity to the atomic problem) measures the component of orbital angular momentum along the axis of the molecule, in multiples of $\hbar$. The total angular momentum of the electrons in the molecule is not quantized, but this component is. The molecular spectroscopists had recognized the existence of this quantum number for a long time, and had adopted Greek letters $\sigma, \pi, \delta, \ldots$, to represent values $m = 0, \pm 1, \pm 2, \ldots$, by analogy with the s, p, d, of the atomic spectroscopists. One realizes that the two states with $m = \pm 1$ will be degenerate with each other, as will the two with $m = \pm 2$, etc., while the σ state will be nondegenerate. The sum of the m's for the various electrons in a molecule would also be quantized, and the molecular spectroscopists had adopted the capital Greek letters, $\Sigma, \Pi, \Delta, \ldots$, to describe these quantum numbers of a whole molecule. Thus, for instance, the states of the hydrogen molecule which Heitler and London had discussed were $^1\Sigma$ for the ground state, and $^3\Sigma$ for the repulsive state, where the multiplicity arises, as in an atom, from the quantized spin angular momentum.

In a symmetrical molecule like H_2, composed of two like nuclei, there is an inversion symmetry with respect to a point midway between the two nuclei. The general properties of Schrödinger's equation when we have a reflection plane of symmetry or a point of inversion, which we commented on in Sec. 6, come into effect: All wave functions of the one-electron problem must be either symmetric or antisymmetric on reflection in the plane or on inversion in the point of inversion. In such a case, the convention arose of denoting the wave function or molecular orbital symmetric on inversion by a subscript g, the antisymmetric wave function by u, from the German terms "gerade" (even) and "ungerade" (odd), which Hund had used in his early papers.

It became the custom through a suggestion of Mulliken in 1932 to refer to the one-electron wave functions in these cases as molecular orbitals. There were no further symmetry properties or quantum numbers aside from those described by the notation σ, π, δ, and the subscript g or u for a symmetric molecule. Thus the various molecular orbitals of one type, such as σ_g, can be labeled $1\sigma_g, 2\sigma_g, \ldots$, in order of ascending energy. In general two curves of energy versus R for molecular orbitals of the same type do not cross, so that this method of labeling the orbitals was ordinarily unique. The problem then was how to find out something about these molecular orbitals and their energies as functions of R, and how to build up pictures of the electronic constitution of the whole molecule by assuming that the lower molecular orbitals, up to a Fermi energy, were occupied, while the higher ones were empty. As I have mentioned earlier, there was practically no quantitative work to go on in the early days, and one had to use rather indirect and approximate methods.

Foremost among the useful practical approximations was that of building up an approximate molecular orbital as a linear combination of atomic orbitals (LCAO). The obvious first example of this was the hydrogen molecule. One can start with the same $1s$ atomic orbitals of the hydrogen atoms which Heitler and London had used, u_a and u_b, centered on the atoms a and b. These $1s$ atomic orbitals fall off exponentially as the distance from the nucleus increases. If then we build up the linear combinations

$$1\sigma_g = \frac{1}{(1+S)^{1/2}}(u_a + u_b) \qquad 1\sigma_u = \frac{1}{(1-S)^{1/2}}(u_a - u_b) \qquad (14\text{-}1)$$

where the orbitals are as in Sec. 13, we shall have normalized symmetric and antisymmetric molecular orbitals of the LCAO type, which can serve as approximations to the lower orbitals of the H_2 problem. Lennard-Jones in 1929 made a beginning toward using this method for discussing the molecular orbitals. However, his results were not yet available when I was in Leipzig in the fall of 1929, and the first piece of calculation I carried out there was a complete investigation of the relation of the molecular orbital method for the hydrogen molecule to the Heitler-London method. This was a sufficiently informing piece of work from the point of view of comparing the two methods so that it will be worthwhile describing it in detail here.

I started by setting up the problem in terms of the molecular orbitals $1\sigma_g$ and $1\sigma_u$ of Eq. 14-1, and of the four spin orbitals $1\sigma_g\uparrow, 1\sigma_g\downarrow, 1\sigma_u\uparrow, 1\sigma_u\downarrow$ which can be formed from them. With the two electrons to be distributed among these four spin orbitals, there are six ways of setting up determinantal functions, as indicated in Table 14-1, similar to Table 9-2. When I started to calculate the coulomb and exchange integrals to use in the general determinantal method, I found that almost all of them involved the same work which Heitler and London and Sugiura had already carried out, and it was easy to evaluate the one integral that was required which had not been worked out previously. In Table 14-1, the two functions with $M_S = 1$ and -1, and one of the combinations of the third and fourth functions in the table, lead to the three components of a $^3\Sigma_u$ molecular state, which proves to be identical with the $^3\Sigma$ of the Heitler-London calculation.

Table 14-1

	M_S	
$1\sigma_g\uparrow 1\sigma_g\downarrow$	0	$^1\Sigma_g$
$1\sigma_g\uparrow 1\sigma_u\uparrow$	1	$^3\Sigma_u$
$1\sigma_g\uparrow 1\sigma_u\downarrow$	0 ⎫	$^3\Sigma_u, {}^1\Sigma_u$
$1\sigma_g\downarrow 1\sigma_u\uparrow$	0 ⎭	
$1\sigma_g\downarrow 1\sigma_u\downarrow$	−1	$^3\Sigma_u$
$1\sigma_u\uparrow 1\sigma_u\downarrow$	0	$^1\Sigma_g$

The other combination of the third and fourth functions in the table proves to be a $^1\Sigma_u$ function, given by

$$\frac{1}{[2(1-S^2)]^{1/2}}[u_a(x_1)u_a(x_2)-u_b(x_1)u_b(x_2)]\frac{1}{\sqrt{2}}[\alpha(1)\beta(2)-\beta(1)\alpha(2)]$$

(14-2)

in the same sort of notation used in Eqs. 9-29 and 13-2. The spin function is the same as for other singlets, and the function of coordinates, as for all two-electron singlets, is a symmetric function of the coordinates of the two electrons. But we se that if we interchange the roles of nuclei and orbitals a and b, which is what is involved in the inversion operation, the wave function changes sign, which is the reason for the $^1\Sigma_u$ notation. But quite contrary to the wave functions used by Heitler and London, both electrons are on the same nucleus, or in the same atomic orbital. Half the time they are both on nucleus a, half the time on nucleus b. These states, composed of a positively and a negatively charged ion, are referred to as ionic states.

We are left with the first and last functions in the table. The determinantal functions are the following:

$1\sigma_g\uparrow 1\sigma_g\downarrow {}^1\Sigma_g$: $\dfrac{1}{2(1+S)}[u_a(x_1)u_a(x_2)+u_b(x_1)u_b(x_2)+u_a(x_1)u_b(x_2)$

$+u_b(x_1)u_a(x_2)]\dfrac{1}{\sqrt{2}}[\alpha(1)\beta(2)-\beta(1)\alpha(2)]$

$1\sigma_u\uparrow 1\sigma_u\downarrow {}^1\Sigma_g$: $\dfrac{1}{2(1-S)}[u_a(x_1)u_a(x_2)+u_b(x_1)u_b(x_2)-u_a(x_1)u_b(x_2)$

$-u_b(x_1)u_a(x_2)]\dfrac{1}{\sqrt{2}}[\alpha(1)\beta(2)-\beta(1)\alpha(2)]$ (14-3)

Here we have two functions of the $^1\Sigma_g$ type, rather than the one such function which Heitler and London found as the ground-state function. If we multiply the first function by $1+S$, the second by $1-S$, and add, the result has an orbital function which is a constant times $u_a(x_1)u_a(x_2) + u_b(x_1)u_b(x_2)$, an ionic state of the $^1\Sigma_g$ symmetry, while if we subtract, the orbital function is a constant times $u_a(x_1)u_b(x_2) + u_b(x_1)u_a(x_2)$, the Heitler-London ground state, from Eq. 13-2. But with the two functions of the same symmetry type, we must make linear combinations of the two in such a way as to minimize the energy, in order to get the best approximation to the ground state which can be obtained from the basis functions of Table 14-1. I computed this ground-state energy as a function of R, with great hopes that it would lead to a significant improvement over the Heitler-London result, but was disappointed; the lowering of energy was very small. Nevertheless, the calculation helped a good deal in showing the relation of the molecular orbital to the Heitler-London method.

This relation is best shown from Fig. 14-1, drawn from the paper I wrote in Leipzig. In this figure we first show the $^3\Sigma_u$ and $^1\Sigma_u$ energy levels as functions of R. Then we show the quantities called H_{11} and H_{22}, diagonal matrix components of energy of the two states $1\sigma_g\uparrow 1\sigma_g\downarrow$ and $1\sigma_u\uparrow 1\sigma_u\downarrow$, $^1\Sigma$ states whose wave functions are given in Eq. 14-3. There is, however, a nondiagonal matrix component between the two, and the secular equation reduces to a quadratic equation, which must be solved for the two eigenvalues. These eigenvalues are the two $^1\Sigma_g$ levels given by the full curves in Fig. 14-1. It is the lower one of these which represents the ground state, and which lies slightly lower and therefore slightly closer to the correct value, than the Heitler-London ground state. It and the $^3\Sigma_u$ states both go to the energy of two neutral hydrogen atoms at infinite R, just as the Heitler-London states do.

We have, however, in addition two higher levels, $^1\Sigma_u$ and $^1\Sigma_g$, which go at infinite R to the energy of a H^+ and a H^- ion, in which both electrons are either in orbital u_a or u_b, with opposite spins. It requires considerable energy to remove an electron from one hydrogen atom and attach it to the other, which is the reason why these go to a higher energy at infinite R. There are such states of the hydrogen molecule, but there are many others as well, going to excited states of one or the other atom at infinite R. These higher levels of the molecule cannot be found at all accurately with the small number of basis functions we have included in Table 14-1. We should need many more determinantal functions to get a proper representation of them.

But it will pay us to look more closely at the energy marked H_{11} in Fig. 14-1. This is the energy of the single determinantal function in which both electrons are assigned to the lowest molecular orbital, $1\sigma_g$, one with each spin. It is this function which would be used to represent the ground state, in the MO-LCAO (molecular orbital, linear combination of atomic orbitals) method. We see that near the equilibrium distance, or minimum of the lowest curve, it lies nearly as low as the Heitler-London curve, but at infinite R it goes to an energy which in fact is half-way between the ground state and the ionized state energies. The reason is that, as one can see from Eq. 14-3, the wave function is a mixture of the two functions, for the ionic and the nonionic state. For instance, suppose electron 1 is on atom a, so that $u_a(x_1)$ is large, while $u_b(x_1)$ is very small. Then the wave function reduces to $u_a(x_1) [u_a(x_2)+u_b(x_2)]$, giving equal probabilities of finding electron 2 on either atom. This behavior is not correct, and it is very widespread with the molecular orbital method. This shortcoming of the MO method had not been properly appreciated before my paper, but it did not affect the fact that it is capable of giving quite good results at values of R near the equilibrium.

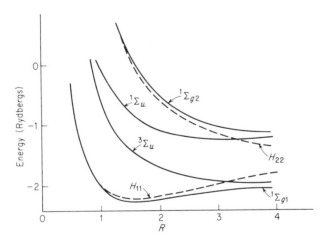

Fig. 14-1. Energy of hydrogen molecule as function of internuclear distance, by molecular orbital method. Full curves, energies as derived from secular equation. Dashed curves, diagonal energies of $^1\Sigma_g$ states constructed from molecular orbital wave functions, without configuration interaction. From *Quantum Theory of Molecules and Solids*, Volume 1, by J. C. Slater. Copyright 1963 by McGraw-Hill Book Company, Inc. Used with permission of McGraw-Hill Book Company.

Most of the work done on diatomic molecules, during the period from 1930 to 1932, was descriptive, with the aim of finding which atomic levels each molecular orbital energy went to at $R=0$ and $R=\infty$. This effort was largely stimulated by the work which O. Burrau, a Danish scientist, had carried out very soon after Schrödinger's equation was first suggested. Burrau noted that the problem of a single electron moving in the field of two bare nuclei was exactly soluble in wave mechanics, as the corresponding problem of a classical particle moving in the field of two attracting centers is exactly soluble in Newtonian mechanics. This meant that the problem of H_2^+ could be exactly solved. In Figs. 14-2 and 14-3 we give the

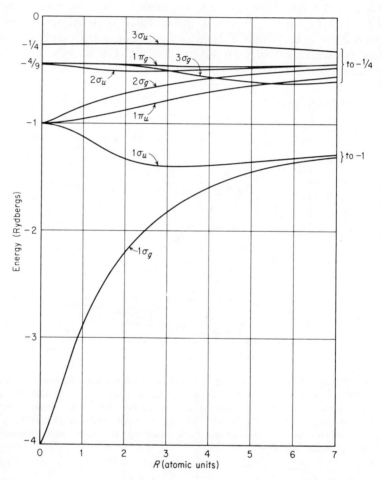

Fig. 14-2. Lowest energy levels of H_2^+ as function of internuclear distance. Internuclear repulsive energy not included. From *Quantum Theory of Molecules and Solids*, Volume 1, by J. C. Slater. Copyright 1963 by McGraw-Hill Book Company, Inc. Used with permission of McGraw-Hill Book Company.

energy levels of this problem partly as worked out by Burrau, partly as filled in much later by Bates, Ledsham, and Stewart. In Fig. 14-2 we give the energy levels omitting the repulsive potential energy $1/R$ between the two hydrogen nuclei, whereas in Fig. 14-3 this repulsion is included. We see that the two lowest molecular orbitals are the $1\sigma_g$ and $1\sigma_u$, as assumed in the H_2 problem. In Fig. 14-2, we see the reason for the large energy separation between these two energy levels: The $1\sigma_g$ goes to the $1s$ orbital of the He^+ ion which is formed when two hydrogen nuclei come together,

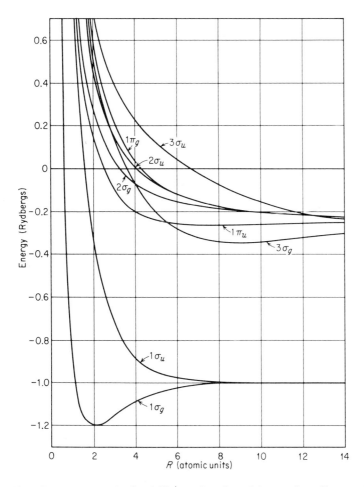

Fig. 14-3. Lowest energy levels of H_2^+ as function of internuclear distance, including internuclear repulsive energy. From *Quantum Theory of Molecules and Solids*, Volume 1, by J. C. Slater. Copyright 1963 by McGraw-Hill Book Company, Inc. Used with permission of McGraw-Hill Book Company.

while the $1\sigma_u$ goes to the much higher $2p$ orbital of He$^+$. Similar information, based on what Mulliken called the "promotion" of some quantum numbers when atoms are brought together, made it possible in diatomic molecules formed from heavier atoms to estimate which energy levels were attractive, which repulsive.

It was also possible to understand why the $1\sigma_g$ orbital in H$_2$ led to a covalent bond between the two atoms, while we have repulsion if one $1\sigma_g$ orbital is replaced by a $1\sigma_u$, by direct study of the electrostatics of the problem. The $1\sigma_g$ orbital, usually called a bonding orbital, has a considerable concentration of electronic charge near the midpoint between the nuclei, where the atomic orbitals u_a and u_b of Eq. 14-1 add. On the other hand, the $1\sigma_u$, called an antibonding orbital, has a charge density which goes to zero at the midpoint. The concentration of charge near the midpoint of a bonding orbital is attracted to both nuclei, and it is this attraction which is largely responsible for the covalent binding. This is seen from Eq. 13-4, where in the Heitler-London method the negative value of the energy H_1, which is what leads to binding, arises from the negative terms $-1/r_{1b}-1/r_{2a}$, the attractions between electrons and nuclei. The charge distribution in the exchange integral H_1, which leads to the binding, is the overlap charge, whose magnitude is the overlap integral S, and which is largely concentrated between the atoms.

Both the Heitler-London and the molecular orbital methods led to about the same amount of charge concentration in the region of the bond, and each one led to the fact that in this region we had two electrons, one of each spin. This was widely held to verify the qualitative ideas which Gilbert N. Lewis, the veteran chemist at Berkeley, had propounded some years earlier, to the effect that an electron pair was required for a covalent bond, the type of bond that one has between uncharged atoms. Lewis had pointed out many cases where such binding occurred. As we can see, it was not essentially the fact that there were two electrons composing the bond, but rather that they were in the region between the atoms, which was the important point. They formed a pair only on account of the exclusion principle, which made it possible to have only two electrons, one of each spin, in the bonding region. The H$_2^+$ ion, shown in Fig. 14-3, has a stable bond, though it has only one bonding electron. But this general interpretation of the nature of the bond led to the concept that an overlap of wave functions from the two atoms involved in the bond was essential in the region between the atoms forming a covalent bond.

Since either the Heitler-London or molecular orbital method seemed to lead to approximately the same picture of bonding, while the Heitler-London method also behaved properly when the atoms were separated, the Heitler-London method appealed to me rather more in the period we are discussing. Pauling, whose thoughts were traveling along much the same direction, felt the same way, and some writers referred to the Heitler-London-Slater-Pauling (HLSP) theory, as opposed to the Hund-Mulliken appellation for the molecular orbital scheme. In 1931, after my return from Leipzig, I wrote a paper describing in more detail the determinantal method of treating the Heitler-London method, with specific applications to a number of molecules. This merely expressed the method of setting up the matrix components of energy between the required determinantal functions, and made no effort to carry out computations for actual cases.

But a difficulty was obvious, once the formulas were written down: The overlap integrals, like the S in the H_2 case, became much more difficult to handle as the number of electrons in the molecule increased. These overlaps were very much simpler to handle, as I pointed out in the paper written in Leipzig, if one used molecular orbitals, for those were automatically orthogonal to each other, if for instance they were derived from a Hartree-Fock type of self-consistent field. It seemed to me that the feeling of partisanship which led many chemists to "take sides," thinking that HLSP was right and MO was wrong, or vice versa, was very unfortunate, and I wrote a short note in 1932, pointing out that just as in the case of H_2, which I have described, one could use either method, provided one included ionized states as well as states of neutral atoms, and could get to the same final results in any problem. And I mentioned this superior feature of the MO method, that it greatly simplified the orthogonality problem, as a strong argument in favor of using that method for any actual calculations. This, I believe, is a point of view that justified itself in postwar days, when there was a great development of numerical calculation, but it did not greatly impress workers in 1932.

In spite of these matters of mathematical technique, the physical fact remained that it was the concentration of electronic charge in a bond which was responsible for holding molecules together. This suggested to both Pauling and to me independently that if we could set up atomic orbitals which projected out in the direction of a bond, the strength of the binding would be increased. Both of us thought of using p orbitals, which have directional properties, rather than the spherically symmetrical s orbitals, to get more charge concentration in the bond, and we both noticed that by making linear combinations of s and p orbitals this concentration could be increased still further. We wrote papers practically simultaneously in 1931, making some applications of these ideas to such molecules as water, ammonia, and methane, where the angles between the bonds were well known, pointing out how such directed bonds would explain many features of the molecular geometry. The work up to that time had largely neglected the importance of dealing with any orbitals other than s orbitals in the explanation of chemical binding. As I have hinted, the combination of orbitals, now called hybridization, which was introduced in this way proved to be equally useful whether one was discussing the Heitler-London or the molecular orbital method. But here, as with the diatomic molecules, work was largely qualitative and descriptive in the prewar period.

This was not true, however, of some general questions of symmetry. I have mentioned the importance of the symmetric or antisymmetric wave functions which one has with a reflection or inversion symmetry. These are only the simplest cases of symmetry operations, whose general treatment comes through the group theory. Mulliken, Van Vleck, and others made great contributions to the study of the molecular symmetry during the period after 1932, contributions which have been of great importance for later work. I ought to say, so that the reader will not get the wrong idea, that the hostility to the group theory which I described in Sec. 8, where I was speaking of the "Gruppenpest", applied specifically to what many people felt was an unnecessary application of the group theory to a problem which could be equally well handled by simple ideas of antisymmetry. The application to molecular symmetry was a completely different case, in which group theory is both appropriate and necessary.

At about the same time that Pauling and I were having the thoughts I described regarding the directional properties of orbitals, a quite independent approach to somewhat similar problems was made by Erich Hückel, a young German whom I had met in Leipzig, and who proceeded to write two very important papers in 1930 and 1931 about the double bond. He treated in particular the molecules of ethylene, C_2H_4, and of benzene, C_6H_6. These are both planar molecules, as shown in Fig. 14-4. And here it proved to be the case that there was a fundamental difference between those orbitals which had a maximum in the plane of the molecule, and those which were antisymmetric in this reflection plane, positive above, negative below, and zero in the plane. It is rather unfortunate that the chemists started calling the orbitals of the first type σ orbitals, those of the second type π orbitals, producing thereby a confusion with the similar but somewhat different use of the notations σ and π orbitals for diatomic molecules.

Let us first think of ethylene, as Hückel did. The carbon atom has six electrons, of which two are in the K shell, leaving four valence electrons. These four from each of the two carbons, plus the four hydrogen electrons, give 12 electrons which might take part in bonding. There are only five bonds, however: four C–H bonds, and one C–C bond. According to the ideas of Lewis, these five bonds would require ten electrons, and this leaves two electrons over. Somehow it has been assumed by the chemists that these two extra electrons formed an additional C–C bond, giving a double

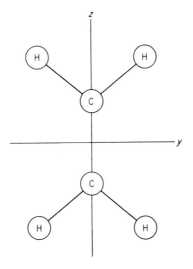

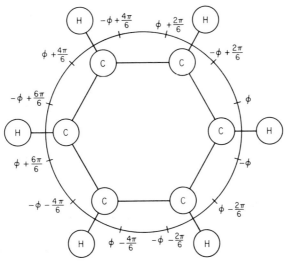

Fig. 14-4. Atomic positions in the (*a*) ethylene molecule, (*b*) benzene molecule. In (*b*), positions of identical potential are indicated. From *Quantum Theory of Molecules and Solids*, Volume 1, by J. C. Slater. Copyright 1963 by McGraw-Hill Book Company, Inc. Used with permission of McGraw-Hill Book Company.

bond between the carbons. Hückel wished to form an idea as to the electronic structure of this extra C–C bond.

Of the four valence orbitals of carbon, three are σ orbitals, and one is a π orbital. The s orbital is obviously of σ type, and the three p orbitals are functions like $xf(r)$, $yf(r)$, $zf(r)$, of which one will have the nodal plane in the plane of the molecule [$xf(r)$, if x is the coordinate at right angles to the plane of the molecule], and the other two will have maxima in the plane of the molecule. Thus we have three σ orbitals, one π orbital. The ordinary single bonds are of σ type, and their molecular orbitals will be linear combinations of the σ-type carbon orbitals and the hydrogen orbitals. Thus we have just ten σ-type atomic orbitals from which we can make five molecular orbitals of bonding type, each of which can be occupied by an electron of each spin, leading to the five single bonds demanded by Lewis's ideas. We can also make five antibonding σ-type molecular orbitals, which are empty. The actual molecular orbitals are fairly complicated combinations of the atomic orbitals on the various atoms. Each molecular orbital must be either even or odd with respect to x, y, and z, to take account of the symmetry of the molecule. We do not need to go into these details to understand the fundamental point of Hükel's work, which is that the additional carbon-carbon bond must be of π type, since that is the only orbital which is not accounted for so far.

The π atomic orbital of carbon, as we have just mentioned, would be of the nature of $xf(r)$, if x is the coordinate at right angles to the plane of the molecule. Let us consider just these two orbitals, one on each of the carbons. If we add them, the wave functions of the two will both be positive and will give an overlap charge on the side of the molecule at positive x. Similarly they will both be negative, and again will lead to overlap charge, on the opposite side of the molecule, with negative x. Hence there will be these two charge concentrations, above and below the plane of the molecule, each of which can be attracted to the carbon nuclei, and can produce bonding. Hückel gave a simplified treatment of the Heitler-London method of considering the bonding of this molecule, disregarding all the overlap integrals S, and estimating the coulomb and exchange integrals by comparison with experiment, and got such a good qualitative explanation of the properties of the double bond that his method has formed the basis of approximations that are still used today.

The main qualitative property of the double bond, which is not found in a single bond, is in its resistance to rotation. It is known experimentally that the restoring force opposing a rotation of half of the molecule with respect to the other half, about the C–C axis, is very large. Since such a rotation would reduce the overlapping between the π-type orbitals on the two carbons, it is qualitatively clear that it should greatly increase the energy, in agreement with observation. A σ-type bond shows no such resistance to rotation, as is known from the case of ethane, C_2H_6, in which only single bonds occur, and rotation of one CH_3 group with respect to the other encounters only a slight opposing force.

The other case considered by Hückel, the benzene molecule, shows additional interesting properties. If we count up σ-type orbitals, as we did for ethylene, we find just enough σ-type orbitals to produce single bonds between each carbon and its adjacent hydrogen, and between each carbon and its carbon neighbors. This would give only three bonds from each carbon, as we had for the σ bonds in ethylene, and there is one π orbital left over from each carbon, which could be used to produce double bonding. Here the situation is not as simple as it seems, however, as the chemists have realized for many years.

The six π orbitals could be used to form three bonding orbitals, with two electrons in each, and this would only be enough to give double bonds between alternating pairs of carbons. One would have to leave single bonds between the intermediate pairs. And yet there is nothing in the chemical behavior of benzene to indicate that half the C–C bonds are stronger than the other half. For instance, Pauling had found from large amounts of empirical evidence, mostly from x-ray crystal structure, that a double C–C bond is always shorter than a single C–C bond, as is natural from the additional attraction resulting from the double bond. But all six C–C bonds in benzene are of the same length, and are just about half-way between the lengths characteristic of single and of double bonds. It seems as if there were an additional one-electron π-type bond between each pair of carbons, rather than a two-electron bond as Lewis's picture would have suggested.

Hückel showed that this is just what we are led to by use of a molecular orbital picture. Let us consider merely the π-type orbitals; the σ orbitals show no new features not found in ethylene. We have six π-type atomic orbitals, one on each of the carbon atoms. What are the six linear combinations of these atomic orbitals which form molecular orbitals? This is a question that can be answered unambiguously by the group theory. The symmetry properties of the regular hexagon, which we have in benzene, resemble those with cylindrical symmetry. If we set up an axis perpendicular to the plane of the molecule, passing through its geometrical center, we note that a rotation through an angle of 60° or any integral multiple of that angle brings the molecule back into coincidence with itself. Such a rotation, according to the group theory, must multiply the molecular orbital by a factor of $\exp(im\phi_0)$, where ϕ_0 is the angle 60° and m is a positive or negative integer. The resemblance of this case to the case of rotational symmetry met in an atom or a diatomic molecule is obvious. In those cases, the orbital will be multiplied by a factor $\exp(im\phi)$ on rotation through an angle ϕ, but in the case of complete rotational symmetry through any angle, ϕ can be arbitrary.

If we have six atomic orbitals, $u_1, \ldots, u_6$, located on the carbon atoms at positions making angles of ϕ_0, $2\phi_0, \ldots, 6\phi_0$ with the axis passing through one carbon, then we can set up by inspection the (unnormalized) linear combinations of these atomic orbitals which have the required properties. They are

$$\psi_m = \sum (k=1,\ldots,6) \exp(imk\phi_0) u_k \qquad (14\text{-}4)$$

We can see at once that it has the required behavior. If we rotate through the angle ϕ_0, this means that we expect the wave function to have the same value near the $(k+1)$th atom that the original wave function has near the kth atom. But we see that the coefficient by which the $(k+1)$th atomic orbital is multiplied is $\exp(im\phi_0)$ times the coefficient by which the kth atomic orbital is multiplied, which is what we need to multiply the molecular orbital by the proper factor.

If we were dealing with a problem of complete rotational symmetry, in which rotation through any angle was allowed, we should have molecular orbitals corresponding to any arbitrary value of m. However, with our finite number of possible atoms, this is no longer the case. If we pick only the values $m = 0$, ± 1, ± 2, 3, or six values in all, we can show that we have included all possible linear combinations. For suppose we add 6 to one of these values of m. Then the factors $\exp(imk\phi_0)$ are all multiplied by $\exp(6ik\phi_0)$. But $\phi_0 = 60°$, $6\phi_0 = 360°$, or 2π. Since $\exp(2\pi i) = 1$, the quantities $\exp(6ik\phi_0)$ are all equal to unity. In other words, adding 6 to the value of m repeats one of the functions we have already found. We have, then, just six independent linear combinations of the six atomic orbitals u_k, and they are the molecular orbitals of the π-electron problem.

The case $m = 0$ corresponds to an addition of the atomic orbitals on all six atomic sites, with coefficients which are all the same. These will then result in overlap charge between each pair of adjacent atoms, and this molecular orbital will result in bonding. On the other hand, the case $m = 3$ gives a coefficient which, in going from one atom to the next, is multiplied by $\exp(3i\phi_0) = \exp(\pi i) = -1$, so that atomic orbitals on adjacent carbon atoms come out with opposite signs in the molecular orbital, and there is no charge density between the atoms. This is then an antibonding orbital. The two degenerate orbitals with $m = \pm 1$ are nearer the bonding case than the antibonding, while the two degenerate orbitals with $m = \pm 2$ are nearer the antibonding case. Hence the molecular orbital with $m = 0$ has the lowest energy, $m = \pm 1$ is higher, $m = \pm 2$ still higher, and $m = 3$ is the highest of all. Since we have six electrons, three of each spin, to be accommodated, we shall have one of each spin in the orbital $m = 0$, one of each spin in $m=1$, one of each spin in $m=-1$, and the others will be empty. This is the essence of Hückel's explanation of the structure of the π bonds in benzene. When we add up the amount of overlap charge in each bond, we find equal amounts in each of the six C–C bonds, half as much as in a complete covalent bond, so that we verify our earlier statement that each C–C bond in benzene has a bonding half-way between a single and a double bond.

15. 1928–1930, Bloch Sums and Brillouin Zones

For some time we have been following the work on molecular orbitals in simple molecules. But at the same time this was going on, there was a great deal of activity devoted to finding the orbitals of electrons in a perfect crystal. The early work was mostly for metals, and the sodium crystal was a standard example. In 1928, Felix Bloch, then a young student working with Heisenberg and Hund in Leipzig, set up the method of making a molecular orbital by the LCAO method, for a simple crystal. The method is very much like the scheme which Hückel used two or three years later for benzene, which we have just described. Hückel in fact was led to the method he used for benzene by analogy with Bloch's work. As in Eq. 14-4, the scheme is to make a linear combination of atomic orbitals, with coefficients which equal the value which a free-electron wave function would have at the nuclear positions of the various atoms. The group theory shows that this is the way to proceed, and we can verify it by simple calculations.

First let us consider these free-electron wave functions. We have written them down in one form, Eq. 11-1 where we were discussing the homogeneous electron gas preparatory to setting up the Fermi statistics. The function we used there, $\sin(n_x \pi x/L) \sin(n_y \pi y/L) \sin(n_z \pi z/L)$, was chosen to go to zero on planes which formed the boundary of a cubic region of side L. For most purposes, however, it is more convenient to use the exponential form of wave functions,

$$u = \exp\left[\frac{\pi i}{L}(n_x x + n_y y + n_z z)\right] = \exp(i\mathbf{k} \cdot \mathbf{r})$$

where $\quad k_x = \dfrac{n_x \pi}{L} \quad k_y = \dfrac{n_y \pi}{L} \quad k_z = \dfrac{n_z \pi}{L}$ (15-1)

This form of function satisfies what are called periodic boundary conditions: If x increases by $2L$, the function is multiplied by $\exp(2\pi i n_x) = 1$ if n_x is an integer, so that the function is periodic, with period $2L$, along each of the three axes. The exponential functions of Eq. 15-1 are traveling waves, in contrast to the standing waves of Eq. 11-1.

Bloch used the exponential functions of Eq. 15-1 as coefficients in setting up a linear combination of atomic orbitals to represent a molecular orbital in a crystal. That is, if u_i is an atomic orbital on the ith atomic site, which is located at the position $\mathbf{r} = \mathbf{R}_i$, Bloch set up a sum

$$u = \sum (i) \exp(i\mathbf{k} \cdot \mathbf{R}_i) u_i \tag{15-2}$$

where $\mathbf{k}$ is ordinarily called the wave vector, to represent the type of LCAO function demanded by the crystal periodicity. Such a sum is called a Bloch sum. These functions have the property that if we find the matrix component of a periodic potential, having the periodicity of the crystal, between two Bloch sums corresponding to different $\mathbf{k}$'s, this matrix component is zero. This means that in setting up a secular equation for finding the eigenvalues of a self-consistent-field problem in a periodic potential, we can consider each value of $\mathbf{k}$ separately. Also we have only a discrete set of possible $\mathbf{k}$ values, with the periodic boundary conditions, much as we found only six values of m, namely $m = 0$, ± 1, ± 2, 3, for the benzene problem. If L is made to equal a large integer times the lattice spacing of the crystal, we shall find a great many of these discrete $\mathbf{k}$ values, with very closely spaced energy levels, similar to our case of the homogeneous electron gas in Sec. 11. There are in fact as many $\mathbf{k}$ values as there are unit cells in the fundamental region of the crystal.

The Bloch sum of Eq. 15-2 has the property that if one goes from one point of space to another point at an equivalent position in another unit cell, the wave function is multiplied by a factor $\exp(i\mathbf{k} \cdot \mathbf{R}_{ij})$, where $\mathbf{R}_{ij}$ is the displacement vector from the ith to the jth atom. This factor is analogous to the factor $\exp(im\phi_0)$ which we met in the benzene molecule, when one rotated through such an angle that one carbon atom was carried into its nearest neighboring carbon atom. We can see why it is that this form of linear combination of atomic orbitals has useful properties: The charge density of the electron around any one of the atomic sites, which would involve $\exp(i\mathbf{k} \cdot \mathbf{R}_{ij})$ times its complex conjugate, or unity, has the same value on each atom of the crystal. If we have satisfied Schrödinger's equation on one atom, we have satisfied it on all of them. This theorem, that a wave function in a periodic potential must be multiplied by a factor $\exp(i\mathbf{k} \cdot \mathbf{R}_{ij})$ in going from one point to an equivalent point in another unit cell, is sometimes called Bloch's theorem.

The diagonal matrix component of the Hamiltonian for a Bloch sum will give a good first-order approximation to the energy level of an electron in a periodic potential. To find this quantity, we must first normalize the function of Eq. 15-2. We have

$$\int u^* u \, dv = \sum (ij) \exp[i\mathbf{k} \cdot (\mathbf{R}_j - \mathbf{R}_i)] \int u_i^* u_j \, dv \tag{15-3}$$

The atoms in the crystal are assumed in this simple calculation to be equivalent to each other, and the orbitals u_i are assumed to be normalized. If there are N atoms in the repeating region of the crystal, we shall have in Eq. 15-3 first a set of N identical terms for $i=j$, each equal to unity; then N terms referring to the overlap of nearest neighbors i and j, with coefficients $\exp[i\mathbf{k} \cdot (\mathbf{R}_j - \mathbf{R}_i)]$; then terms for second nearest neighbors; and so on for more distant neighbors. The normalized function will be the quantity u of Eq. 15-2, divided by the square root of this sum. Similarly we find the matrix component $\int u^* Hu \, dv$, where H is the periodic potential Hamiltonian met within a crystalline problem. Here again we shall have terms for which $i=j$, then terms from nearest neighbors, and so on. The final result is

$$(H)_{av} = \frac{\sum (j) \exp(i\mathbf{k} \cdot \mathbf{R}_j) \int u_0^* H u_j \, dv}{\sum (j) \exp(i\mathbf{k} \cdot \mathbf{R}_j) \int u_0^* u_j \, dv} \qquad (15\text{-}4)$$

where u_0 stands for the atomic orbital of an atom at the origin, and the sum is over the j neighbors at distance $\mathbf{R}_j$ from this atom. Ordinarily only nearest neighbors will be important, both in the numerator and denominator, so that the energy is a simple function of the wave vector $\mathbf{k}$. The integrals $\int u_0^* H u_j \, dv$ are ordinarily negative, for they contain terms coming from the attraction of the overlap charge $u_0^* u_j$ with the nuclei. Thus for $\mathbf{k}=0$ we have a negative energy term coming from the interactions with nearest neighbors. On the other hand, the maximum value of $\mathbf{k}$ will lead to $\exp(i\mathbf{k} \cdot \mathbf{R}_j) = -1$, as in the benzene case, and this will give a positive energy term. Hence we find a band of energy levels, the bottom of the band coming for $\mathbf{k}=0$, corresponding to bonding orbitals, while the top of the band corresponds to antibonding orbitals.

Each of the atomic energy levels will be broadened in this way into a band, capable of holding one electron per atom of each spin. This is entirely analogous to our π orbitals in benzene, where we had six atoms, six molecular orbitals, of which only the lower three were occupied, each with an electron of each spin. The broadening depends on the overlap of the orbitals on neighboring atoms. Thus the bands arising from inner, x-ray electrons, whose orbitals are very small and do not overlap their neighbors appreciably, will hardly be broadened at all. On the other hand, outer orbitals will lead to quite broad energy bands, and it is even possible to have them wide enough to overlap their neighboring bands. In such a case, it is not legitimate to use only the diagonal matrix component of energy, as in Eq. 15-4. Rather, one would have to take linear combinations of the levels of the same $\mathbf{k}$ value in the two overlapping bands, leading to a secular equation for the energy.

Bloch in his paper looked into the behavior of these wave functions, as far as their conduction of electricity was concerned. He was able to show that a filled band of electrons will not conduct any current, so that electrical conductivity comes only from outer energy bands, in which on account of the Fermi statistics we might have the lower part of a band filled, the upper part empty. In such a case there can be conduction, much as in the classical theory of electronic conduction of Lorentz and Drude, except that the electrons will be accelerated like particles with an effective mass which can be much greater than the actual mass: the narrower the band, the greater the mass.

We see in these results the beginning of the theory of insulators, metallic conductors, and semiconductors. In an insulator, certain lower bands are filled, upper bands are empty, and there is a considerable energy gap between the top of the highest occupied band, the valence band, and the bottom of the lowest empty band, the conduction band. In a semiconductor, the gap is narrow enough so that one can get appreciable excitation of electrons to the conduction band, leaving gaps in the valence band. The electrons in the conduction band behave like negative electrons with appropriate effective masses, while the holes in the valence band act like positive carriers of charge, but both types of carriers take part in conduction. In a metal, the Fermi energy comes within a band, and conduction is possible without thermal excitation of electrons. It is not our aim in this volume to go into the theory of electrical conductivity, but it is of course a topic on which a good start was made during the decade we are considering, by Bloch, Peierls, Lothar Nordheim, Alan H. Wilson, and many others, and which developed greatly in postwar days, when solid-state electronics became a topic of great practical importance. Sommerfeld and Bethe wrote a very useful handbook article on the electron theory of metals in 1933, which consolidated this whole field.

The LCAO method was not the only one used to discuss molecular orbitals in crystals in the 1920s and early 1930s. In addition, there was a good deal of development of the idea that the whole wave function could be expanded as a sum of plane waves, essentially a three-dimensional Fourier series. One of the first suggestions of this type of calculation was made in 1928 by Hans Bethe, a young student of Sommerfeld. He was considering the problem of electron diffraction. He started with an electron beam, described as a plane wave, $\exp(i\mathbf{k} \cdot \mathbf{r})$. He allowed it to be scattered by a periodic potential in a crystal, which can be expanded as a sum of terms of the nature of $\exp(i\mathbf{K} \cdot \mathbf{r})$. He assumed that the scattered wave would be of the form $\exp(i\mathbf{k}' \cdot \mathbf{r})$. The amplitude of the scattered wave depends on the matrix component of the Hamiltonian between the incident and the scattered wave. This involves

$$\int \exp(i\mathbf{k} \cdot \mathbf{r})^* \exp(i\mathbf{K} \cdot \mathbf{r}) \exp(i\mathbf{k}' \cdot \mathbf{r}) \, dv = \int \exp[i(-\mathbf{k}+\mathbf{K}+\mathbf{k}') \cdot \mathbf{r}] \, dv$$

Such an integral of an exponential function over the whole repeating volume of the crystal is zero, unless the quantity $-\mathbf{k}+\mathbf{K}+\mathbf{k}'=0$, or

$$\mathbf{k}'=\mathbf{k}-\mathbf{K} \tag{15-5}$$

This wave vector **k** is analogous to a momentum, and in fact is sometimes called the crystal momentum, and Eq. 15-5 expresses a conservation law for this crystal momentum. When Bethe looked into the problem of superposing the incident and the scattered waves, with a given nondiagonal matrix component of the Hamiltonian between them, he was led to a secular equation between the two basis functions, which took the form of a quadratic equation. When this was solved, an energy gap opened up between otherwise continuous energy bands. This is illustrated in Fig. 15-1. These gaps come just at the positions where simultaneously one has

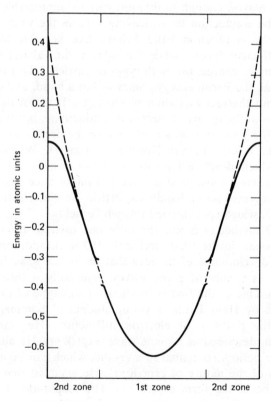

Fig. 15-1. Energy as function of wave vector **k**, showing gaps at boundaries of Brillouin zones.

conservation of crystal momentum and of energy. These are closely tied with the electron diffraction. Electron beams approaching the crystal with such energy and at such directions that the corresponding electron wave within the crystal would fall in these forbidden gaps are the ones that are diffracted. The electrons remain outside the crystal, since they are not able to be propagated inside.

Leon Brillouin, a French scientist some years older than most of those active in wave mechanics, followed these ideas further in important papers in 1930 and 1931. Brillouin made use of some of the techniques which the crystallographers had worked out many years earlier, in the study of crystalline properties in general, and of x-ray diffraction in particular. In a crystal, the vector $\mathbf{R}_n$ from a point in one unit cell to an equivalent point in another unit cell can always be written as a linear combination of three vectors, $\mathbf{t}_1$, $\mathbf{t}_2$, $\mathbf{t}_3$, with integral coefficients:

$$\mathbf{R}_n = n_1 \mathbf{t}_1 + n_2 \mathbf{t}_2 + n_3 \mathbf{t}_3 \tag{15-6}$$

It then is convenient to set up three other vectors, $\mathbf{b}_1$, $\mathbf{b}_2$, $\mathbf{b}_3$, which have the property that

$$\mathbf{t}_i \cdot \mathbf{b}_j = \delta_{ij} \tag{15-7}$$

Then one can set up what is called a reciprocal lattice, in which the vectors $\mathbf{b}_j$ form fundamental vectors the same way that the $\mathbf{t}_i$'s do in the ordinary space lattice. In terms of the reciprocal lattice, we can set up vectors $\mathbf{K}_h$ by the equation

$$\mathbf{K}_h = 2\pi (h_1 \mathbf{b}_1 + h_2 \mathbf{b}_2 + h_3 \mathbf{b}_3) \tag{15-8}$$

where the h's are integers. Then one can prove very easily that the exponential function $\exp(i\mathbf{K}_h \cdot \mathbf{r})$ has identical values at corresponding points of each unit cell of the crystal.

To prove this theorem, we need only show that when $\mathbf{r}$ is increased by one of the vectors $\mathbf{R}_n$, this exponential function is unchanged, which demands that $\exp(i\mathbf{K}_h \cdot \mathbf{R}_n) = 1$. But this follows from the fact that, from Eqs. 15-6, 15-7, and 15-8,

$$\mathbf{K}_h \cdot \mathbf{R}_n = 2\pi (h_1 n_1 + h_2 n_2 + h_3 n_3) = 2\pi \text{ (integer)} \tag{15-9}$$

Therefore any such exponential function can be used in expanding the periodic potential in a crystal in a three-dimensional Fourier series:

$$V(\mathbf{r}) = \sum (\mathbf{K}_h) W(\mathbf{K}_h) \exp(i\mathbf{K}_h \cdot \mathbf{r}) \tag{15-10}$$

It is easy to set up the equations for expanding the wave function in the form

$$u(\mathbf{k},\mathbf{r}) = \sum (\mathbf{K}_h) v(\mathbf{k}+\mathbf{K}_h) \exp[i(\mathbf{k}+\mathbf{K}_h) \cdot \mathbf{r}] \qquad (15\text{-}11)$$

provided the electron is moving in the potential given by Eq. 15-10. This form for u satisfies Bloch's theorem: If r is increased by R_n, the exponentials are all multiplied by the same factor $\exp[i(\mathbf{k} \cdot \mathbf{R}_n)]$, in consequence of Eq. 15-9. Then we find

$$(\mathbf{k}+\mathbf{K}_j)^2 \frac{\hbar^2}{2m} v(\mathbf{k}+\mathbf{K}_j) + \sum (\mathbf{K}_l) W(\mathbf{K}_l) v(\mathbf{k}+\mathbf{K}_j - \mathbf{K}_l) = E(\mathbf{k}) v(\mathbf{k}+\mathbf{K}_j)$$

$$(15\text{-}12)$$

The first term is the kinetic energy $-\hbar^2 \nabla^2 / 2m$, the next is the potential energy, times the appropriate values of v. If one could use a finite number of v's and still get a good expansion, one could set up a secular equation from the coefficients of Eq. 15-12, which are of the same form as Eq. 9-7, and solve it for the energy levels E, after which we could solve for the coefficients $v(\mathbf{k}+\mathbf{K}_j)$. But unfortunately if one looks into the convergence of this expansion of the wave function in plane waves, one finds that the convergence is so slow that an impossibly large number of plane waves would be required to get satisfactory representations of the wave function, so that this method is not practical for computation.

Nevertheless, this type of analysis is very useful for proving some general results, as Brillouin showed. In particular, let us set up all the vectors $\mathbf{K}_h$, from Eq. 15-8, in a $\mathbf{K}$ space. These are all to be regarded as vectors pointing out from the origin. Then set up the plane which is the perpendicular bisector of each of these vectors $\mathbf{K}_h$. These planes will bound certain polyhedra. Take the smallest polyhedron, including the origin. This is called the first Brillouin zone. It was shown by Brillouin that the wave vectors $\mathbf{k}$ lying inside this first Brillouin zone are just those which give distinct wave functions for the problem. That is, the interior of the first Brillouin zone plays the same role in the study of crystals that the range of m from 0 to ± 1, ± 2, and 3 did in the benzene problem. The reason this is so is simple. To go from the center of one face of the Brillouin zone to the opposite face must involve a vector displacement $\mathbf{K}_i$ in the $\mathbf{K}$ space. If we add $\mathbf{K}_i$ to the $\mathbf{k}$ in the expression of Eq. 15-11 for the wave function, we have exactly the same set of wave vectors $\mathbf{k}+\mathbf{K}_h+\mathbf{K}_i$ as before, only they are labeled differently. Hence the resulting wave function must be the same one we had before $\mathbf{K}_i$ was added. We thus see that the generalization of the benzene situation, where the lowest energy came for $m=0$, the highest for $m=3$, is that the lowest energy in the Brillouin zone should come at the center of the zone, the highest values somewhere along the surface of the zone.

These simple statements are oversimplifications, however. The full study of the shape and symmetry properties of the Brillouin zones is a complicated one. In Fig. 15-2 we show the familiar zones for the body-centered and face-centered cubic structures found in many metals. One ordinarily finds energy as a function of **k** along certain lines in this zone. In the period of 1928–1932, which we are talking about, no method which had been proposed for computing the energy was very practical, and no experiments told very much about its detailed behavior. This was one of the features of the problem which remained in an unsatisfactory state for many years.

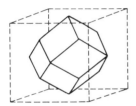 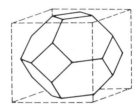

Fig. 15-2. Brillouin zones for body-centered and face-centerfed cubic structures.

16. 1927–1930, Ferromagnetism

In Sec. 6, we mentioned the two very important papers which Heisenberg published in 1926, regarding the many-body problem in wave mechanics. The third in this series, which came out in 1927, was devoted to ferromagnetism. To give a little background, let us go back to an account of some of the work on this subject which had been done in the early 1900s.

Back in the late 1890s, experimental work of Pierre Curie and others had shown that there were two quite different types of magnetic behavior which were common. One was ferromagnetism, which was known in only a few metals—iron, cobalt, nickel, and some alloys. A sample of one of these metals, put in a magnetic field, acquires a large magnetic moment, part of which, because of the effect known as hysteresis, remains after the external field is removed. If the external magnetic field is increased, the magnetic moment of the sample increases, but only to a certain limit, described as saturation. We now know, though it was not known at that time, that the cause of these effects is that the substance is made up of microscopic domains, each of which is magnetized to saturation all the time, whether in an external field or not. The effect of the field is to rotate the directions of magnetization of the various domains, lining them up with the field. Thus the hysteresis and the process of magnetization are rather secondary effects, while the primary fact is the permanent magnetization of a domain. A permanent magnet is formed from some particular alloy which has the property that once the domains are lined up, they stay that way, as a result of extreme hysteresis, rather than reverting to random orientations as in soft iron.

The paramagnetic substances, which are much more common than ferromagnetics, have a very much smaller magnetic moment in the presence of an external magnetic field, and the magnetic moment goes back to zero as soon as the external field is removed. Langevin, in the early 1900s, produced a theory of these substances. He assumed that each individual atom had a magnetic moment, but that in the absence of a field, they were oriented at random. The field tended to line them up, but this effect was opposed by thermal agitation. A very simple treatment showed that on account of this thermal effect, the magnetic moment in the presence of a field should not only be proportional to the external field, but also should be inversely proportional to the absolute temperature. This inverse proportionality to the temperature had been noticed by Curie, and is often called Curie's law. Langevin's theory was able to lead to approximate orders of magnitude for the magnetic moments of the individual atoms, and was one of the early indications of the Amperian currents, the circulating currents in an atom leading to magnetic moments, which were later incorporated into quantum theory and wave mechanics.

As low-temperature experimentation became more widespread in the early 1900s, it was found that paramagnetic substances frequently did not have a magnetic moment exactly proportional to $1/T$, but more often to $1/(T-\theta)$, where θ was a low temperature which came to be known as the Curie temperature, on account of Curie's activity in this field. If this law were to be extended down to sufficiently low temperature, it would predict an infinite magnetic moment at the temperature $T=\theta$, but experiments at low enough temperature to verify this were not practicable at the time. However, high-temperature experiments on the ferromagnetic metals, iron, cobalt, and nickel, began to exhibit a behavior which suggested that the two types of magnetism might be related.

It had been known since very early days that the magnetization of a permanent magnet disappeared at a sufficiently high temperature. Experiments then began to be made on the saturation magnetization of the ferromagnetic metals, as a function of temperature, and it was found that the saturation moment decreased to zero at a temperature of several hundred degrees Kelvin, well above room temperature. Above this temperature the substances became paramagnetic, following approximately the law $1/(T-\theta)$, with a θ which was approximately equal to the temperature at which the ferromagnetism disappeared. It began to seem, in other words, that the infinite magnetic moment predicted at $T=\theta$ was merely the permanent moment already found in ferromagnetics. Naturally the suspicion arose that a paramagnetic material below the Curie temperature would become ferromagnetic, and when experimentation became practical at sufficiently low temperatures, this was often found to be the case.

It was at this time, in 1907, that Pierre Weiss, a French physicist then in his forties, proposed a theory to explain both types of magnetization in a unified way. He started with Langevin's theory of paramagnetism, but he assumed that the magnetic field tending to line up the magnetic atoms was the sum of the external field applied from outside, and of what he called the inner field, proportional to the magnetization itself. In other words, he assumed what is often described as a cooperative effect, by which the magnetization of the crystal tends to pull other atoms into alignment along the same direction as the magnetization. When he put this assumption into the equations, they proved to lead very simply to results in close agreement with experiment.

He found the saturation magnetization at very low temperatures, with the saturation moment decreasing with increasing temperature up to a temperature θ, which could be predicted in terms of the inner field. The predicted curve agreed satisfactorily with observation. Above the temperature θ, the theory predicted a paramagnetic behavior, with moment proportional to $1/(T-\theta)$, again in agreement with experiment. The theory even predicted the coefficient of the equation giving paramagnetic moment as a function of field and temperature, with the right order of magnitude. Seldom has a simple theory worked better. And yet seldom has the fundamental basis of the theory been more unreasonable, as Weiss himself realized. For if one looked into the magnitude of the inner field, and assumed, as would seem plausible, that it arose from magnetic interactions between the various magnetized atoms, the fact immediately emerged that the observed inner fields were tens of thousands of times greater than could possibly be explained.

This left physicists in a dilemma for 20 years. Naturally they accepted the Weiss theory as a working hypothesis, but they were left completely in the dark as to the nature of the large forces which must be acting to line up the magnetic moments of neighboring atoms. What Heisenberg was able to do in his 1927 paper was to tie up this difficulty with another one whose explanation he had already given in his 1926 papers: the large energy differences between the singlet and triplet states in the spectra of two-electron atoms. We have mentioned that any magnetic effects of interaction between the magnetic moments of the two electrons were enormously too small to explain the large observed energy differences. And we have seen how the exchange integral which Heisenberg had introduced in his 1926 papers, an electrostatic effect and hence of a very much larger order of magnitude than magnetic effects, was responsible for the effect. Heisenberg opened a whole new era in the theory of magnetism by suggesting in 1927 that the interactions resulting in Weiss' inner field were likewise a result of exchange integrals.

Heisenberg's suggestion was not so successful, however, when it came to predicting details of experiment. He based his arguments on a theory similar to the Heitler-London theory for the hydrogen molecule. If two neighboring atoms had lower energy when their spins were parallel than when they were opposite, he would have the material for explaining Weiss' inner field. In fact, such interactions almost certainly would fall off so rapidly with distance that only nearest neighboring atoms would be important, and one could find the energy difference between the state of a cluster containing a central atom and its nearest neighbors, when all atoms had their magnetic moments parallel, and the state when the central atom had reversed its magnetization. This energy difference could be written in terms of the exchange integrals of the central atom and its nearest neighbors, and alternatively in terms of the Weiss field. Thus one could get a straightforward interpretation of the Weiss field in terms of exchange integrals.

But as we have seen in Sec. 13, in our discussion of the Heitler-London method, the exchange integral H_1 of Eq. 13-4 is ordinarily negative, leading to a lower energy for the singlet state of the molecule, with the spins antiparallel, than for the triplet, with parallel spins. To explain ferromagnetism, Heisenberg had to assume that for some reason the exchange integrals for the ferromagnetic metals should be positive rather than negative. In Eq. 13-4 for the exchange integral, this of course is possible. The negative terms in that integral, $-1/r_{1b}$ and $-1/r_{2a}$, are present only on account of the lack of orthogonality of the two orbitals. If we do not have overlap, we have only the term $1/r_{12}$, which is present in an atom, and which necessarily gives a positive exchange integral. We have pointed out that this is what leads to Hund's law, making the atom have the lowest energy if the spins are parallel.

Something like this must be going on in the ferromagnetic case. And yet one could not see why just the ferromagnetic crystals, iron, cobalt, and nickel, should stand out from other substances, and be ferromagnetic. This difficulty with Heisenberg's theory was really never cleared up. No calculation of exchange integrals led directly to integrals which gave any explanation of which substances should be ferromagnetic, which should not. This rather stopped fundamental work with Heisenberg's theory, but left it as a very useful semiempirical theory. If one was ready to accept a positive exchange integral as an experimental fact, one could go on from there, and derive a theory from Heisenberg's hypotheses which went considerably beyond the Weiss theory, and gave many useful results in accord with experiment. Van Vleck in particular took up the Heisenberg theory, developed it to the point where he became a leading figure in the field of magnetism, and in his book *Electric and Magnetic Susceptibilities*, published in 1932, gave a treatment so complete and fundamental that it has remained the leading book in the field ever since. But this fundamental difficulty with the Heisenberg theory was not touched. It is largely for that reason that the study of magnetism has remained a field rather outside the main line of solid-state theory. It was not clear how to tie it in with the energy-band theory we have sketched in the preceding section.

Quite a different approach to magnetic theory was taken by Bloch in 1929. I have mentioned that on my visit to Zurich in 1929, I first met Bloch, and showed him the preprint of my paper on the determinantal method. He was at a point where he wanted to get into the theory of magnetism, and he saw that the method gave him a way of introducing magnetism into the free-electron theory of the type which Sommerfeld and his students had been using. He carried this through very rapidly, and within a few weeks wrote a very useful paper embodying this approach to magnetism. It is so important, on account of its relation to the $X\alpha$ method, that I shall describe it in some detail.

Bloch started by assuming a free-electron gas with different numbers of spin-up and spin-down electrons, so that it would have a net magnetization, and he computed the Hartree-Fock energy of such a gas, as a function of magnetization. His idea was to find whether the unmagnetized state had a higher or a lower energy than a magnetized state. We shall state Bloch's results in terms of hartree atomic units, rather than in terms of the ordinary units which he used. Furthermore, we shall describe the number of spin-up and spin-down electrons in a form which will fit in with applications we shall later make of the method.

Let us define a density $\rho\uparrow$ of electrons with spin up, $\rho\downarrow$ of electrons with spin down, the unit being the number of electrons per unit volume a_0^3, where a_0 is the bohr unit of length. We let $\rho = \rho\uparrow + \rho\downarrow$, the total density of electrons. We shall let

$$x = \frac{\rho\uparrow - \rho\downarrow}{\rho} \tag{16-1}$$

so that x goes from zero for the unmagnetized material, where $\rho\uparrow = \rho\downarrow$, to unity for the completely magnetized case, where $\rho\uparrow = \rho$, $\rho\downarrow = 0$. We can then give the energy of a sample of the material as a function of x and ρ, or of x and the Fermi energy ϵ_F, which is connected with ρ.

The first term in the energy is the kinetic energy. In Eq. 11-4 we gave the number of stationary states with energy less than E, in a perfect Fermi gas, occupying a given volume. This number of states is proportional to $E^{3/2}$, so that the maximum energy, or Fermi energy if we are using the Fermi statistics, is proportional to the two-thirds power of the density. In a given volume, the total kinetic energy is proportional to $\rho^{2/3}$ times the integral of the density over the volume, which is the total number of electrons. The exact value worked out on this model is

$$\text{kinetic energy} = \frac{6}{5}\pi^2 \left(\frac{3}{4\pi}\right)^{2/3} \int \left[(\rho\uparrow)^{5/3} + (\rho\downarrow)^{5/3}\right] dv$$

$$= \frac{3}{5}\epsilon_F \frac{(1+x)^{5/3} + (1-x)^{5/3}}{2} \text{ per electron} \tag{16-2}$$

where in the second form of Eq. 16-2, the Fermi energy ϵ_F is computed for the case $x=0$, or the unmagnetized case, and is proportional to $\rho^{2/3}$.

We are finding the diagonal matrix component of energy for a determinantal function, and this, according to Eq. 9-13, includes first the kinetic energy, which we have just computed. Next there is a coulomb energy, arising from the sum of electrostatic energies of all electrons in the field of the nuclei, plus the coulomb part of

$$\Sigma(\text{pairs } i,j)[(ij/g/ij) - (ij/g/ji)] \tag{16-3}$$

There is one point about this summation that we have not yet mentioned. We could perfectly well include the case $i=j$ in the summation over pairs, since for this case the coulomb integral $(ij/g/ij)$ equals and cancels the exchange integral $(ij/g/ji)$. Then the quantity of Eq. 16-3 can be rewritten as half the double sum over i and j of the quantity written above, or

$$\frac{1}{2}\Sigma(i)\Sigma(j)[(ij/g/ij) - (ij/g/ji)] \tag{16-4}$$

It is this form which we wish to use, because then $\Sigma(j)(ij/g/ij)$ equals the potential energy acting on the ith electron of all electrons including itself. To get our homogeneous electron gas, we must assume that the positive charge, instead of being concentrated in discrete nuclei, is uniformly distributed over the volume of the electron gas, and furthermore that the charge density of the electron gas precisely balances the charge density of the nuclei. Hence in this case the total coulomb interaction, equaling the interaction with the nuclei from the term $\Sigma(i/f/i)$ of Eq. 9-13, and the first term of Eq. 16-4 precisely cancel, and give nothing.

We are left, then, with an exchange term in the total energy.

$$-\frac{1}{2}\Sigma(i)\Sigma(j)(ij/g/ji) \tag{16-5}$$

which includes the term $i=j$, the self-interaction energy of the ith spin orbital. This quantity of Eq. 16-5 is nonvanishing only if the spin orbitals u_i and u_j have the same spin. It is a simple problem to compute this integral for the Fermi distribution and the perfect gas. We shall have much more to say later about this term, for it is met in the $X\alpha$ method which we describe later. For the moment, however, we shall merely state its value, as Bloch computed it, but expressed in the units we are using. It is

$$\text{exchange energy} = -\frac{3}{2}\left(\frac{3}{4\pi}\right)^{1/3} \int \left[(\rho\uparrow)^{4/3} + (\rho\downarrow)^{4/3}\right] dv$$

$$= -\frac{3}{2\pi}\frac{\sqrt{\epsilon_F}}{\sqrt{2}} \frac{(1+x)^{4/3} + (1-x)^{4/3}}{2} \text{per electron} \tag{16-6}$$

Now let us apply these results to answering the question which Bloch asked: Will the electron gas have a lower energy, at fixed density, if it is nonmagnetized, with $x=0$, or totally magnetized, with $x=1$? It can be easily shown that the lowest energy will come from one or the other of these cases. We find, using the second form of Eqs. 16-2 and 16-6,

$$\text{energy }(x=1) - \text{energy }(x=0) = \frac{3}{5}(2^{2/3}-1)\epsilon_F - \frac{3}{2\pi\sqrt{2}}(2^{1/3}-1)\sqrt{\epsilon_F}$$

(16-7)

For small ϵ_F, which occurs at low density, the term in $\sqrt{\epsilon_F}$ outweighs that in ϵ_F, and the energy difference is negative. This means that in this case the magnetized state is more stable than the nonmagnetized, and the gas would spontaneously magnetize. For large ϵ_F, larger than about 0.062 hartree units, the positive term, the kinetic energy term, outweighs the negative term arising from exchange, and the nonmagnetized state would be the stable one. When Bloch examined the electron density at which the change would occur, he found that it corresponded to a lower density, or smaller ϵ_F, than would be found using the free-electron model of any of the alkali metals with which he was particularly concerned. Hence he concluded that none of them should be ferromagnetic.

I received a preprint of Bloch's paper while I was working in Leipzig in the fall of 1929. The main problem I was exploring there was the relation between the two approaches to the theory of a solid, and in particular of an alkali metal, namely the Heisenberg and the Bloch methods. I recognized that these were analogous to the Heitler-London and the molecular orbital approaches, respectively, to the molecular problem. It was to clarify this relation that I worked out the hydrogen molecule problem by the two approaches, as described in Sec. 14. The general conclusion of that study was that while the Heitler-London ground-state solution for H_2 was fairly reliable over the whole range of internuclear separations, the molecular orbital method required a linear combination of more than one determinantal function to give results at large separation which were at all reliable. The diagonal energy of a single determinantal function made from molecular orbitals went to the wrong energy at infinity. In particular, at infinite separation, from Fig. 14-1, we see that the molecular orbital energy H_{11} of the determinant in which both electrons are in the lowest molecular orbital state rises too high, above the triplet state energy, whereas the triplet state, which would correspond to a magnetic crystal, goes to the right energy.

But this was just what Bloch was doing. He was finding the energy of the single determinant made from Bloch functions, or plane waves, the analog of the molecular orbitals, and he was finding a lower energy for the magnetic state than for the nonmagnetic state, in the limit of small ϵ_F, which is the limit of large internuclear separation. In other words, I saw that while Bloch's conclusion was correct that for wide bands, or a large ϵ_F, the nonmagnetic state should be stable, there was no reason to believe his conclusion that for small ϵ_F, or large internuclear separation, the magnetic state was stable. One would have to use something equivalent to the Heisenberg or Heitler-London method to get any valid results for the limit of large separations.

Ferromagnetism thus clearly could not come for substances with all their electrons in broad bands, as we had with the alkali metals. And yet such broad bands were necessary for explaining metallic cohesion, which was the particular topic of my work in Leipzig. As an energy band broadens out, when the atoms approach each other, only the lower part of the band will be occupied in an alkali metal, corresponding to the bonding orbitals in the molecular problem. The upper parts of the band, the antibonding orbitals, will be empty. It is the broadening of the band which lowers the energy of the occupied levels enough to explain the cohesive energy.

But this is definitely contrary to the requirements for ferromagnetism. The only possible place to look for ferromagnetic states was in narrow bands. In H_2, when we took the interaction of the ionic and nonionic states into account, we saw in Fig. 14-1 that the ground-state singlet, analogous to the nonmagnetic case, lay below the triplet, analogous to the magnetic case, at all distances, no matter how great. But there was nothing in the theory which made it impossible that in some cases the magnetic energy should be below the nonmagnetic at large enough distances, or small enough band widths. The only conclusion, then, was that since we actually had ferromagnetic crystals in nature, they must have two separate energy bands. One, a broad band, would be responsible for cohesive energy. The other, a much narrower one, might in some circumstances lead to ferromagnetism. To quote from the paper I wrote at the time of leaving Leipzig, "It is a very attractive hypothesis to suppose that in the iron group the existence of the $3d$ and $4s$ electrons provides in this way the two electron groups apparently necessary for ferromagnetism; for it is only in the transition groups that we have two such sets of electrons, and this criterion would go far toward limiting ferromagnetism to the metals actually showing it."

This conclusion was directly responsible for some work on atomic orbitals which I carried out soon after returning to Harvard, and then MIT, in 1930. Clarence Zener, one of our Harvard graduate students in physics, who took his Ph.D. in 1929, was in Leipzig at the same time that I was, and spent several additional years in Europe. He did a good deal of fundamental work on analytic approximations to atomic orbitals during this period, following the general lines which Kellner in 1927 had applied to helium. That is, he set up analytic functions similar to hydrogenic functions, but with effective atomic numbers to be adjusted to minimize the energy, and by varying the energy, he obtained quite good approximations to wave functions for a number of the lighter atoms. This suggested to me that one could use his results as far as they went, and could fit analytic functions to some of Hartree's self-consistent-field orbitals for heavier atoms, thereby getting analytic formulas which one might be able to interpolate for other atoms that had not yet been computed. In 1930 I wrote a paper outlining simple ways to set up such analytic functions, which to my astonishment proved to be so popular with the chemists that they are still often used.

I hoped to be able to make from these analytic orbitals some estimate of the widths of actual energy bands, to see if this would throw some light on why it was just iron, cobalt, and nickel which were ferromagnetic. Naturally an energy band will be narrow if the atomic orbitals concerned in it are far apart compared to some reasonable measure of atomic radius. It was easy to set up approximate atomic radii from my wave functions, or in fact to set up radii for the various shells of the atoms. What I found was striking: Of all the unfilled shells in the atoms in the periodic system for which I could make calculations, the partly filled $3d$ shells of iron, cobalt, and nickel were furthest apart in proportion to the dimensions of the shells. They should, in other words, be the most likely candidates for showing ferromagnetism. This was a clear indication of the strength of the energy-band approach to the problem of ferromagnetism.

There was another likely use for these analytic atomic orbitals: One could use them to calculate exchange integrals. This was one of the reasons I had set them up. There was some half-hearted effort to compute exchange integrals for the $3d$ orbitals in the transition elements, to see if perhaps Heisenberg's hypothesis that they should be positive was justified. I did none of this calculation myself, for I did not feel it was profitable: All of Heisenberg's work had been for orbitals like s orbitals, whereas the d orbitals had directional effects which were completely disregarded in most of the magnetic literature. In any case, none of the work trying to relate Heisenberg exchange integrals to integrals between actual atomic orbitals had any success.

During the latter part of my stay in Leipzig, however, after I had come to the general conclusions I have been describing, it became pretty obvious to me that for some reason which was not yet clearly understood, the Heisenberg approach was working better than one might expect. It was not until some work in 1937, which I describe in Sec. 25, that I really understood what was going on. But at least I explored the information which one could get by assuming that his method was reasonable for large interatomic distances. The state of maximum multiplicity, like the triplet state of H_2, was represented by a single determinant, with all spin orbitals having α spins. If only one electron had its spin reversed, to a β spin, one obtained only as many determinants as one had atoms in the crystal: Any one of the N atoms in the crystal could have its spin reversed. If one made Bloch sums of these determinantal functions, each one would have a different wave vector, and it was easy to show that there were no nondiagonal matrix components of the Hamiltonian between such functions with different wave vectors. Thus their energies could be found from their diagonal matrix components, which were easy to compute. These functions are now called spin waves; Bloch independently came upon them at the same time that I did, and he carried their theory further than I did, in their application to ferromagnetic problems. For a ferromagnetic crystal, if one uses the Heisenberg exchange picture, they form the lowest excited states of the crystal, most important at low temperatures, and I was able to derive conclusions about the probable distribution of energy levels in such a case.

With the negative exchange integrals characteristic of a nonferromagnetic crystal, however, these states were at the top rather than the bottom of the distribution of states, just as the $^3\Sigma_u$ of H_2 lies far above the $^1\Sigma_g$ ground state of the molecule. The interesting states for the problem of cohesion of an alkali crystal were the ones with small magnetic moment, and I wished to consider these. There were certain special states which had no magnetic moment, and yet which could be shown to have a very low evergy. In a body-centered cubic crystal such as an alkali metal, one could put electrons of α spin on the atomic sites at the centers of the cubes, and β-spin electrons on the corners of the cubes. Such an arrangement of spins is what we now would call antiferromagnetic. I brought out some reasons for thinking that such states should be considered in the theory of a nonferromagnetic metal. The idea is one which later led Per-Olov Löwdin and Reuben Pauncz to their theory of alternant molecular orbitals; but it did not lead to anything useful at the time, and we shall postpone discussion of it until Sec. 35, when we come to antiferromagnetism.

17. 1926–1932, Quantum Electrodynamics

Let us now return to the problem where wave mechanics started in the first place, the electromagnetic radiation field. It was natural that all of us who had been concerned with photons before 1926 should try out Schrödinger's methods as soon as they were suggested, to see if they gave the answers to the problems for which they were devised. The outcome, in a word, was that some results came out without trouble, but others were not so easy, and needed further work. The first and simplest problem was to derive the Kramers-Heisenberg dispersion formula. If one included an external electric field, a sinusoidal function of time, in the Hamiltonian as a perturbing field acting on the electronic system, an oscillating charge density with the same frequency as the external field was automatically set up, and it proved to satisfy the Kramers-Heisenberg formulas exactly. Schrödinger showed this in one of his series of 1926 papers, and it indicated that the wave mechanics was on the right track here, as well as in the problem of atomic and molecular structure.

It was also easy not only to derive the dispersion formula, but to consider transitions between stationary states induced by external radiation, and thus get the Einstein B coefficients. Schrödinger's equations for a time-dependent problem are easy to handle by assuming that the wave function ψ is a linear combination of the wave functions of an unperturbed problem in which the time-dependent perturbation is missing, of the form $\psi = \Sigma(i) C_i \psi_i$. With the time-dependent perturbation, the C_i's become functions of time. One can interpret $C_i^* C_i$ as the probability of finding the system in the ith stationary state, and can find the time rate of change of this quantity, and hence the transition probability. If one starts at $t=0$ with the assumption that the system is in one particular stationary state, then one finds that the other $C_i^* C_i$'s are zero at $t=0$ by hypothesis, but they start to build up proportionally to the time. The time rate of change gives the desired transition probability. This was the same type of calculation which I had made in my 1925 work on the quantum theory of optical phenomena. In the fall of 1926, I worked out such methods for the wave-mechanical problem, and on November 26, 1926, sent in the first of my papers involving wave-mechanical calculations to the *Proceedings of the National Academy* for publication. Before I read proof on it, Dirac's first paper using wave mechanics for this same problem came out in the *Proceedings of the Royal Society*; he had sent his paper in for publication just three months earlier, August 26, 1926. This was one of several cases when Dirac was just a few weeks or months ahead of me. I did not withdraw my paper, for reasons which I shall now mention.

The calculations of the Einstein B coefficients which Dirac and I had made were identical. We both used the same time-dependent form of Schrödinger's equation, and both found the same term $B\rho$ for the time rate of increase of the number of atoms in an excited state, under the action of a radiation field of energy density ρ. The difficulty came with the spontaneous radiation probability A, arising according to Einstein merely from the fact of being in an excited state. I had tried to incorporate this probability into the theory by including a perturbation term to represent what is called in classical electrodynamics the radiation resistance. This is a resisting force proportional to the time derivative of the acceleration, which classical theory leads to in order to give the rate at which the energy of a radiating system of charges loses energy to the radiation field. When I put it in, I found a change of the $C_i^* C_i$'s arising from it, but the dependence on the occupancies of the various states was not correct, though the A factor came out correctly. The error showed itself in that the sum of $C_i^* C_i$ over all states did not remain constant, which was obviously incorrect since this sum was required to be normalized. The mathematical difficulty arose because the assumed perturbing potential was not Hermitean, a requirement for terms in the Schrödinger Hamiltonian. I tried to modify the relations in such a way as to correct this error, but the modification was not very convincing.

Dirac must have tried the same thing, and come to the same conclusion; he did not try in his paper to include the radiation resistance, though of course he knew of it. I am sure that he knew of it for an amusing reason: I found years later that he and I had both attended the same course of lectures on electromagnetic theory given by E. Cunningham, at St. John's College, Cambridge, in the fall of 1923. Cunningham, an expert on electromagnetic theory, was one of the best informed men on such questions as radiation resistance, and he took it up in detail in his lectures. It was characteristic of the remoteness of students from each other in the English universities that Dirac and I did not meet at the time.

My difficulty in handling the spontaneous transitions immediately suggested to me that the trouble was that we were handling the radiation field as an external force, rather than as part of a closed system. In any closed system, Schrödinger's equation automatically results in having the sum of the $C_i^* C_i$'s remain constant in time. I therefore wrote another note for the *Proceedings of the National Academy*, sent in two months after the first, January 29, 1927, entitled "Action of Radiation and Perturbations on Atoms." After describing the time-dependent perturbation method in detail, I went on to describe a succession of cases which would suggest the type of closed system we should be considering. First I took the case of the two-electron system for which Heisenberg's paper dealing with the symmetric and antisymmetric states had just appeared. My arguments were really rather general, merely referring to two coupled systems with only a small possibility of passing from one to the other. We may as well describe the argument in terms of a single electron moving in a potential which has two identical wells with a rather high barrier between, as shown in Fig. 17-1, which represents the potential found in the H_2^+ problem. The same symmetry situation holds here that we have with any reflection plane: All solutions must be symmetric or antisymmetric with respect to the center of symmetry of the problem, and if the barrier is high enough, the antisymmetric wave function will have an eigenvalue only slightly higher than the symmetric function.

If we had only a single potential well, there would be certain eigenvalues and eigenfunctions, the atomic orbitals. The sum and difference of these functions are the symmetric and antisymmetric molecular orbitals of the problem, in the LCAO approximation. Each energy level of the single well goes into two closely spaced levels in the two-well problem. Then I asked the question, suppose one set up the problem in terms of the atomic orbitals, and assumed that at $t=0$ the electron is necessarily found in the first well. What will happen as time goes on? If we forget the nonorthogonality of the two sets of atomic orbitals, the symmetric molecular orbital can be written as $(u_a + u_b)/\sqrt{2}$, where u_a and u_b are the atomic orbitals, and the antisymmetric orbital can be written as $(u_a - u_b)/\sqrt{2}$. If E_1 and E_2 are the energies of the symmetric and antisymmetric states, and u_1 and u_2 are the corresponding molecular orbitals, then the function

$$\psi = \frac{1}{\sqrt{2}} \left[u_1 \exp\left(\frac{-iE_1 t}{\hbar}\right) + u_2 \exp\left(\frac{-iE_2 t}{\hbar}\right) \right] \quad (17\text{-}1)$$

is a solution of Schrödinger's equation which behaves properly at $t=0$, since at that time the function will reduce to $(u_1 + u_2)/\sqrt{2} = u_a$. As time goes on, however, there will be a beat phenomenon, by which the electron will gradually pass from the first well to the second and back again. This is the phenomenon discussed at length by Condon and Gurney and by George Gamow in 1928, and now generally known as the tunnel effect.

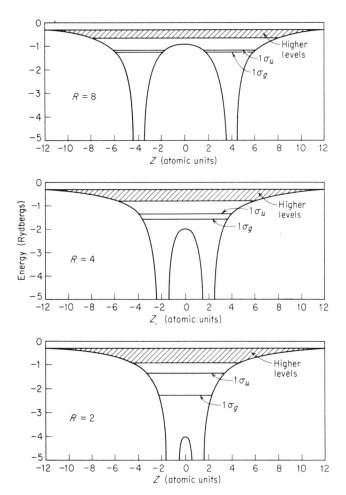

Fig. 17-1. Potential energy of electron along internuclear axis in H_2^+, for various internuclear separations. Energy levels of $1\sigma_g$ and $1\sigma_u$ are indicated, shaded area indicates region occupied by higher levels shown in Figs. 14-2 and 14-3. From *Quantum Theory of Matter*, 2nd ed., by J. C. Slater. Copyright 1968 by McGraw-Hill, Inc. Used with permission of McGraw-Hill Book Company.

We can see this phenomenon by forming $\psi^*\psi$. We have $\psi^*\psi = C_a^* C_a u_a^* u_a + C_b^* C_b u_b^* u_b$, where

$$C_a^* C_a = \frac{1 + \cos[(E_1 - E_2)t/\hbar]}{2}$$

$$C_b^* C_b = \frac{1 - \cos[(E_1 - E_2)t/\hbar]}{2}$$

$$C_a^* C_a + C_b^* C_b = 1 \tag{17-2}$$

We see a pulsation, with angular frequency $(E_1 - E_2)/\hbar$. The more the two wells are separated from each other, the smaller is the energy separation $E_1 - E_2$ between the two eigenvalues, and hence the slower is the beat frequency. In my paper I suggested next that we consider many different coupled systems, which in our simple picture would correspond to an electron moving among many identical potential wells. Each energy level would be split into as many levels as there were wells, and this pulsation phenomenon would become a complicated motion with many beat frequencies. If we started at $t = 0$ with the electron in well a, as in the preceding example, it would be a long time, perhaps infinity, before the complete probability would be located on atom a again.

Finally, as I pointed out in the paper, the mathematical characteristic of a radiation field is that it contains essentially an infinite number of natural frequencies. I mentioned the well-known treatments of the electromagnetic radiation in a perfectly reflecting cavity, which are usually used as the first step in discussing blackbody radiation. This has a distribution of energy levels somewhat resembling that of the homogeneous electron gas which we discussed in Sec. 11. It is in the case of such a continuous distribution of energy levels, coupled with a system like an atom, that the superposition of all the pulsations adds up to a C^*C for the atom which varies linearly with the time. This had been proved in my paper of two months earlier. The interaction of only two identical systems, like our case of Eq. 17-2, looks like a case of resonance, and it was presumably from such a way of looking at it that Heisenberg was led to refer to his two-electron problem as a resonance phenomenon, thus introducing that term into the study of atoms and molecules. But as we see, there is no very clear distinction between this case and the one with a great many degrees of freedom, like the radiation field, in which we speak of quantum transitions rather than of perturbations of the atoms.

It would have been an obvious next step after this paper for me to have gone ahead and set up the radiation in the cavity, and have considered the interaction of this radiation field with an atom. But here again Dirac got ahead of me. I had sent my paper in to the *Proceedings of the National Academy* on January 29, 1927. On February 2, 1927, just four days later, Dirac sent the paper to the *Proceedings of the Royal Society* in which he had handled the interaction of atoms with radiation by just this method. I had already sensed that Dirac and I were pursuing the same line of thought and that he was several months ahead of me, and I had already decided to switch to some other problem. It was obvious that I would never catch up with Dirac to the point of being clearly ahead of him. Thus at this point I shifted my interest to the problem of the helium atom, where I felt one should make every effort to go beyond the very simple approximations that Heisenberg had suggested. I shall postpone discussion of that problem until Sec. 19, and instead shall continue with Dirac's work, and with the development of the theory of the electromagnetic field and of the electron.

It was obvious from reading Dirac's paper, "The Quantum Theory of the Emission and Absorption of Radiation," that he had succeeded in doing what he and I had hoped, namely to include the spontaneous radiation along with the absorption in a consistent scheme. But it was also obvious that his paper was a typical example of what I very much distrusted, namely one in which a great deal of seemingly unnecessary mathematical formalism is introduced. It was not until Fermi published a paper in the *Reviews of Modern Physics* for January 1932, on the quantum theory of radiation, that I felt that I understood what it was all about, even though I had been on the point of working out the same type of theory myself. I was very much heartened to hear some time later that Fermi had written his paper, which was a set of lecture notes for some lectures he had given at the University of Rome, because he also had not been able to get anything out of Dirac's paper, and felt he would never understand it until he had worked it out for himself. Fermi's paper was very much better, as was Heitler's treatment of the problem in his book *The Quantum Theory of Radiation*, whose first edition came out in 1936. But, like Fermi, I never felt that I really understood the theory until I worked it out for myself, and for that reason I refer the reader to the treatment in my book, *Quantum Theory of Atomic Structure*, Vol. 1, Chap. 6 and Appendix 12, McGraw-Hill, 1960. A somewhat abbreviated version of the same treatment is given in my *Quantum Theory of Matter*, second edition, Chap. 13, McGraw-Hill, 1968. It is the discussion in these books that I shall sketch here.

The important step which must be taken if the radiation field is to be included as part of an atomic system is to find a Hamiltonian function to describe the whole combined system. For it is only through a Hamiltonian function that Schrödinger's method leads to a wave function and wave equation. Fortunately it was well known how to set up this Hamiltonian for the electromagnetic problem. First, one must describe the electromagnetic field in terms of scalar and vector potentials, ϕ and $\mathbf{A}$. From them, one defines the electric and magnetic fields, $\mathbf{E}$ and $\mathbf{B}$, by the equations

$$\mathbf{E} = -\operatorname{grad}\phi - \frac{\partial \mathbf{A}}{\partial t} \qquad \mathbf{B} = \operatorname{curl}\mathbf{A} \qquad (17\text{-}3)$$

We shall first take up the problem of a radiation field in a cavity which contains no electrons. For this case, the scalar potential ϕ is zero. If the cavity is taken to have perfectly reflecting walls, it can be shown in classical theory that the vector potential can be expanded in terms of an infinite set of vector functions $\mathbf{E}_a(x,y,z)$ that satisfy the equation

$$\nabla^2 \mathbf{E}_a = -\mathbf{k}_a{}^2 \mathbf{E}_a \qquad (17\text{-}4)$$

and that also satisfy suitable boundary conditions over the surface of the cavity. The quantity $\mathbf{k}_a$ is of the nature of a wave vector, similar to the wave vector we met in Sec. 15 in connection with the wave functions of free electrons in a cavity. If our cavity is cubical, the functions $\mathbf{E}_a$ are similar to the functions u of Eq. 15-1, except that they are vector quantities, not scalars as the wave function of Schrödinger's equation is. The $\mathbf{E}_a$'s can be normalized, and satisfy certain orthogonality relations.

In terms of these functions $\mathbf{E}_a$, we can then expand the vector potential in the form

$$\mathbf{A} = \Sigma(a) Q_a \mathbf{E}_a \qquad (17\text{-}5)$$

where the Q_a's are scalar functions of time. We must now remember that the vectors $\mathbf{E}$ and $\mathbf{B}$ of Eq. 17-3 must satisfy Maxwell's equations, in order that they may describe the radiation field according to classical electromagnetic theory. It is easily shown that this requirement simplifies to the condition that $\mathbf{A}$ satisfy the wave equation

$$\nabla^2 \mathbf{A} - \frac{1}{c^2} \frac{\partial^2 \mathbf{A}}{\partial t^2} = 0 \qquad (17\text{-}6)$$

where c is the velocity of light. If we substitute the expression of Eq. 17-5 for the vector potential and use Eq. 17-4, we find that Eq. 17-6 is reduced to a simple differential equation for Q_a,

$$\frac{d^2 Q_a}{dt^2} = -k_a^2 c^2 Q_a = -\omega_a^2 Q_a \qquad (17\text{-}7)$$

where $\omega_a = k_a c$, and this is the angular frequency of the particular electromagnetic standing wave within the cavity whose space dependence is given by E_a.

In Eq. 17-7 we see that we have just the differential equation for a sinusoidal oscillation with angular frequency ω_a. Our general solution for the vector potential is a superposition of such standing waves with arbitrary coefficients. Let us now find the electromagnetic energy of these standing waves. It is well known that, if we use the mks units for describing the electric and magnetic fields, the electromagnetic energy per unit volume in the cavity is

$$\text{energy} = \frac{\epsilon_0}{2} E^2 + \frac{1}{2\mu_0} B^2 \qquad (17\text{-}8)$$

where ϵ_0 and μ_0 are fundamental constants in these units. The first term of Eq. 17-8 is the electric energy, the second is the magnetic energy. In terms of the expansion of Eq. 17-5, it can be shown that this energy when integrated over the volume of the cavity is

$$\text{energy} = \frac{\epsilon_0}{2} \Sigma(a) \left[\left(\frac{dQ_a}{dt} \right)^2 + \omega_a^2 Q^2 \right] \qquad (17\text{-}9)$$

Here again the first term is the electric energy, the second the magnetic energy.

This expression is just like the sum of the energies of an infinite set of linear oscillators. The term in $(dQ_a/dt)^2$ is like the term $\frac{1}{2}mv^2$, or the kinetic energy, of an oscillator, and the term in Q_a^2 is like the potential energy, where Q_a is the coordinate giving the position of the oscillating particle. The energy can be very simply converted into a Hamiltonian form, by introducing momenta P_a proportional to the velocities dQ_a/dt, and in turn this Hamiltonian function can be converted into an operator, leading to a wave function and a Schrödinger equation. The energy is like that of a superposition of independent linear oscillators, and the corresponding Schrödinger equation can be separated into equations for each oscillator. The energy then proves to be the sum of the energies of the oscillators,

$$\text{energy} = \Sigma(a)(n_a + \tfrac{1}{2})\hbar\omega_a \qquad (17\text{-}10)$$

As we have mentioned in Sec. 5, only the energy differences are of physical importance, and one can disregard the half-quantum numbers. The number n_a is interpreted in Dirac's theory as the number of photons in the ath radiation mode in the cavity, each photon having the energy $\hbar\omega$ or $h\nu$, agreeing with the information which we had had about photons for many years. For the radiation field in an empty cavity, without electrons, photons would neither be emitted or absorbed, and the stationary state we get in this way would persist indefinitely.

We can see what the physical meaning of the wave function would be, by use of the analogy with a system of linear oscillators. Since these oscillators are independent of each other, and there is no requirement of antisymmetry, the wave function can be taken merely as a product of wave functions for each individual oscillator. More accurately, since photons are indistinguishable, one uses the symmetric combination of these products, and this can be shown to lead to the Bose-Einstein statistics for the photons, as the use of the antisymmetric combination would have led to the Fermi-Dirac statistics. Each of the wave functions representing an individual oscillator would be like a Schrödinger wave function for a linear oscillator in a particular quantum state, with quantum number n_a. The whole wave function would then represent the situation where there were n_a photons in the ath radiation mode, where the numbers n_a, analogous to quantum numbers, could be adjusted at will to describe any assignment of photons to modes.

To go further, one must next have a Hamiltonian for the electrons in the presence of the electromagnetic field. This was known to be, in the nonrelativistic limit,

$$(H)_{\text{op}} = \Sigma(i)\left\{ \frac{[\mathbf{p}_i - e_i\mathbf{A}(x_i)]^2}{2m_i} + e_i\phi(x_i) \right\} + \Sigma(\text{pairs } i,j) g_{ij} \quad (17\text{-}11)$$

where we are summing over electrons or other charges e_i with mass m_i (some of the particles could be nuclei), $\mathbf{A}(x_i)$ and $\phi(x_i)$ are the vector and scalar potential at the position of the ith particle, and g_{ij} is the coulomb interaction between the ith and jth particles. The scalar potential $\phi(x_i)$ would arise from nuclei, in case they were not included as particles in the system, but instead were assumed to be fixed in the spirit of the Born-Oppenheimer approximation.

The only term in Eq. 17-11 which is unfamiliar is the one in the vector potential. It can be shown in classical electrodynamics that if one applies Hamilton's equations to this Hamiltonian, it results in the Newtonian equations of motion, and the force acting on the ith particle is the sum of the electrostatic force and the magnetic force, $e_i(\mathbf{v}_i \times \mathbf{B})$, where $\mathbf{v}_i$ is the velocity of the ith particle. This magnetic force expresses the ordinary motor rule, incorporated in the Lorentz force which had been familiar in electron theory since the early 1900s. The terms in $\mathbf{A}(x_i)$ in Eq. 17-11, which include both a linear term in $\mathbf{p}_i \cdot \mathbf{A}(x_i)$ and a quadratic term in $|\mathbf{A}(x_i)|^2$, represent the additional terms which enter on account of the radiation field described by the vector potential, according to this simple formulation.

To get the Hamiltonian for the complete system, one then has to do two things. First, one adds to the Hamiltonian of Eq. 17-11 the energy of the radiation field, which came from Eq. 17-9 if the radiation energy were expressed in Hamiltonian form,

$$\text{radiation energy} = \Sigma(a)\left(\frac{P_a^2}{2\epsilon_0} + \frac{\epsilon_0}{2\omega_a^2}Q_a^2\right) \qquad (17\text{-}12)$$

Second, one expresses the vector potential $\mathbf{A}$ as it appears in Eq. 17-11 in terms of the Q_a's, by Eq. 17-5. When this is done, it can be shown that Hamilton's equations for the electrons still describe their motion in the combined electrostatic and radiation field. But it can also be shown that Hamilton's equations for the components of the radiation field are modified just as they should be to lead to Maxwell's equations in the presence of the electronic charge. In other words, they describe the creation of a radiation field by an oscillating charge.

When we then convert this Hamiltonian into an operator, and write a Schrödinger equation for the combined system of radiation and electrons, we have introduced the terms in $\mathbf{p}_i \cdot \mathbf{A}(x_i)$ and in $|\mathbf{A}(x_i)|^2$ as perturbations. Ordinarily for radiation fields of moderate intensity, it is only the linear term $\mathbf{p}_i \cdot \mathbf{A}(x_i)$ which is important. We convert $\mathbf{p}_i$ into an operator $-i\hbar \partial/\partial q_i$ in the ordinary way, and $\mathbf{A}(x_i)$ into an expression involving the Q_a's, by use of Eq. 17-5. We then apply the time-dependent form of Schrödinger's equation to the resulting wave function, regarding this linear term as a perturbation, and are led to expressions for the time rate of change of the numbers of systems in each stationary state. The definition of stationary state now involves both the electronic stationary state and the numbers of photons in each radiation mode. We have many degenerate modes, one corresponding to the electronic system in an excited state and all radiation modes in an unexcited state, the others to the electronic system in the ground state and certain radiation modes whose frequencies satisfy the Bohr frequency condition having one photon in them. The net result, as Dirac showed, is that the time rate of change of the number of electronic systems with time is given not only by the Einstein $B\rho$ terms, but also by the Einstein A's, which were not properly handled in the earlier theory. This description of the situation is very sketchy, but it is an involved situation, which really cannot be described in such simplified language. We have already suggested a reference in which the reader who wishes to follow it further can look it up, expressed in the same language we have used here.

Up to this point, the theory as I have described it seemed to me reasonable. But there were two further developments to which I took serious exception. The first is a matter of taste, not of principle. Jordan and various associates (Pauli, Oskar Klein, Wigner, John von Neumann) started with the mathematical methods that Dirac had used, and developed a method that is usually called second quantization. It is not hard to see what this term means. In treating the electromagnetic vibrations in a cavity, we have first set up electromagnetic waves $\mathbf{E}_a$ satisfying certain boundary conditions, which contain what are essentially quantum numbers, like the n_x, n_y, n_z of the Bloch sum of Eq. 15-1. Second, we have just been handling each of these oscillators $\mathbf{E}_a$ and their coefficients Q_a as quantities similar to harmonic oscillators, which were quantized with quantum numbers n_a of Eq. 17-10. Here are two successive quantizations. A similar situation occurs in a more straightforward case, though one that we have not taken up in this volume. In the elastic vibrations of the atoms of a crystal, one can set up normal modes, as was done many years ago by Born and von Kármán. Each of these normal modes is characterized by numbers like the n_x, n_y, n_z we have just mentioned. Then in quantum theory each of these normal modes acts like a harmonic oscillator, and is quantized with energy $(n_a + \frac{1}{2})h\nu$. This type of second quantization was used in the Debye theory of the specific heat of a crystal, many years before there was any thought of wave mechanics.

This idea, then, is nothing new or out of the way. But Jordan and his associates, following the lead of Dirac, proceeded to set up quantities $a_r^\dagger$ and a_s, such that $a_r^\dagger a_s$ formed a matrix whose diagonal component $a_r^\dagger a_r$ represented the quantum number n_r giving the number of quanta in the rth oscillator. Being matrices, these quantities did not necessarily commute with each other. Jordan and his associates showed that if one postulated one type of commutation rule, the resulting n_r's satisfied the Bose-Einstein statistics, which is the right type to describe the photons. But if one postulated another type of commutation rule, the n_r's satisfied the Fermi statistics: Each n_r could take on only the value zero or unity. It then became formally possible to express the theory of a many-electron problem in terms of these operators, using the commutation rule for the Fermi statistics, and throwing the whole theory into a matrix form. I felt at the time, the early 1930s, that this method was just as objectionable for handling the many-electron problem as the group-theoretical methods of the "Gruppenpest" had been in the period before 1929. It meant essentially the construction of a great framework of mathematical theory that did not contribute at all to the physical understanding of the problem, and should appeal only to those theorists who are more fascinated by pure mathematics than they are by physical results.

This method appealed particularly to those users of the field theory who left chemical physics in the 1930s and took up nuclear and high-energy theory instead. There were many of these theorists, and they more than anyone else set the style for what came to be called many-body theory. The result is that even now, a great fraction of the world of physicists feels that if one is to work on many-body effects, this means that one must use the $a_r^\dagger$ and a_s operators, often now called creation and annihilation operators, must use a Hamiltonian set up in terms of them, and must base one's work on the diagram technique which has become so popular. I do not hold with this view at all, any more than I was willing to accept the "Gruppenpest" in 1929.

There is a parallelism between these methods of second quantization and the equally fundamental and powerful method of building up linear combinations of determinantal functions to form approximate solutions of Schrödinger's equation for the many-body problem. The diagram techniques are methods of handling very high-order perturbation treatments with such a configuration interaction. The attention of the theorists who are devoted to the techniques of double quantization is attracted almost completely to the formalism of making these perturbation treatments, and is deflected practically completely away from what is the really significant problem, that of getting the best orbitals possible. In many cases they even proceed as our simplified treatment would indicate, using plane waves as the basis functions in terms of which everything is expanded. I have already mentioned in Sec. 15 that such a plane-wave expansion, though useful for proving general theorems, is so poorly convergent that it is of no use for any practical purposes of getting wave functions for molecules or solids.

We shall make no use of these techniques of second quantization in the present volume, and my personal feeling is that the ordinary student of the electronic structure of atoms, molecules, and solids would do much better to avoid them completely. By using the simple methods of determinantal functions, or even more the use of the charge density, as we describe in Secs. 30, 33, and 35, we focus our attention on the problems that are physically important, and avoid those that appeal only to the abstract mathematician.

The other point in which I took exception to some of the further development of field theory is concerned with the electrostatic self-interaction of the electron. This problem comes up particularly clearly in the formulation which Fermi gave the theory, in his paper in 1932. Fermi expanded the vector potential in a set of orthogonal vector functions, similar to the E_a we have used, but he also introduced a set of orthogonal scalar functions and expanded the scalar potential in terms of them. The resulting electrostatic energy, of the nature of the integral of the product of charge density and scalar potential, could then be transformed into a form looking like the coulomb interaction in the familiar Hamiltonian, but this energy came out to be

$$\frac{1}{2}\Sigma(i,j)\frac{1}{r_{ij}} \qquad (17\text{-}13)$$

if we use atomic units. This includes the infinite self-energy, coming from the case $i=j$. There was nothing in the theory to exclude this infinite self-energy. And yet we can be as sure as we can of any part of the theory, that no such term should be included in the Hamiltonian. The summation of Eq. 17-13, excluding the term $i=j$, is the one that has yielded the Schrödinger equation which has given very accurately correct solutions for the problems of the helium atom and hydrogen molecule, and which gives every indication of being correct.

There is a great deal of history behind this problem of the self-energy of an electron. H. A. Lorentz and Max Abraham, in the early 1900s, were working in a period when they were already familiar with the fact, emphasized by Einstein, that there was a relation between the mass m of a particle and its energy mc^2. The electrostatic energy of a point charge is infinite. It is the integral over all space of the quantity $(\epsilon_0/2)E^2$, where E is the electrostatic field, which is proportional to $1/r^2$ near a point charge. If we integrate this quantity over all space, there is an infinite amount of energy concentrated at infinitesimal values of r. If one assumes a finite electron, such as a charge e placed on the surface of a sphere of radius r_0, then there is no field inside this sphere, and the remaining electrostatic energy, which comes from the space outside the sphere, is finite. One can then choose r_0 so that the mass would equal the electronic mass, and the energy mc^2 would equal this electrostatic energy in the field. The radius comes out in the range of 10^{-11} centimeter, and there is no evidence whatever that the electron has such dimensions. In fact, electrons of this size would be too big to fit into the inner shells of the heavier atoms. But Lorentz and Abraham were familiar with this type of calculation.

They had made further developments of the theory. They considered the fact that if one put a whole electron's charge on the surface of a sphere of the indicated size, the electrostatic forces tending to blow the charge apart would be enormous. There would have to be some form of external pressure to keep the hypothetical electron from blowing up. This would contribute some additional energy, and hence should be considered in finding the mass. Then another point was that it is known that any distribution of matter set in motion will contract along the direction of motion. This contraction, predicted by the theory of relativity, had already been postulated by Lorentz, and is generally called the Lorentz-Fitzgerald contraction: Fitzgerald was a man of the last century who had anticipated both Lorentz and Einstein in suggesting it. What sort of forces would make a complicated finite electron of the type we are discussing contract in the proper way?

All these problems of the structure of the electron failed to receive a satisfying answer in the early 1900s, when they were being actively discussed. Somehow they suggested that the self-energy should be finite, not infinite, and that it should appear in the theory as the explanation of why the electron had the mass it did. But there was no suggestion of this in the sort of quantum electrodynamics that Dirac and Fermi set up. To my way of thinking, this indicates a fundamental shortcoming of the standard versions of quantum electrodynamics, which must be corrected in the future. Fermi had the same opinion. The last sentence of his 1932 paper is "In conclusion we may therefore say that practically all the problems in radiation theory which do not involve the structure of the electron have their satisfactory application; while the problems connected with the internal properties of the electron are still very far from their solution."

What the Schrödinger theory and the Hamiltonian it uses seem to be telling us is that a single scalar potential simply does not exist: The potential that should appear in the Hamiltonian, and in Schrödinger's equation, is different for each electron, or more generally for each charged particle. It is the potential computed by electrostatics from all *other* charges in the universe on the charge in question. A consequence of this is that the type of theory which Dirac and Fermi used, in which we start from the outset with the preconception that a single scalar potential exists, is the wrong starting point, and the theory cannot possibly be right. In a similar way, I suspect that the assumption that a single vector potential exists is equally wrong, since I do not believe that an electron or other charged particle acts on itself through magnetic forces any more than it does through electrostatic forces.

Hartree of course was perfectly aware of this situation in setting up his self-consistent field: He assumed that each electron was acted on by all nuclei, and all electrons aside from itself. The Hartree-Fock method also gives a very reasonable way of correcting for this fact. It is just this process of correcting for the fact that the electron does not act on itself which is the point of the $X\alpha$ method, which we are discussing in this volume. In some fundamental way, the scalar potential ϕ which should appear in electrodynamics is not that arising from all charges in the system, but only from all charges except the one whose motion we are considering.

I do not pretend to be able to say what the right theory should be. I suspect that when the answer is finally found, it will be some more inclusive theory which predicts the masses not only of the electron, but of all the elementary particles. But failing this right theory, the main lesson to be learned is that we should not consider the quantum electrodynamics as it exists today to be the last word. We should treat it as something still to be corrected, as in 1923 we treated the form of quantum theory we had then as something tentative, which had to be greatly modified. The moral, in a word is "physics isn't all done yet."

18. 1928–1932, Dirac and the Theory of the Electron

The theory of quantum electrodynamics which we discussed in the preceding section is one that could probably have been produced by any one of a number of physicists who were working in the area at the time. In contrast, the theory of the behavior of the electron given by Dirac in 1928 is one that we can hardly conceive of anyone else having thought of. It shows the peculiar power of the sort of intuitive genius which he has possessed more than perhaps any of the other scientists of the period. Dirac was not satisfied with the mathematical beauty of the theory of the electron as we had possessed it up to that time. He started out to make it more symmetrical. He ended up with a theory which completely explained the electron spin and its properties, but which stubbornly predicted also a positively charged particle with properties like those of the electron, which had never been observed. This prediction puzzled him and the scientific world for four years, until August 1932, when Carl Anderson of Pasadena observed the first appearance of a positron in a cloud chamber. Once positrons were discovered, they were seen often enough to confirm that they were the particles predicted four years earlier by Dirac.

Dirac's theory was concerned with a single electron of charge $-e$ in an external field described by a scalar and vector potential. From Eq. 17-11 its Schrödinger equation, in the nonrelativistic form we were using there, would be

$$\left[\frac{(\mathbf{p}+e\mathbf{A})^2}{2m} - (E+e\phi)\right]\psi = 0 \qquad (18\text{-}1)$$

in which we are to replace $\mathbf{p}$ by $-i\hbar\nabla$ and E by $i\hbar\partial/\partial t$. Here we have the three components of a vector $\mathbf{p}+e\mathbf{A}$ appearing as a square, in an equation in which the scalar $E+e\phi$ is appearing to the first power. But in relativity theory, $\mathbf{p}$ and E appear as the four components of a four-vector, as $\mathbf{A}$ and ϕ also do. Dirac's aesthetic sense was outraged at seeing an equation which purported to be one of the fundamental equations of physics, in which we did not have the symmetry which Einstein had believed all physical laws should possess when expressed in four-dimensional form.

I mentioned in Sec. 5 that the ordinary wave equation of physics has a second time derivative, which would give the term $E + e\phi$ coming in as a square rather than as a first power. This would have the sort of symmetry that Dirac wished. But we have also seen that this could not be the correct equation, since it did not lead to Bohr's frequency condition. It was to correct that difficulty that the term in the first derivative was inserted. Very well, said Dirac to himself, the only way to get the symmetry we want must be by having a Hamiltonian function which is linear in $\mathbf{p} + e\mathbf{A}$, as well as in $E + e\phi$. That is what he proceeded to do.

Naturally he did not work with this nonrelativistic Hamiltonian, however, for his endeavor was to get a proper relativistic theory of the electron. One can set up a proper Hamiltonian to describe the motion of an electron in relativistic but non-quantum-theoretical mechanics. If it is represented by H, it is given by the equation

$$(H + e\phi)^2 - (\mathbf{p} + e\mathbf{A})^2 c^2 - m_0^2 c^4 = 0 \tag{18-2}$$

If one writes this in the form

$$(H + e\phi) = \left[m_0^2 c^4 + (\mathbf{p} + e\mathbf{A})^2 c^2 \right]^{1/2} \tag{18-3}$$

and expands the square root by the binomial theorem, one finds

$$(H + e\phi) = m_0 c^2 + \frac{(\mathbf{p} + e\mathbf{A})^2}{2m_0} - \frac{1}{8 m_0^3 c^2} (\mathbf{p} + e\mathbf{A})^4 + \cdots \tag{18-4}$$

Here the first term on the right is $m_0 c^2$, the self-energy of the electron, if m_0 is the rest mass. The next term is the one given in Eq. 18-1, and the last term is the beginning of an expansion in power series of the relativistic corrections arising on account of the change of mass with velocity given by the relativity theory. Substantially this derivation of a relativistic Hamiltonian was given by Schrödinger in one of his 1926 papers.

In Eq. 18-2 one has the sort of symmetry between the term $H + e\phi$ and $\mathbf{p} + e\mathbf{A}$ which Dirac wanted, but we still have the quadratic dependence on these quantities. It was at this point that Dirac's ingenuity showed itself. He said, let us replace the quadratic expression of Eq. 18-2 by a linear wave equation, which he wrote in the form

$$\left[\frac{1}{c}(H + e\phi) + \alpha_1(p_x + eA_x) + \alpha_2(p_y + eA_y) + \alpha_3(p_z + eA_z) + \alpha_4 m_0 c \right] \psi = 0 \tag{18-5}$$

In this equation, Dirac's equation for the electron, he wished to have the quantities α_1, α_2, α_3, α_4 determined as dimensionless quantities not involving the coordinates or time, and p_x, p_y, p_z, to be replaced by $-i\hbar \partial/\partial x$, etc., and H by $i\hbar \partial/\partial t$. In order to establish a relation between this equation linear in the derivatives and the Hamiltonian of Eq. 18-2, Dirac next allowed the operator

$$\frac{1}{c}(H+e\phi) - \alpha_1(p_x + eA_x) - \alpha_2(p_y + eA_y) - \alpha_3(p_z + eA_z) - \alpha_4 m_0 c \quad (18\text{-}6)$$

to operate on Eq. 18-5, where again H and p are to be replaced by operators. He assumed that the σ's would commute with the expressions $(p_x + eA_x)$, etc., since they were assumed not to depend on the coordinates. Then he found that the resulting Hamiltonian would agree as closely as possible with that of Eq. 18-2, provided that we have the relations

$$\alpha_i^2 = 1 \quad \alpha_i \alpha_j + \alpha_j \alpha_i = 0 \quad \text{if} \quad i \neq j \quad (18\text{-}7)$$

As a next step, Dirac examined the quantities α_i, satisfying the relations of Eq. 18-7. It is not possible to satisfy these relations if the α's are ordinary numbers. He found, however, that the relations can be satisfied if the α_i's are matrices with four rows and columns, and if the multiplications of Eq. 18-7 are regarded as matrix multiplications. This is as far as we shall try to describe his procedure in detail, and I shall now describe Dirac's further procedure in general language.

In the first place, if Eq. 18-5 is to have any meaning when the α's are matrices with four rows and columns, we must assume that ψ is a quantity with four components, and that the multiplications are to be interpreted as

$$(\alpha_k \psi)_i = \Sigma(j)(\alpha_k)_{ij} \psi_j \quad (18\text{-}8)$$

where the α_{ij} are the matrix components of the four-by-four matrices. Then Eq. 18-5 reduces to four simultaneous linear equations for the four function $\psi_1, \psi_2, \psi_3, \psi_4$. It is easy to get solutions of these four simultaneous equations for the case where the scalar and vector potentials are zero, or the particle is freely moving in space. These solutions are of the form of plane waves for each of the four components of the wave function, of the form

$$\psi_i = c_i \exp\left[\frac{i(p_x x + p_y y + p_z z - Et)}{\hbar}\right] \quad (18\text{-}9)$$

where p_x, p_y, p_z, E are constants. When these functions are inserted into Eq. 18-5, they become four simultaneous linear equations for the four constants c_i. They cannot be satisfied unless the determinant of coefficients is zero, and this condition, which determines the eigenvalues, is

$$E = \pm \left[m_0^2 c^4 + \left(p_x^2 + p_y^2 + p_z^2 \right) c^2 \right]^{1/2} \qquad (18\text{-}10)$$

which is just what we should expect from Eq. 18-3, if **A** and ϕ were set equal to zero, except for the $\pm$ sign.

The method, in other words, leads not only to positive eigenvalues but to negative eigenvalues as well. The positive energies start with the rest energy $m_0 c^2$, the next term is the ordinary kinetic energy, as in Eq. 18-4, and the further terms represent the relativistic corrections. But the negative term represents a negative rest mass and a negative kinetic energy. There was great confusion at first as to what these negative energies meant. They were not a characteristically quantum-theoretical feature; the Hamiltonian of Eq. 18-3 has nothing to do with quantum theory, and there is no reason why we should not have used the negative square root there as well as the positive value.

There is a pleasant story, which may or may not be true, but which I think is very plausible, that Dirac was in a great quandary about what these negative energies meant, at a time when he was to take the long ride on the Transsiberian Railway from Siberia to Europe. Those were the days when a west European could still travel from Japan to western Europe by this means. The long and tedious trip across the frozen north was said, according to this story, to have given Dirac the uninterrupted time to think through the problem. Whether this was the case or not, he had thought it out at the time of his first paper, but the answer when he got it was quite fascinating.

Briefly, his explanation was that the negative-energy states, an infinite number of them, really existed, but there were enough electrons in the universe so that they were all filled with an electron each, the maximum number allowed by the Pauli exclusion principle. Thus they acted like closed shells, and it was only the relatively few electrons in positive-energy states that we ever saw. However, there was the possibility that if one started with all the negative states filled, all the positive ones empty, one could absorb a photon of energy $2m_0 c^2$, which could raise an electron from the highest occupied negative-energy state, whose energy was $-m_0 c^2$, up to the lowest positive-energy state, of energy $m_0 c^2$. This would produce an electron and a hole in the negative-energy states. The analogy to electrons and holes in a semiconductor was obvious, and as in that problem, one can show that the hole in the negative-energy states should move under the action of an external field like a positive electron. It was this process of creation of a positive hole and negative electron out of empty space which Anderson observed in 1932. And more careful experiments proved that in fact the threshold for this process, under the stimulation of gamma-ray absorption, was just the amount $2m_0c^2$ predicted by Dirac's theory. This was only in 1932, however. Until then physicists, including Dirac himself, were puzzled.

Once we accept the possibility of the negative- as well as the positive-energy states, we can go back to Dirac's equations for the force-free case, and find the relations between the c_i's. It turns out that for the positive-energy states the first two constants, c_1 and c_2, are very small, while the last two are large. For the negative-energy states we have the opposite situation. In the ordinary problems we meet in chemical physics, we do not have to worry about the negative-energy states, since we do not ordinarily have positrons present in our experiments. Consequently the large components are ψ_3 and ψ_4. Dirac's equations prove able to give the values of ψ_1 and ψ_2, the small components, in terms of the large ones, and when we make this elimination, we come to equations for ψ_3 and ψ_4, which look like ordinary Schrödinger equations, with the second derivatives we are used to. This process can be carried through even in case we have an external vector and scalar potential. And when that is done, we find the answer to the question as to the significance of ψ_3 and ψ_4.

The meaning is very simple. Automatically, out of the various terms contained in Dirac's equations resulting from the σ's and the potentials, one finds that the electron has a spin angular momentum and magnetic moment of just the amount predicted by Uhlenbeck and Goudsmit and by Pauli. All of the terms giving the interaction between these magnetic moments and the external magnetic field which we should expect are there, as well as the anomalous g value of 2 for the gyromagnetic ratio we mentioned in Sec. 4. It is the fact that all of these results come out of Dirac's idea, with really nothing more in the way of fundamental ideas than we have hinted, which makes it such a wonderful feat of imagination. And in addition to that, the existence of the positron came from the same simple postulates.

But now what are ψ_3 and ψ_4? It is very simple: They are the wave functions for the spin-up and the spin-down electrons, respectively. The rules for operating with these spins, which Pauli had worked out rather intuitively in 1927, immediately fell into place from Dirac's theory. There are also all the various terms in the Hamiltonian which are required to discuss spin-orbit interaction and various features of the experimental spectra. For the behavior of a single electron in an external field, Dirac's theory seems to have given the final answer.

All this, however, is only for one electron. The question of the interaction of electrons, taking account of the electron spin as well as relativistic effects, is much more complicated. Gregory Breit, whom I had known at Harvard during my last graduate year when he was a National Research Fellow there, was the first one to work this out. The general approach was the direct one: to find the electric and magnetic fields produced by one electron at the position of the other, and to introduce their scalar and vector potentials into the Dirac equation for the other electron. Breit's paper, in the *Physical Review* for 1929, was substantially a complete treatment, but there has been a good deal of work since in making minor improvements on it. It is notable that in these treatments, it is simply assumed from the beginning that an electron does not act magnetically on itself, any more than it works electrostatically on itself. In other words, all terms of the nature of $1/r_{ij}$, where $i=j$, are omitted from the first. There seems, as I mentioned in the preceding section, to be no doubt that the vector potential, in a many-electron problem, just like the scalar potential, is to be found from the fields of all other electrons, without any self-interaction term.

19. 1927–1932, The Helium Atom

I mentioned in Sec. 17, that in 1927, after it became obvious that Dirac was going to work out the problem of the radiation field, I shifted my interest to the problem of the helium atom. This in a sense was going back to the question that had led me to wave mechanics in the first place: the compressibility of the alkali halides. I mentioned in Sec. 1 that when I was working on that problem for my Ph.D. thesis, I concluded that nothing in the quantum mechanics as known in that period would explain the repulsive forces felt at short distances between the atoms or ions of an alkali halide. Obviously, since the ions were closed shells, an equivalent problem was the repulsions between inert gas atoms, helium, neon, argon, and so on. The simplest case was helium. Why were they relatively hard and impenetrable, as kinetic theory showed them to be? I felt that if one could first investigate the wave functions of a helium atom, and then their interaction, one should be able to throw light on this problem. It was with this in mind that I started work on the helium atom.

The reader should remember that at that time Heisenberg's paper concerning the symmetric and antisymmetric electronic wave functions for the helium atom had already come out, but not yet the Hartree self-consistent field. However, Heisenberg, as I indicated in Sec. 6, was working with a wave function that was a product, or more generally a symmetrized or antisymmetrized product. This incorporated the general idea of the self-consistent field, which was based on the idea that the repulsive interactions between the electrons would not be handled in detail, but only as a sort of average over the electronic charges. I felt that if I were to make a real advance over that type of treatment, I should include the $1/r_{12}$ term explicitly in my calculations. During the next year I tried out several schemes for doing this, none of which proved to be very successful, but which on account of their inherent interest deserve to be mentioned. They played a fairly important part, in fact, in the development of the ideas which later led to the $X\alpha$ method.

After considerable experimentation, I found that it was useful to find a function which behaved properly in the limits where r_1, r_2, r_{12}, the distances of the two electrons from the nucleus and from each other, became very small. I noticed that a very simple function, namely $\exp(-2r_1 - 2r_2 + \frac{1}{2}r_{12})$, behaves properly in each of these limits. I tried a function that approached this function when all r's were small, but behaved quite differently when r_1 was small, r_2 large. In that limit, electron 2 was moving around a nucleus shielded by electron 1, so that its wave function was approximately hydrogenic, but with a different energy from the ordinary hydrogenic energy. I described the behavior of electron 2 in this limit in terms of a solution of the hydrogenic equation which behaved properly at infinity, and would diverge at the origin, but I was not using it except at large r, where I used an interpolation scheme to make it join smoothly onto the function I was using at small r. An equivalent function was set up in the limit where r_2 was small, r_1 large. I tested the function in several ways. First I integrated the square of the function over dv_2 to get the probability that electron 1 lay within given limits, or to give a charge density. By the time I had arrived at this part of the calculation, Hartree's first calculation of the helium atom by the self-consistent field had come out, and I could compare my calculated density with Hartree's, and obtained very satisfactory agreement. Next I computed the diamagnetic susceptibility, a very sensitive test of the accuracy of a wave function. The agreement with experiment was again excellent.

At this point I felt it worthwhile to make the tedious calculation of the repulsion of two helium atoms, using the techniques of Heitler and London, which by then had been published. I therefore built up the four-electron wave function, using a determinant for the two electrons with spin up multiplied by a determinant for the two with spin down, and the wave functions I was using. I carried through the various integrations which one runs into with the method of Heitler and London, and ended up with a repulsive potential which could be adequately approximated by an exponential repulsion. Here was the repulsive potential I had been looking for for five years.

In order to test this potential, I wished to compare it with potentials of interaction between helium atoms, as the experts in the kinetic theory of gases had determined them. This demanded the van der Waals attraction as well as the repulsion. A value for this quantity was determined by S. C. Wang in 1927, anticipating the better known work of Eisenschitz and London on this problem in 1930. Wang was a young Chinese student who had taken his master's degree at Harvard in 1926, and then had gone on to Columbia to finish his Ph.D. Debye had suggested some time earlier that the origin of the van der Waals attraction came from a mutual polarization effect of one atom for the other, and had shown that the attractive energy should be proportional to the inverse sixth power of the interatomic distance. Wang worked out this idea for helium, obtaining a value that could be related to the polarizability of the atom. A combination of my exponential repulsion with Wang's inverse sixth power attraction was the final outcome of the paper I wrote in 1928, about the normal state of helium.

Calculations that I made in 1931, in collaboration with John G. Kirkwood, then a student of chemistry at MIT who later became a leading theoretical chemist, verified the accuracy of the van der Waals attraction which Wang had calculated. Of course, there was no good reason for supposing that the simple superposition of the two terms in the energy should be accurate, but it proved to be a fact that the potential suggested in the 1928 paper gave such good agreement with experiment for the properties of gaseous and liquid helium that it was a good many years before anyone had produced a more accurate calculation of this interaction. At any rate, the result was good enough to convince me that we understood these repulsions. When one looks at the problem from the point of view of molecular orbitals, one finds that the ground state of the molecule consisting of two helium atoms has two occupied bonding σ_g orbitals and two occupied antibonding σ_u orbitals. The repulsion arising from the antibonding orbitals slightly outweighs the attraction arising from the bonding orbitals. It is this effect that is responsible in all cases for the repulsion of closed shells of electrons in atoms or ions.

By the time I wrote up this helium work for publication, in 1928, Hartree's papers about the self-consistent field were coming out, and it was obvious that that method worked so much better than one might have anticipated that I felt that the added refinements I was trying to introduce with my helium work might not be very necessary. I started the work relating to the self-consistent field which resulted in my determinantal wave function papers in 1929. But another reason for dropping the helium problem was that this was when Hylleraas started his very important series of papers. Egil Hylleraas was a young Norwegian, a year or two older than I, whom I unfortunately never had the pleasure of meeting until 1963, two years before his death. At that time, his contributions to physics were being celebrated at one of the Sanibel symposia organized by Per-Olov Löwdin in Florida. This was, in fact, the first one of those symposia which I visited, and it was the one at which the negotiations were started which eventually induced me to move to Florida myself. Hylleraas's introductory talk at that symposium gives a delightful account of his early days with the helium problem.

After some studies in Norway, he had received a fellowship to work in Göttingen, where Born was the head of the Institute for Theoretical Physics, and where Heisenberg and Jordan had been working with Born on the development of quantum mechanics. As soon as Schrödinger's equation was proposed, Born suggested to Hylleraas that he try to make a real calculation of the energy of the ground state of that atom, to give a crucial test of the correctness of wave mechanics. Heisenberg's paper with the symmetric and antisymmetric wave functions had already come out.

The problem appealed to Hylleraas, for much the same reasons that it appealed to me: Hylleraas's earlier work had been on the theory of ionic crystals, the same problem that led me toward a study of helium as the simplest example of a closed-shell atom.

Hylleraas's first attack on the problem was by the method of configuration interaction. For this purpose he had to have a complete set of excited orbitals, and rather than using the hydrogenic ones he had the good idea of using an alternative set which does not have a continuous spectrum. The convergence of a configuration interaction with such a basis set is much better than with the excited orbitals of a Hartree calculation. He carried through a quite ambitious calculation of this type, and got a remarkably good value for the ionization potential, good enough to make everyone believe that wave mechanics was probably exactly right for such a problem.

But for a second step, he decided, as I had done, to include r_{12} explicitly in his calculations. Rather than putting r_{12} in an exponential, as I had done, he decided to include it in a power series term. He tried a number of different forms of approximation function, of which a fairly simple but very good one is, in unnormalized form,

$$u = \exp(-1.82s)(1 - 0.100828s + 0.353808u + 0.033124s^2$$
$$+ 0.128521t^2 - 0.031799u^2) \tag{19-1}$$

where $\quad s = r_1 + r_2 \quad t = -r_1 + r_2 \quad u = r_{12}$

The average energy he obtained for this function was -2.90324 hartrees, which compares very well with one of the most recent values, -2.9037245 hartrees, which is extremely close to experiment. For comparison, the energy of the Hartree-Fock solution is -2.861679 hartrees. Hylleraas went on and used some power series solutions longer than that of Eq. 19-1, getting results in even closer agreement with experiment.

It is difficult to see what such a wave function as that of Eq. 19-1 really means. How well does Hylleraas's solution describe the cusp, the behavior of the function when r_{12} goes to zero? In Fig. 19-1, we show the wave function as a function of r_{12}, for two cases. One is for the case $r_1 = r_2$, so that both electrons are on the surface of the same sphere. The other is the case where both electrons are on the same radius. In each case, $r_1 + r_2 = 1$ atomic unit. The dashed curve is $1 + \frac{1}{2}r_{12}$, the first two terms of the expansion of $\exp(\frac{1}{2}r_{12})$, the form that I had been using. We see from the figure that Hylleraas's function leads to the correct behavior as r_{12} goes to zero, and shows that this behavior is quite soon lost as r_{12} becomes larger. Hylleraas pointed out that it was in fact rather more convenient to describe this cusp in terms of a linear term in r_{12} in the power series multiplying the exponential, as in Eq. 19-1, than in terms of an exponential, as I had done.

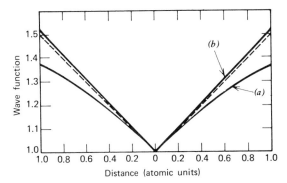

Fig. 19-1. Helium wave function as function of interelectronic distance, as given by Eq. 19-1. Abscissa, r_{12} in atomic units: ordinate, quantity proportional to wave function. For curve a, $r_1 = r_2$, so that both electrons are on the same sphere, for curve b, both electrons are on the same radius. In each case, $r_1 + r_2 = 1$ atomic unit. Dashed curve, wave function $= 1 + \frac{1}{2} r_{12}$.

As I have mentioned several times, this problem of the helium atom has been a celebrated one in quantum mechanics, and later more extensive solutions than that of Eq. 19-1 give results which are indistinguishable from experiment. Equally good solutions have been found for excited states of the atom. The other equally important two-electron problem is that of the hydrogen molecule, and here solutions of the accuracy of Hylleraas's treatment of helium were somewhat longer in coming. We may anticipate by saying that in 1933 two Harvard workers, Hubert M. James, then a graduate student, and Albert Sprague Coolidge, a faculty member in chemistry, working under Kemble's direction, applied methods similar to those which Hylleraas had used to the molecular problem. The results were equally satisfactory, and later more elaborate treatments, as with helium, have converged more and more closely to the experimental values.

20. 1929–1931, Bethe and Frenkel, Localized States in Crystals

In the latter 1920s and early 1930s, experimental evidence was piling up to indicate that the absorption spectra of solids had much more structure in them than had been believed earlier. Thus, Hilsch and Pohl in Göttingen had made studies of the absorption spectra of the alkali halide crystals, and had found sharp absorption lines near the main ultraviolet absorption edge. The energy-band theory was far enough advanced even at that time so that it was generally accepted that the valence electrons of these crystals fell in bands that were fully occupied, while above an energy gap there was a conduction band that was empty. Naturally it had been supposed that the smallest photon which could be absorbed would be the one to raise an electron from the top of the valence band to the bottom of the conduction band. The electron thus placed in the conduction band, and the hole left in the valence band, would both be potential carriers of electricity, so that such absorption would lead to electrical conductivity, called photoconductivity, which would persist until the excited electrons fell back into the holes in the valence band.

What Hilsch and Pohl found, on the contrary, was that the sharp absorptions did not lead to photoconductivity. It was only somewhat larger energy absorptions, coming from absorption further in the ultraviolet, which led to photoconductivity, and which formed the beginning of a continuous absorption band. Clearly the sharp-line absorption must have some different mechanism. These were not the only sharp-line absorptions not involving photoconductivity which began to be observed. It was found that not only crystals but also solutions of compounds containing rare earth ions showed sharp-line absorption in the visible part of the spectrum. It was these absorptions that were responsible for the brilliant colors which some of these substances showed. Some crystals and solutions of the $3d$ and $4d$ transition elements showed similar behavior, though the lines were not so sharp. It was naturally supposed that the sharpness of the rare earth absorptions came from the fact that they arose from $4f$ electrons, far down inside the atom and hence well protected from a broadening of their bands on account of the action of neighboring atoms. On the other hand, the orbitals of the $3d$ and $4d$ electrons were more spread out, and one would expect more interaction between the orbitals in neighboring atoms or ions. This would be expected to broaden the discrete levels of the atom or ion into bands.

Naturally these phenomena attracted the attention of the theorists. It looked as if the absorption were more like that of free atoms than like that arising from energy bands. One of the first pieces of theoretical work relating to this was that of Bethe, in 1929. He had already made the study of electron diffraction, which I mentioned in Sec. 15. He wished to investigate the perturbation of atomic levels by the most obvious effect arising from a crystalline environment: an electrostatic crystal field which did not have spherical symmetry. His 1929 paper on this topic, in which he worked out by group theory the various levels into which an atomic level was split, has been a classic ever since. For instance, in a cubic field, an atomic p level remains threefold degenerate, or is not split, whereas an atomic d is split into a twofold and a threefold degenerate set of levels.

The theory was soon taken up by Van Vleck and his students, since the splitting of levels was fundamental to the study of the magnetic properties of crystals containing $3d$ or $4d$ transition elements or rare earth ions. A famous set of papers was written by Penney and Schlapp in 1932 and 1933, under the direction of Van Vleck. It is significant that Penney and Schlapp had come from England to work with Van Vleck in Madison. This was one of the early signs of a reversal of the trend by which Americans went to Europe, but Europeans seldom came to America. W. G. Penney, then a young student who had just taken his Ph.D. at Imperial College, later rose in England to the point where, as Lord Penney, he headed the U.K. Atomic Energy Authority, and was Rector of Imperial College.

Another contribution to the localized states in crystals came from J. Frenkel, in 1931. Here again it is significant that this contribution, "On the Transformation of Light into Heat in Solids," by a distinguished Soviet scientist was published in the *Physical Review* in 1931, while Frenkel was a visiting scientist at the University of Minnesota. Frenkel was the first to set up a theory of the exciton, a localized excitation in a solid. He assumed that just one atom, at position $\mathbf{R}_m$ in the lattice, was excited, the rest being in their ground states. He then made a linear combination of such many-electron functions, of the nature of a Bloch sum. If $A(\mathbf{R}_m)$ is a many-electron function, ordinarily a determinantal function, with excitation on the atom at $\mathbf{R}_m$, this Bloch sum was

$$\Sigma(\mathbf{R}_m)\exp(i\mathbf{K}\cdot\mathbf{R}_m)A(\mathbf{R}_m) \qquad (20\text{-}1)$$

Such excited states are now generally called exciton states. Frenkel went into the problem of the interchange of energy between such an optically excited state and lattice vibration states, showing how the energy of optical excitation could gradually be dissipated into heat. We shall not go further into Frenkel's theory here, for we mention it in more detail in Sec. 25 in connection with further work on the exciton.

It is interesting, however, to think that Frenkel was freely visiting the University of Minnesota, talking at meetings of the American Physical Society, and publishing in American journals. It is hard for readers of the present age to realize how relaxed the international scene was in the early thirties. There we were in the middle of Herbert Hoover's presidency, a time which according to present misconceptions was one of political reaction. And yet there has never been a later period when the world was as easy to travel in. For example, when I was in Leipzig in 1929, I also visited Prague, in Czechoslovakia, to attend a meeting of the Deutsche Physikalische Gesellschaft, the German Physical Society, which out of a spirit of good feeling was holding a meeting there. I visited not only Vienna but also Budapest. Leipzig, which is now in East Germany, was a short trip on the train from Berlin, and one traveled perfectly freely from one to the other and to all other parts of Europe. I am very glad to have had this chance to see Germany, and to visit the eastern European countries when it was still easy. I did not, however, visit the Soviet Union, though it would have been perfectly simple. My own sentiments, ever since the time of the Russian revolution in 1917, had been strongly anti-Communist, and I had no desire to visit their country. In this, I was quite different from most scientists and in fact most university people, among whom there were very great numbers of Communist sympathizers.

But in spite of any political feelings, Frenkel was a great favorite, with me as well as with Americans in general. He was a physicist of the type I preferred, writing clearly and comprehensibly. His books on wave mechanics, written in English, were among the most usable of all those that were published in the early days. He made many distinguished contributions to physics. It was a great pity that he died in the 1950s; he would have been a great help to his country in the difficult days that followed.

These relaxed days of the early 1930s were really the end of an era, not only for physics but for the world. Quite arbitrarily, I shall regard March 4, 1933, the day of Franklin Roosevelt's inauguration as president, as a turning point between two ages. That was politically a dramatic period in the history of the United States. Deepening financial troubles in the European banking system had caused a financial panic in the United States, undoing all the improvement in the economic situation which had helped greatly to overcome the disastrous stock market crash of 1929. Roosevelt declined all efforts of Hoover to soften the impact of the change of administrations in the spring of 1933. During the week before the presidential inauguration, every bank in the country had closed. The stage was set for Roosevelt to come in as the saviour of the country, to change the government as far as he was able, and to make it over in a new form. The United States was never the same again.

But before March 1933 was over, Adolf Hitler was voted dictatorial powers in Germany, and the Nazi regime really started. Things had been turning in this direction for several years. Even when I was in Leipzig in 1929, our German landlady had been dreadfully afraid of the "Roten," the young Communists who were trying to terrorize the city. She, like so many Germans, felt that the only way to counter such violence on the left was with equal violence on the right. I do not doubt that when the Nazis came in, she welcomed them. And when they in turn were driven out by the Communists, I do not doubt that she felt that her worst dreams had come true. By 1932, when we made a short summer trip to the Austrian Tyrol purely as a vacation, and made a short stop in Germany to visit Nuremburg and Rothenburg, stresses were much worse, and international finance was so complicated that it was hard to transport even one's ordinary travel money from one country to the other. After 1933, few people would have thought of trying to visit Germany. I crossed the ocean again in 1934, but only to Britain. I have never been back to Germany, or to eastern Europe, since those days.

Before we leave this rather ideal but vanished age of the decade from 1923 to 1932, let us run over some of the main centers of quantum science in the countries that were so soon to lose their scientists. Copenhagen, of course, had remained a leading center, with Bohr, though I was not invited back during this prewar period. I had made myself sufficiently unpopular with them so that apparently they did not care to see me again. The only one of the centers which I knew personally was Leipzig, with Debye, Heisenberg, and Hund, where as I have said I had a very pleasant welcome in 1929. But the two leading centers aside from these and Berlin were Munich, where Sommerfeld had been the professor for many years, and Göttingen, where they had the fabulous situation, for Germany, of three physical institutes. It was the usual custom in Germany for the professor to be the boss of the whole show, and for him to expect everyone in the department to work for him. But in Göttingen, there were three professors, each with his own institute: Max Born in theoretical physics; James Franck, of the Franck-Hertz experiments, in electronic experimental physics; and Pohl in the optical properties of solids. It was in Pohl's laboratory that the Hilsch and Pohl work I mentioned was carried out. Many Americans visited these two centers, including many who were later colleagues of mine at MIT, so that we did not lose the spirit of these great laboratories after they were dismantled in the early Hitler days. They formed a model which we perhaps emulated more than we realized, when we began to build up the American universities to take their place, in the years that followed.

BOOK

II

Transitional Years,

1933–1940

21. MIT and Princeton, 1930–1940

The 1930s were the years in which the initiative in quantum physics shifted from Europe to America. I have somewhat arbitrarily chosen inauguration day 1933 as a transition date, but I shall start by going back to 1930, the year I moved from Harvard to MIT. I have already covered fairly well the developments in quantum theory up to 1933, but it will be worthwhile to give a picture of the situation in American universities at the time. Fortunately we were advancing, allowing us to take up the leading position which we acquired by the end of the 1940s. MIT and Princeton were particularly close during that period, and for that reason I shall pay particular attention to what was going on there. But many other places come into the picture: obviously Harvard, where I continued to teach one course each term for a couple of years after 1930; but also Berkeley, where I spent the summer of 1931; Pasadena, which I visited during that summer; Chicago where I had taught in the summer of 1928; Wisconsin and Minnesota and Michigan; as well as numerous other leading centers. The particular advances in quantum theory which I discuss in later sections, however, all happened to come from MIT or Princeton.

The close ties between MIT and Princeton were an outcome of the fact that Compton had come from Princeton to MIT, and he did not want to break relations between the two institutions, any more than I wanted to break off between MIT and Harvard. My position as a new department head under a new president who had come to the institute with a mandate to build up the physics department was obviously ideal, except for the fact that, as a young man 30 years of age, I obviously had to avoid rushing in brashly and changing everything. Fortunately that was not necessary anyway. I realized from the outset that there already was a very strong foundation to build on at MIT. The move to improve the physics department, then one of the smallest in point of numbers at the institute, came from the more forward-looking members of that department itself, so that I came into a sympathetic environment. The department members had chosen me, just as much as Compton had. In the course of the negotiations leading to my going to MIT, Compton had confided to me the fact that four other people had been approached for the position, and had turned him down, before they spoke to me. But when I heard who the four were—W. L. Bragg, G. P. Thomson, W. V. Houston, and A. W. Hull—and when Compton said that I was in fact the first choice, but that they felt they should try first to get someone from outside Cambridge, I felt that I must have been well thought of by those who had chosen me.

Two activities kept me and Compton busy from the first, and he and I worked very closely together on them. First, there was the recruiting of new staff. He had listed for me a number of possibilities in the first long letter he wrote me while we were in negotiations. Second was a decision about a number of people, some on temporary appointment and some more senior, who, we felt, should look for positions elsewhere. I need say no more about the second activity, except that it was handled without any disruption of the department. But the first was important, and we had the cooperation of everyone in our efforts. All during the decade of the 1920s efforts had been going on at MIT to improve physics. Leading professors in other departments, particularly electrical engineering and mathematics, as well as those in physics, had been agitating for this, and were very helpful.

Among the department members whom I found already there when I arrived in 1930, and who proved to be among the strongest members of the department as it was finally formed, were several theorists. Manuel S. Vallarta, a brilliant Mexican, had taken his doctorate at the Institute in 1924. He was interested in relativity, relativistic quantum mechanics, cosmic rays, cosmology, and various related topics. He was up to date on quantum theory, as he showed during the year 1930–1931 when he gave a Harvard colloquium on Dirac's electron theory. Nathaniel H. Frank, a native of Boston, had taken his doctorate at MIT in 1927. He spent the year 1929–1930 with Sommerfeld at Munich, working on the theory of magnetoresistance and related topics. He had also specialized in classical mechanics, and was regarded as one who could well revolutionize the teaching of undergraduate mechanics at the institute. It should be pointed out that in those days, all undergraduates at MIT were required to take two years of physics, during their first two years, so that they had a very thorough course. Frank took a leading part in bringing the undergraduate training up to date, and he served as department head for several years in the 1950s. William P. Allis had also been with Sommerfeld at Munich during 1929–1930, and remained there during 1930–1931, coming back to MIT in September 1931. Allis, a son of wealthy parents, half-American and half-French, had been brought up in practically a palace on the Riviera, and had then taken his degree at the University of Nancy in 1924, before coming to MIT. I shall speak later of some of his research.

In addition to these very strong members of the theoretical part of the department, there was Julius A. Stratton, who had been a member of the electrical engineering department, but who transferred to physics in 1930. Stratton's work on electromagnetic theory, culminating in the classical book on that subject which he wrote in 1941, is known to every physicist. He had taken his degree at the Federal Institute of Technology in Zurich in 1927, after doing his undergraduate work at MIT. Even as a very young man, his good judgment and grasp of the fundamentals of administration were obvious to everyone, and no one was surprised when he eventually become president of MIT, and later chairman of the board of the Ford Foundation. From the time of my first acquaintance with him, he was one of my closest friends, and an inestimable help in the effort to build up the physics department and the institute.

In spite of this strength in the field of theoretical physics, which of course was my own particular interest, Compton and I both felt that we needed additional staff members. We were anxious to build up a group second to none. The first step we took was to appoint Philip M. Morse to the staff. He had taken his undergraduate degree at the Case Institute in Cleveland, and then in 1929 took his Ph.D. at Princeton, where Compton had formed a very high opinion of his qualities. During the two years 1929–1931 he, like Frank and Allis, was with Sommerfeld in Munich, and he joined the MIT staff in 1931. I shall speak later of joint work which he and Allis had done in Munich. But he had also made other important contributions before coming to MIT. Everyone has heard of the Morse curve, a curve giving a good approximation to the actual interatomic potential energy in the diatomic molecule. He had suggested this, not merely as a useful approximation, but because it permitted an exact integration of the Schrödinger equation for the molecular vibration. He had also, in the course of a summer spent at Bell Telephone Laboratories, done some of the very early work on energy bands. And in collaboration with Condon during the year 1929 when he was taking his degree at Princeton, he produced a text on quantum mechanics which was one of the very first American books on wave mechanics, and which still is a valuable treatment of the subject. Morse's greatest interest has always been the application of rather sophisticated mathematical methods to problems in physics, and his joint book with his student Herman Feshbach, *Methods of Theoretical Physics,* has become a classic.

Compton and I still felt that we would like more strength in the quantum-theoretical field, however, During the first two years I was at MIT, we made strong efforts to get both Condon and Pauling to join the department. We were not successful, since each one felt that he was already well off where he was, but each one visited MIT frequently, and gave extended series of lectures in the department during those first years. They were the last ones whom we tried to get from outside, but as I mentioned, Feshbach, who took his degree some years later in the department, stayed on as a faculty member, and at the time of this writing is the head of the department.

In addition to this strength in theoretical physics, Compton and I both felt that we wanted to have very strong experimental groups. There were already several excellent department members in experimental fields. Two of these were results of previous visiting professors in the department. It had been the department policy for some years to have eminent lecturers from abroad, and I have already mentioned the lectures which Max Born had given in 1925, lectures which led to his useful book on crystal theory and quantum mechanics. But in addition, in the preceding year, 1924, Debye had lectured for a considerable period in the department. He was at that time a department member of the Federal Institute of Technology in Zurich, and he brought with him his student Hans Mueller, to help him with his demonstration lectures. To everyone's delight, Mueller decided to stay on, and he was a very valuable department member, full of ideas on properties of dielectrics and similar topics.

Then in 1927, W. L. Bragg, then of the University of Manchester, before becoming Cavendish Professor at Cambridge, was a visiting lecturer. One of the young MIT students, Bertram E. Warren, so impressed him that he arranged for Warren to spend some time with him in England, and also to visit other European x-ray centers. After returning to MIT, Warren took his degree in 1929, and started a lifetime of most useful work at MIT. He became universally respected among x-ray crystallographers, and always had one of the most productive experimental groups in the department. My interest in directional properties of covalent bonds was an outgrowth of the great deal I learned from him about the geometrical arrangement of atoms in crystals. During the first days of the 1930s, his interest was in the silicates, but it turned later to amorphous compounds and glasses, a subject in which he made great contributions.

I should mention in the field of crystalline properties, in addition to Warren, another department member, Donald C. Stockbarger, who was more interested in techniques than in fundamentals, but who was the originator of some of the best methods for making single crystals. Another department member with a very good experimental program was Arthur C. Hardy, in the field of optics. His interests were of the rather practical sort —geometric optics as concerned in the design of optical instruments, and color measurements. He devised very useful instruments for getting the color properties of all types of samples of materials. He had particularly close relations to industry.

But in addition to these fields, Compton and I both had our ideas as to how to build up further experimental branches of physics. First, we both were interested in spectroscopy, and wanted to have strong groups in that field. Compton had had with him at Princeton Joseph C. Boyce, whose interest was in the spectroscopy of the vacuum ultraviolet. Boyce joined our staff in 1931, and remained through the 1930s. But in addition to him,

I had come to know George R. Harrison at Harvard during the years 1924–1925, when he was a postdoctoral fellow working with Lyman on far-ultraviolet spectroscopy, and when I was an instructor. Harrison, who had taken his doctorate at Stanford and had returned there after his Harvard year, was particularly interested in the measurement of the intensities of spectral lines, and he and I had collaborated on working out the practical methods of applying my optical theory of 1925. He was a man of particular genius in the laboratory, an expert at the design of equipment, but interested also in applying his methods. I had known him again in the summer of 1926, when at his suggestion I was given a summer teaching position at Stanford.

My first thought, when I was appointed to the MIT department in 1930, was that Harrison would be invaluable in seeing to it that the laboratory ran well. I felt that as a theorist, I could hardly undertake to manage a laboratory by myself, and Compton had suggested in our very early negotiation that if I wished, I could appoint some department member as director of the laboratory. I immediately suggested Harrison to Compton, who had not met him, and he agreed to offer Harrison an appointment in the department, with the additional title of director of the laboratory. Harrison accepted promptly, and was on hand by September 1930.

At the time I was appointed, plans were already well underway for a new laboratory, jointly for physics and chemistry: the George Eastman Research Laboratory, to be financed from a fund which Eastman had given to the Institute after the main buildings had been completed. It was by then fairly well known that the "mysterious Mr. X" who had contributed many millions to the Institute to help with the construction of their monumental building complex in Cambridge was Eastman, who modestly refused to let his name be mentioned, but this additional fund which he had left led to the Eastman Laboratory. If there had not been such a fund available, it is not likely that the Institute would have been able to go ahead with the new building so soon after the 1929 stock market crash. There were very active negotiations with the architects during the summer of 1930, and Harrison helped at long distance with these. We enlarged the building plans by constructing an additional building, hidden in a courtyard, to house the gratings and vacuum spectrographs which Harrison wished to install.

The spectroscopic program was an important feature of the department all during the 1930s. When Francis Bitter, the distinguished specialist on magnetism, joined the Institute staff in 1934, first in the metallurgy department and later in physics, he constructed a magnet in the spectroscopy laboratory able to reach the then unprecedented steady-state field of 100,000 gauss. We thus had unequaled facilities for studying the

Zeeman effect. Harrison's interest then turned to the spectra of the transition group and rare earth elements, whose spectra could now be interpreted, following the pioneer work of Russell and Saunders. These were spectra so complicated that mechanized methods of measuring the enormous number of lines were needed, and Harrison constructed equipment for this. During the years when the WPA (Work Projects Administration) was a method of helping out with unemployment, he managed to secure a WPA project to get unemployed scientists to help use this equipment, and a rather wholesale effort produced measurements of the spectra of many of these elements. One of the younger colleagues whom Harrison brought with him, Walter Albertson, proved to be quite a genius at the analysis of such spectra, and MIT became a center for the spectroscopy of these complex atoms. From our point of view in physics, it was a sad day when Harrison was appointed dean of science in 1942, and had to reduce his connection with the departmental program. One of the valuable enterprises which he started was an annual spectroscopy conference, held each summer beginning in 1933, which attracted many participants from all over the world.

Another interesting spectroscopic program went on during the early 1930s. For several years following 1931, Henry M. O'Bryan was an instructor in physics. His interest was in the field of soft x-ray spectroscopy. During the year 1932–1933 he was joined by a visitor from England, H. W. B. Skinner. The two combined to do some of the first serious work on the soft x-ray spectra of metals. This method gave early indications of the energy width of the energy bands in metals, and of the sharp Fermi surface. Unfortunately, though it was a popular technique for a number of years, both here and overseas, there were constant difficulties arising from contamination of the surface of the sample, and in the long run it has not turned out to be as useful for studying energy levels of crystals as the photoemission technique which is more in vogue at present.

Another experimental field which Compton was very interested in establishing was electronics, which had been his own main field of interest. Two of his former students whom he wanted to invite to the Institute were Wayne B. Nottingham and Edward S. Lamar. Both of them came, Lamar working particularly closely with Compton and Boyce on gas discharge problems. Nottingham, who had spent some years at the Bartol Laboratory after taking his doctorate at Princeton, soon induced Dr. Erik Rudberg, from Sweden, whom he had known at the Bartol, to come, and Rudberg was a department member for several years. Nottingham's interest was varied, but centered mostly on thermionic emission, phosphors, and related fields. He had a particularly strong feeling for coordinating his experiments with theory, and I learned a great deal from him about the practical use of the Fermi statistics in studying such problems as field emission, surface effects on various faces of a single crystal of such a material as tungsten, and so on.

One of the students who worked with him on this was William Shockley, who later took his degree under me on a theoretical problem which I shall describe in Sec. 25. He then went on to the Bell Laboratories. As one of the inventors of the transistor, he received a Nobel Prize. But in addition to him, many other leading scientists in later years at the Bell Laboratories and in other industrial laboratories obtained much of their training under Nottingham. One of his enterprises was to establish annual electronics conferences, meeting every spring at the institute. These were attended by several hundred participants, some from universities but more from industry, who felt that they led to a more scientific approach to electronic technology than could be obtained elsewhere.

These early years of the 1930s were of course the ones when high-energy and nuclear physics were rapidly becoming of surpassing interest. I well remember sitting in the lunchroom of the Union Station in Washington for breakfast one morning in the early 1930s, when the physicists were congregating to attend the spring meeting of the American Physical Society, and having Ernest Lawrence of Berkeley come and sit beside me, breathless with excitement at the new idea he was working on. He drew for me a diagram of a cyclotron, which was still unnamed and unbuilt, but was merely a gleam in his eye. By the summer of 1931, when I was in Berkeley, the first cyclotron was running under the hands of Lawrence and Livingston, and produced 600-kilovolt protons. This was the beginning of an exponential growth, and before the decade was over, we had our own cyclotron, constructed by Stanley Livingston, who became a member of our staff. But there were several steps before that.

One of the most ingenious of the young experimenters whom Compton wanted to bring from Princeton was Robert Van de Graaff. He came in 1931, becoming an associate professor in 1934. He never took any interest in teaching, and was regarded as essentially a research man. When he came, he had only small models of the accelerator which bears his name. But he started at once building a mammoth accelerator, at a surprising place: the estate of Colonel E. H. R. Green, the eccentric son of a more eccentric mother, Hetty Green. Colonel Green had inherited Hetty's money, and had a large estate by the side of the water down near New Bedford. At this estate, called Round Hill, he carried out all sorts of scientific experiments. At one time he had an airstrip and landing field, and it was the airplane hangar in which Van de Graaff and his group worked. The MIT people were working with the Colonel, hoping (vainly, as it turned out) that he would leave them some money in his will.

Actually, no will was found, and the entire estate went to his relatives. To show the variety of interests he had, the old whaling ship Charles W. Morgan, the last extant member of the whaling fleet, was tied up at his dock, where it remained until the maritime museum at Mystic, Connecticut, was established.

The first accelerator which Van de Graaff and his considerable group of associates built at Round Hill was an enormous affair, for it used the ordinary atmosphere as an insulator. To get even a couple of million electron volts (abbreviated mev), one had to have electrodes 20 feet high, with spheres at the top big enough to walk into. The belts ran inside the vertical columns up to the spheres, and fine sparks were emitted when the 2-mev voltage was built up between them. It was soon realized that this was not a hopeful direction for future progress, and further accelerators were built to use air at high pressure as an insulator, so that the whole thing was enclosed in a tank, and could be much smaller. Successive versions of this were built, until finally one was set up which reached a good many mev, and which proved to be a fine research instrument. In the meantime, the old original machine had enough historical interest so that it has been set up at the Boston Museum of Science, to make sparks for the schoolchildren.

A number of colleagues gathered around Van de Graaff to work on various aspects of the program. John G. Trump of the electrical engineering department developed engineering applications, principally for medical purposes. The leading person interested in using the accelerator for nuclear physics research was a student named William W. Buechner, who got his start before the war, continued with the largest instrument after the war, studying nuclear energy levels, and eventually served for a number of years as department head. One of the young students who was much interested in the program in its early days was James B. Fisk, who went on to Bell Laboratories, after spending time in the Society of Fellows at Harvard. He eventually rose to be director of research for the Atomic Energy Commission, president of the Bell Telephone Laboratories, and a life member of the MIT corporation. He has been the corporation member most closely associated with picking out the last several presidents of MIT.

Compton and I never felt that we wanted to have the department go overboard in the direction of nuclear and high-energy physics, as so many university departments were doing. We felt that at a great institution like MIT, in which engineering and technology were associated closely with science, we should develop all the branches of physics together, emphasizing the way in which a development in pure science soon takes its place in technology. The history of the physics department had emphasized this. The departments of electrical engineering and aeronautical engineering at

MIT had been outgrowths of work first started in physics. We had no intention of forsaking electrical engineering, metallurgy, and many other fields which were dependent on physics for their future development. I felt this as strongly as Compton did, and my interests in this direction were one of the things which had impelled me to leave Harvard for MIT. Thus it was natural that we had no intention of dropping the more standard branches of physics to go completely into the new fields that were opening up.

Fortunately, however, the Institute was growing enough, and particularly the physics department was expanding enough, so that we could make new appointments and keep up with the development of the new fields. We decided by 1934 that we should be doing some work in the field of radioactivity, and Compton combed the country to try to find a suitable man. At Millikan's suggestion, we approached Robley D. Evans, who had taken his earlier training at Cal. Tech. and at Berkeley in the field of applied radioactivity, and he joined the department in the fall of 1934. He built up what he called a radioactivity center, gathered an impressive group around him, and started an excellent course in nuclear physics. His text on the atomic nucleus was one of the first and best texts on the subject, and he had many students. It was at his suggestion that we decided, in 1938, that we had to have a cyclotron. We raised money from grants to build it, and induced Stanley Livingston to come from Cornell, where he was teaching after leaving Berkeley, first to build it, and then to be a faculty member. Livingston, who as Lawrence's graduate student had in fact built the first cyclotron ever constructed, was a master of his field, and by the time the war started we had an excellent cyclotron in running condition. It was one of the few cyclotrons which were operated during the war for the production of radioactive isotopes, not only for the use of our radioactivity center, but for universities around the country.

I have touched on most of the aspects of the department's work during the 1930s, as far as they concerned research. There were others; for instance, we felt that there should be work in acoustics, and since Morse had been interested in this field while at Case, I asked him to teach a course in the subject. With his usual enthusiasm, he became an expert in the field, produced one of the best texts, and started an acoustic laboratory which was very active for a number of years. And there are still other enterprises which I could mention. But the thing that should be said above all is that along with the research, the whole department entered very actively and enthusiastically into teaching, both on the undergraduate and the graduate level. It was a natural consequence of this that many students decided to go into physics. From one of the smallest departments at the Institute, the number of students and faculty rose so that by the time we were back in operation after the war, we were one of the three or four largest departments.

In the introduction to this section, I mentioned the close ties that existed between MIT and Princeton during this period. One of the ways this showed itself was through annual visits, either of a group of the MIT graduate students and staff to Princeton, or vice versa. In this way our younger members got to know Wigner, Condon, von Neumann, and others of the famous group who were gathered at Princeton. Then there were a number of students of the one department who went to the other for postdoctoral work. The National Research Council Fellowships in those days furnished very much needed support for postdoctoral work, and had a great deal to do with the development of physics in this country. Among those from Princeton who worked with us for periods were George Shortley, who collaborated with Condon on the book on atomic spectra; John Bardeen, who divided his time between Harvard and MIT, and who is now one of the few men who have won two Nobel Prizes; and Conyers Herring, who was with us for two years after taking his Ph.D. at Princeton. In 1937, I had the good fortune to be invited by the Institute for Advanced Study at Princeton to spend several months there. It was during the early years of that organization, while they were still occupying quarters on the Princeton campus, so I had an excellent chance to get to know not only the members of the Institute, but also the whole Princeton scientific community. As I shall describe in Secs. 23 and 25, I was able to get more research done that spring than during most of the rest of my time in the 1930s.

In addition to all of this, there were many foreign visitors, and many meetings were being held. The Lyman Laboratory at Harvard was dedicated in 1932, the Eastman Laboratory at MIT in 1933, in each case with meetings or symposia and visiting lecturers. At various times during those years, we had Debye, Scherrer, Sommerfeld, Bragg, Hartree, Pauling, Condon, and many others as visitors or lecturers for extended periods. All of this led to a feeling that physics was very active at MIT, and it was not surprising that the number of students was growing as rapidly as it was. Hartree particularly was interested in what was going on at MIT, on account of the work of Vannevar Bush with his differential analyzer, an analog computer which he was continually perfecting, which we shall return to in Sec. 38. This worked completely with mechanical components, and it fascinated Hartree, who was shortly using it for solving the self-consistent-field problem. Hartree soon had constructed one of his own in Manchester. But unfortunately it had only a three-figure accuracy, and this was not enough for our main work. It was only in the 1950s, with the electronic digital computers, that quantum mechanics was revolutionized by the development.

22. 1933–1934, The Cellular Method and Energy Bands

Let us go back to physics. The year 1933 was a turning point in the development of the theory of solids, and from this point on the advances we talk of are those bearing directly on our main problem. The first step came from Princeton: the first of several papers by Wigner and Frederick Seitz, his student, on the constitution of metallic sodium. Seitz, a young man ten years younger than I, who had come from Stanford, was taking his doctorate under Wigner, and was headed for a distinguished career: many universities, including Rochester, Pennsylvania, Carnegie Tech., and Illinois, a period at General Electric, then the presidency of the National Academy of Sciences, and now president of Rockefeller University. His book, *The Modern Theory of Solids,* published in 1940, was a landmark in the history of the field, and still remains the leading textbook on the subject. His genius, as well as Wigner's, made the series of 1933–1934 papers a memorable one.

These papers were the first ones which broke definitely out of the pattern of LCAO versus plane waves which had been the direction in which solid-state theory had been traveling. They represented the first attempt to apply something very much like the self-consistent field to a crystal. They began by suggesting what has come to be called the Wigner-Seitz cell. The sodium crystal, which they chose to discuss, has the body-centered cubic structure, an atom at the center and at each corner of a cubic lattice. They noted that if one sets up a plane that is the perpendicular bisector of the line joining a sodium atom to its nearest neighbors, these planes will enclose a polyhedron like that shown in Fig. 22-1, containing one atom, which departs only slightly from a sphere. We recognize in this construction a similarity to that which we have described in Sec. 15 for the Brillouin zone. The Wigner-Seitz cell is in fact the equivalent in ordinary space of the Brillouin zone in reciprocal space, and it can be proved that the Wigner-Seitz cell for the body-centered cubic structure is identical in shape with the Brillouin zone for the face-centered cubic structure, and vice versa. Thus we have seen the polyhedron of Fig. 22-1 before, in Fig. 15-2.

Wigner and Seitz then noted that a self-consistent field acting on an electron in a sodium crystal would be very nearly spherically symmetrical within each of the cells. It would be appreciably different from the potential in an isolated sodium atom, for the valence or conduction

173

electron in the crystal had to be squeezed into a single cell, and this led to a considerably higher density near the atom than one would have in an isolated atom. But it was as reasonable as with an atom to assume spherical symmetry, and this led to the possibility of solving the one-electron Schrödinger equation in spherical polar coordinates, just as Hartree did for his isolated atoms. Wigner and Seitz applied this type of numerical calculation to the particular wave function which was identical in each cell. We recall that on account of Bloch's theorem, the wave function had to be multiplied by a factor $\exp(i\mathbf{k}\cdot\mathbf{R})$ on going from a point in one cell to a corresponding point in a cell distant by the vector $\mathbf{R}$, and for the special case $\mathbf{k}=0$ this led to a periodic wave function.

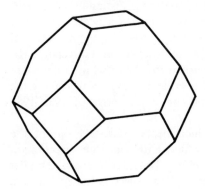

Fig. 22-1. Wigner-Seitz cell for the body-centered cubic sodium crystal.

The actual potential which Wigner and Seitz used was not derived from a self-consistent field, but it closely approximated it. They first noted that if a valence electron were to be found within a certain cell, it was highly unlikely that another valence electron would be in the same cell. The presence of a second valence electron would have led to an ionic state of the sort which we have seen does not occur in the hydrogen molecule to any extent, and it would be just as unlikely to find it in the crystal. Consequently they assumed that if an electron were in one cell, the other conduction electrons would be found in the other cells, one to a cell, so that they would be electrically neutral, while the electron in question would be acted on by the inner shells of the atom in its cell, but not by an outer electron. Thus they effectively assumed that the electron in question was acted on by all the electronic charge in the crystal, as well as by the nuclei, except for the removal of just enough electronic charge from the cell it was in to account for one electron.

Even with this interpretation, they did not literally compute a self-consistent field. For there had been a study of the field required to produce one-electron wave functions in sodium to agree with the experiments on the spectrum of a sodium atom, carried out in 1929 by Prokofjew, essentially a continuation of the sort of calculations made by Schrödinger, Hartree, and others in the early 1920s. What Wigner and Seitz did was to use this potential of Prokofjew's, and to use numerical integration of the radial wave equation to get the solutions of Schrödinger's equation for the spherical field.

They then wished to find a solution of Schrödinger's equation, with the spherically symmetrical potential of Prokofjew within each Wigner-Seitz cell, but joining smoothly and periodically from one cell to the next. If this condition were applied rigorously, the wave function should go to a maximum or minimum at every point of the surface of the cell, so that it would be periodic, and would have zero normal derivative at the bounding surface. This condition was too hard to apply, so they approximated by replacing the cell by a sphere of equal volume, and demanding that the wave function go to a maximum (or minimum) at the surface of this sphere. This was easy to handle by numerical integration of the radial differential equation, just as Hartree had done for atoms, only now the boundary conditions were different. The wave function still had to remain finite at the origin (they were using an s-like function, which is finite at the origin) but would have zero slope at the sphere radius. When they worked this out, they found that the one-electron energy of this state was that shown in Fig. 22-2, as a function of the sphere radius, or of the lattice spacing in the crystal.

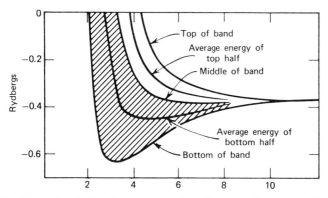

Fig. 22-2. Energy band of sodium crystal arising from atomic $3s$ state. Bottom of band indicates state calculated by Wigner and Seitz. Shaded area, up to middle of band, is occupied. Average energy of bottom half indicates Wigner-Seitz energy including average kinetic energy.

This curve has the right form to represent the energy per atom in a crystal, showing a minimum and an energy decrease on going from the isolated atoms to the crystal. The energy for the isolated atom of course came out right, since Prokofjew had devised his potential to lead to the correct energy levels for the atom. However, Wigner and Seitz noted that the electrons actually would partially fill an energy band, which would broaden as the lattice spacing decreased, and whose lowest energy was for the state $\mathbf{k}=0$, which they had calculated. They assumed that the electrons acted like a homogeneous electron gas filling the same volume as the crystal, and computed the average energy of this electron gas as a function of lattice spacing. This energy was taken from Bloch's work on the homogeneous electron gas, as described in Sec. 16, and is proportional to the two-thirds power of the density, as in Eq. 16-2. When this energy was added to the energy of the lowest state, which has $\mathbf{k}=0$, they found the additional curve shown in Fig. 22-2. This gave a binding energy, equilibrium separation of the atoms, and compressibility of the right order of magnitude to agree with experiment. They discussed various possible corrections to this curve, which were believed to bring it into better agreement with experiment, but there were so many approximations that it is hard to accept the numerical results very seriously. However, for the first time they had given a usable method for estimating energy bands in actual crystals.

In their later papers, and in other work, this first start was improved in two major directions. First, their use of the homogeneous electron gas for estimating the Fermi energy was obviously open to a great deal of suspicion. As soon as the first paper came out, it occurred to me, and independently to Wigner and Seitz, that it should be possible to make a real calculation of energy bands based on the cellular method. One could give an exact solution of Schrödinger's equation within a cell by expanding as a linear combination of radial functions $R_l(r)$ times spherical harmonics $Y_{lm}(\theta,\phi)$. The radial functions would have to be calculated for a given energy. For any arbitrary energy, one has solutions of this type for all possible values of l and m, which behave regularly at the origin. In general, if one continued the functions outward to infinity, they would go infinite at infinite r. It is only for the eigenvalues or stationary states of the atomic problem that they go to zero at infinity. But this infinite behavior at infinity does not interfere with the use of these functions inside the finite Wigner-Seitz cell. Thus one can rigorously write the solution of Schrödinger's equation within a given cell in the form

$$\psi = \Sigma(lm) C_{lm} R_l(r) Y_{lm}(\theta,\phi) \qquad (22\text{-}1)$$

This sum, however, would not in general satisfy proper boundary conditions at the surface of the cell.

We know that solutions of the wave equation for the periodic potential in the crystal must satisfy Bloch's condition that the wave function be multiplied by the factor $\exp(i\mathbf{k}\cdot\mathbf{R})$ when one goes from one point in a cell to a corresponding point in another cell distant by a vector displacement $\mathbf{R}$ from the first. This condition can be set up, for a given $\mathbf{k}$, by making a statement entirely within a given cell. If one goes from a point on one face of the Wigner-Seitz cell, along a line normal to the face, to the corresponding point on the opposite face, the displacement $\mathbf{R}$ is one of the vectors mentioned above. Consequently the value of the function, and of its derivative along the direction of the vector, must be multiplied by $\exp(i\mathbf{k}\cdot\mathbf{R})$ in making such a displacement. If one wishes to satisfy this condition at many points on the surface, obviously one must have many constants C_{lm} to be adjusted, and the problem is reduced to one of many simultaneous equations for the C's.

As a first attempt to get a solution, Wigner and Seitz and I independently decided to satisfy the conditions simply at the center of each of the faces of the Wigner-Seitz cell. This required a finite set of l's and m's, which could easily be determined for the various faces. The conditions then led to a finite number of simultaneous equations, whose secular equation had to be satisfied if the equations were to be compatible. This led to an equation for the energy, which for special symmetry directions in the Brillouin zone broke down into equations simple enough so that they could be easily solved. I carried this calculation further than Wigner and Seitz did in their later paper, and my paper was written before I knew of that later paper, so I shall describe my results; theirs were consistent with mine. First, in Fig. 22-3, I give the limits of the energy bands as a function of lattice spacing. The band going to the atomic $3s$ level at infinite separation is the one which is partly filled, containing one electron per atom though the band is capable of holding two electrons. The bottom of this band and the average energy of the occupied levels are shown in Fig. 22-2.

Next, in Fig. 22-4, I show the energy as a function of the magnitude of $\mathbf{k}$, for propagation along a particular symmetry direction in the Brillouin zone. As had been shown by Brillouin and others, and as was mentioned in Sec. 15, this curve would be a parabola for a free electron, the energy being $k^2\hbar^2/2m$, but when the periodic potential is taken into account, it shows a break at each plane separating one Brillouin zone from another. The calculations by the cellular method show both the approximately parabolic nature of the curve and the breaks. The occupied levels lying below the

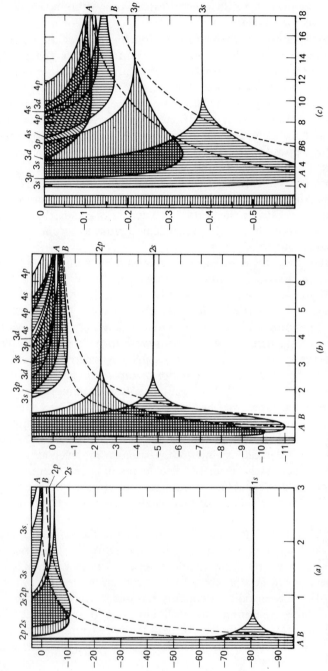

Fig. 22-3. Energy plotted against interatomic distance, for sodium crystal. Different scales are used in each graph. The dashed curve A represents the maximum potential energy, at edge of the cell, and B represents the mean potential energy through the cell, as functions of the distance. Some states are omitted in the upper right-hand corners of all graphs, where the overlapping is really much more than indicated. In the actual crystal, the half-distance of separation is between 3.5 and 4 units. Energies in rydbergs. From Slater, *Rev. Mod. Phys.*, **6**, 209 (1934).

Fermi energy come only about half-way up to the first break, and show that it is a quite accurate assumption to use the free-electron parabola for the occupied states of sodium. Hence the calculations of Wigner and Seitz were justified, as far as the energy-band states were concerned.

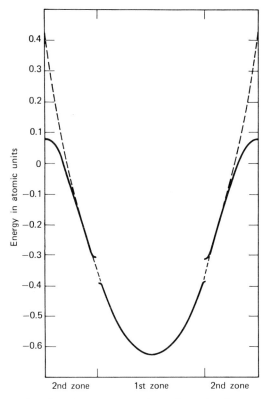

Fig. 22-4. Energy as function of wave vector **k** for sodium, showing gaps at boundaries of Brillouin zones. From Slater, *Rev. Mod. Phys.*, **6**, 209 (1934).

Sodium turned out later to be a special case, and there are few other crystals in which the energy levels follow the free-electron parabola so closely. This rather confused the issue as far as later work was concerned. The workers in the field were entirely too inclined to assume that a single plane wave formed a good approximation to the wave function. The true situation is shown much better in Fig. 22-5, which gives my calculations for the wave function of a conduction electron, as a function of distance along a line in the 111 direction through the crystal, for orbitals whose wave vector **k** was also in this 111 direction. The case a represents the bottom of the band, Wigner and Seitz's case $k=0$. We see that between the atoms, the wave function is quite accurately constant, but in the neighborhood of each nucleus the wave function goes through oscillations much like those of a $2s$ orbital. As we go up in the band, to cases $b, c, d, \ldots,$ the long-range behavior of the wave function is like a sine curve of shorter and shorter wavelength, but still the wave function behaves near each atom like a $2s$ orbital, or like a $2p$ orbital near the nodes of the sine curve. We thus have a compromise between a plane wave and an atomic function. It is important to bear in mind the general nature of the wave function when one is thinking of techniques for getting more accurate approximations to the solution of Schrödinger's equation in the periodic potential, which we come to in later sections.

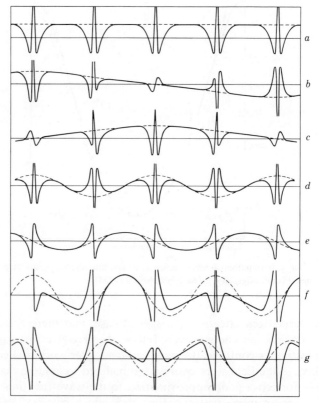

Fig. 22-5. Wave functions for conduction electrons, for sodium, as function of distance along a line in 111 direction. From *Quantum Theory of Matter*, 2nd ed., by J. C. Slater. Copyright 1968 by McGraw-Hill, Inc. Used with permission of McGraw-Hill Book Company.

In addition to this improvement in the calculation of energy bands, Wigner and Seitz in their various papers considered further the nature of the potential which should be used in their calculation. We have stated that their general approach was to assume than an electron in one cell acted as if it were in the field of the inner shells forming the ion in that cell, but of neutral atoms in all other cells. They made this conclusion plausible by an argument that we shall find important in our future work. They started with a homogeneous electron gas, such as Bloch had considered in his argument on ferromagnetism, which we discussed in Sec. 16. They set up a determinantal function composed of plane waves, half with spin up, half with spin down, filling all states up to the Fermi energy. If this many-electron determinantal wave function is ψ, they computed the quantity

$$\int \cdots \int \psi^*(1,2,\ldots,N)\psi(1,2,\ldots,N)dv_3 dv_4 \cdots dv_N \qquad (22\text{-}2)$$

where $1, 2,\ldots,N$ represent the coordinates and spins of the N electrons, and where the integrations over coordinates include summation over the spins. This quantity of Eq. 22-2 gives the probability that electron 1 should be found at coordinates and spin symbolized by 1, and electron 2 at coordinates and spin 2. But on account of the antisymmetry of the wave function, electrons 1 and 2 are completely typical of any pair of electrons in the system.

We can then ask the question, if electron 1 is at a particular point, which we may take to be the origin, with a particular spin, which we may take to be spin up, what will be the density of other electrons both of spin up and of spin down, as functions of position in space? It is a straightforward mathematical problem to answer this question, and Wigner and Seitz gave the answer. Because of the way in which the spin enters the antisymmetric wave function, the distribution of density will be quite different for spin up and spin down. There are supposed to be $N/2$ electrons of spin up, $N/2$ of spin down, in the system. Since the electron which is known to be at the origin has spin up, there are only $N/2-1$ other electrons of spin up, but $N/2$ of spin down.

As for the spin-down electrons, the Hartree-Fock method, which is what we are considering, has no correlation between the motions of electrons of opposite spin. Consequently the density of the electrons of spin down comes out to be just half the total density of all electrons in the crystal. On the other hand, for electrons of spin up, the integral of density over all space must add to only $N/2-1$ electrons. What happens is that the density of electrons of spin up becomes equal to that of spin down at some distance from the electron at the origin. But in the neighborhood of the origin, there is a deficiency of charge of spin up, which Wigner and Seitz christened the Fermi hole, just big enough to include one electron. In Fig. 22-6 we show the density as a function of distance from the electron at the origin, clearly showing the Fermi hole.

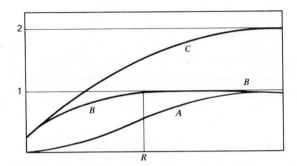

Fig. 22-6. Density of charge near an electron, plotted against distance from a given electron. Curve A for another electron of same spin, B of opposite spin, C for both spins combined. One unit of density represents maximum allowable value for electron of one spin. Integrated deficiency of charge, for curves A and C, 1 electron unit; for B, zero. From Slater, *Rev. Mod. Phys.*, **6**, 209 (1934).

As Wigner and Seitz showed, it is easy to find the approximate size of this Fermi hole. If it were a real spherical hole, such that no other charge of spin up were to be found inside it while outside it there would be no modification of the density of spin-up electrons, its radius R_s would be such that the volume of the sphere, $\frac{4}{3}\pi R_s^3$, times the number of spin-up electrons per unit volume, which is $N/2V$, where V is the volume of the crystal, would equal one electron. That is, we should have

$$R_s = \left(\frac{3}{2\pi}\right)^{1/3}\left(\frac{V}{N}\right)^{1/3} \qquad (22\text{-}3)$$

The function of Fig. 22-6, which Wigner and Seitz called $P(r)$, rises to half its final value at a radius $1.175 R_s$. This is such a radius, for the actual crystal, that the probability of finding another electron of spin up inside the Wigner-Seitz cell would be very small.

There was nothing in the original Wigner-Seitz postulates, however, which would lead to correlation between the spin-up and spin-down electrons. One would expect to find half an electron of spin down in the central Wigner-Seitz cell, as in every other cell. This is the same situation we find in the molecular orbital treatment of the hydrogen molecule, which we discussed in Sec. 14. There we had only two electrons, one of spin up and the other of spin down, and each moved independently of the other, so that on the average an electron of spin up would find half an electron of spin down on the same atom that it was on. But we know from the work of Heitler and London that this was not actually the case. And we know how to avoid this situation according to the configuration-interaction technique of handling the many-electron problem. One makes a linear combination of various determinantal functions (two, in the case of H_2) which has the effect of reducing the probability of finding electrons close together, even if they have opposite spins. This effect is generally called the correlation effect, to distinguish it from the effect of the Fermi hole on keeping two electrons of the same spin apart, which is generally called the exchange effect.

We have seen the effect of the correlation in our discussion of the helium atom, in Sec. 19. The accurate treatments of the hydrogen molecule, such as that of James and Coolidge, show the same sort of effect there: The probability of finding an electron of spin down near an electron of spin up falls to a lower value than the average, as the two electrons approach, but does not vanish, as the antisymmetry principle requires for electrons of the same spin. We showed such an effect in Fig. 19-1, giving the wave function of the helium atom as a function of r_{12}. In the second paper of Wigner and Seitz, and in a later paper by Wigner, this same problem was attacked for the homogeneous electron gas, as found in the sodium crystal. The approach was much the same as I had used in my first helium paper, which I did not describe in detail in Sec. 19. Namely, the electrons of spin down were assumed to be located at fixed positions in the crystal, and the behavior of an electron of spin up was investigated in the neighborhood of the fixed electron of spin down. Account was taken of the fact that one had to use the reduced mass, so that the cusp which Wigner and Seitz deduced was of the same nature as the one I had found for helium, and which was illustrated in Fig. 19-1.

This led to a decreased probability of finding two electrons close together and an additional hole in the distribution function near $r_{12}=0$. The correlation effect, however, is different from the exchange effect, in that the total number of spin-down electrons is still $N/2$, so that if one is less likely to find a second electron very close to the first, one must be more likely to find it at a somewhat greater distance. In Fig. 22-6, we show schematically the probabilities of finding a second electron of either the same spin or the opposite spin in the neighborhood of a given electron of one spin. The effect of deepening the hole near an electron, and making it more shallow at a somewhat greater distance, but of leaving the total deficiency equal to one electron, is equivalent to narrowing the hole, which can be called the exchange-correlation hole. We shall find in later sections, mainly Sec. 35, that this deepened exchange-correlation hole is what is used in the $X\alpha$-SCF method to describe the fact that one is unlikely to find a second electron close to a given electron.

This was as far as Wigner and Seitz carried the discussion of exchange and correlation. They worked out a possible behavior of the correlation effect at large interatomic distances, which I have never been able to accept as a real approach to the properties of crystals. The reason is that they based their discussion on a uniformly distributed positive charge and a homogeneous electron gas, whereas it is obvious that at large interatomic distances we must have the formation of individual atoms behaving as they do when isolated from each other. This is entirely lost in the discussion of Wigner and Seitz, and yet a great deal of the theory of the electronic structure of solids, by those who call themselves many-body physicists, is based on this unrealistic assumption. For this reason I distrust most of what they do concerning correlation.

23. Further Energy Bands, 1934–1940

The method that Wigner and Seitz and I had used for the sodium calculation was not perfect, but it was immensely better than anything that had been suggested earlier, and it started a considerable series of calculations of energy bands, carried out in a number of laboratories. It was the first time that we had had energy bands accurate enough to verify the general conclusion that metals should come from partly filled bands, insulators from filled bands with a large gap above, and then empty bands, and semiconductors from a similar situation with a much narrower gap. Consequently there was a great deal of interest in calculating examples of these various sorts of crystals, to see if the energy bands had the right qualitative behavior.

Seitz continued for a short length of time after the sodium calculations, getting energy bands for the lithium crystal, which showed close resemblance to sodium. He soon shifted his interests, however, to more general properties of crystals. I began to get students interested in energy-band calculations. Up to that time, I had never had graduate students working with me. One of the first who went in for it was H. M. Krutter, who worked out the energy bands of the copper crystal in 1935. I had been particularly interested in getting energy bands for the $3d$ transition elements, to see if my hypothesis, suggested in Sec. 16, that the $3d$ bands were narrow enough to show ferromagnetism in iron, cobalt, and nickel was actually justified. I felt that copper, the first element beyond these, and yet with a single valence electron like sodium, would be a good one to start with. As I mention in Sec. 25, Krutter's bands were good enough so that I was able to use them during the next year for a first study of the ferromagnetism of nickel, which showed that the general concept was entirely correct.

Then to take a very different substance, George E. Kimball in 1935 computed the energy bands of diamond. He had taken his Ph.D. in chemistry at Princeton in 1932, and spent the years 1933–1935 at MIT as a postdoctoral fellow. He later had a distinguished career, first in university work at Columbia, then in industry. The energy bands which he found for diamond were completely different from those of either sodium or copper, but fitted in perfectly with the known properties of diamond. Hund, in Leipzig, had favored the cellular method from the time it was first proposed, and he and his student Mrowka also studied diamond, getting results in agreement with those of Kimball. They later went on to study various refractory compounds.

Hund and Mrowka and Kimball did not try to make actual calculations of the cohesive energy in these tightly bound elements and compounds, as Wigner and Seitz did for the sodium crystal, but it was very clear from their results why the bonds were so strong. The energy bands were split into two groups, the lower ones consisting of bonding orbitals, the upper ones of antibonding orbitals, with a very wide separation between them. The energy bands of Kimball are shown in Fig. 23-1, and we see the interesting way in which the $2s$ and $2p$ levels of carbon, which are separated at large distances, coalesce, and then split again into the lower, bonding energy band and the very much higher antibonding band. The Fermi energy of course lies between these two bands, and the fact that diamond is very transparent not only through the visible but way into the ultraviolet is a result of the wide gap. Postwar calculations on diamond were generally not made for a wide range of interatomic distance, as these early calculations of Kimball were, and hence do not show the behavior so clearly.

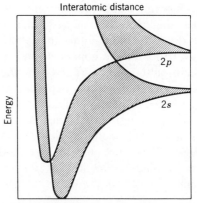

Fig. 23-1. Boundaries of energy bands of diamond, as functions of interatomic distance, as computed by Kimball. From *Quantum Theory of Molecules and Solids*, Volume 2, by J. C. Slater. Copyright 1965 by McGraw-Hill, Inc. Used with permission of McGraw-Hill Book Company.

The first compound we tried at MIT was sodium chloride, for which Shockley calculated the energy bands in 1925. His energy bands in general were what we expected, with a large gap between the highest occupied band, the valence band, and the lowest empty band, the conduction band. He and I published an accompanying paper dealing with optical excitation in NaCl, which I shall describe in Sec. 25. But Shockley was always an ingenious man with an inquiring mind, and he was not ready to accept the cellular method as we were using it. He had a feeling that the procedure of matching the boundary conditions only at the centers of the faces of the cell was pretty poor. To test it, he invented what is called the "empty lattice test."

He simply took a constant potential and applied the cellular method to it, using cells inside each of which the potential was constant. The answer of course should have been a plane wave, and the energy should have been $k^2\hbar^2/2m$. Instead, he obtained gaps in his diagram of energy versus k, that were almost as large as those which I had found for the sodium crystal in Fig. 22-4. It was clear, in other words, that this threw doubt on the numerical accuracy of all the cellular calculations that we had made up to that point. The obvious thing to do was to match boundary conditions at more points, but with the primitive computational methods available at that time, it did not seem practical. Now, with digital computers, it is possible, as von der Lage and Bethe (Sec. 28) and Altmann and Bradley have shown, to get very high accuracy from the cellular method. But at the time, it seemed desirable to look for methods which might be simpler computationally.

During the spring of 1937, as I have mentioned, I had the opportunity of spending several months at the Institute for Advanced Study at Princeton, and one of the things I wanted to try was to look for an improved method of computing energy bands. Shockley's paper about the empty lattice test had not yet come out, but I was aware of his conclusions. I was influenced in my line of thought by work which my colleagues Allis and Morse had done during their period together in Munich, in 1929–1931. They had been studying the scattering of electrons by neutral atoms. An experimentalist named Ramsauer, in the early 1920s, had discovered that very slow electrons, with energies of only a few electron volts, are hardly scattered at all by some atoms such as neon; they act as if they went right through the atoms without interference.

It seemed to Allis and Morse, stimulated by Sommerfeld, that this might be explained in terms of the quantum theory of scattering of an electron wave by a spherical obstacle. They replaced the atom by a spherical symmetrical potential inside a finite sphere, and assumed that the potential was zero outside the sphere. They then considered the wave-mechanical problem of a plane de Broglie wave striking the sphere, and being scattered. The method for handling this is well known. One expands the incident plane wave, the scattered waves, and the wave function inside the sphere, in linear combinations of radial wave functions of the radius r, and spherical harmonics of the angle. In other words, one uses just the type of expansion used in Eq. 22-1 for the cellular method.

Inside the atomic sphere, the radial functions $R_l(r)$ are the same sort used in the cellular method, radial solutions of Schrödinger's equation, though Allis and Morse used a rather crude assumption for the potential within the sphere and so did not use the real radial functions. Outside the atomic sphere, where the potential is constant, one can expand the radial function in terms of the spherical Bessel and Neumann functions. My first real training in using spherical Bessel and Neumann functions came from studying their work.

One had to build up the wave function outside as a linear combination of a plane wave (expanded in spherical coordinates) and of an outwardly traveling scattered wave. Then one had to apply the conditions that the function and its derivative were continuous over the sphere separating the two regions. This can be done easily, and the type of theory Allis and Morse used is now the standard method for handling scattering problems. It is treated exhaustively, for example, in Leonard Schiff's *Quantum Mechanics*. Schiff was another of our graduate students at MIT during the 1930s, who went on to a distinguished career at Stanford, serving for years as head of the physics department. The results obtained by Allis and Morse showed that in fact the Ramsauer effect follows directly from this sort of analysis. Furthermore, the fact that neon shows the effect, and that the neon wave function is very much like that of Na^+, the inner shells of the sodium atom, suggested that the very free-electron-like nature of the sodium energy bands might be tied up with this same Ramsauer effect. The waves, as we see in Fig. 22-5, seem to be going through the atoms with very little disturbance.

The mathematical simplicity of the scheme used by Allis and Morse arose because they had a sphere over which to apply boundary conditions. This is a very much simpler surface over which to impose boundary conditions than the very awkward polyhedral cell of the Wigner-Seitz method. It seemed to me quite plausible that in a crystal such as sodium, the potential inside the Wigner-Seitz cell, while spherically symmetrical over most of the volume, might well be approximately constant in the corners of the cell. The potential falls to an infinite negative value at each nucleus and rises to a maximum in the region between atoms, and this maximum might well be flat enough so that it could be approximated by a constant. I proposed, then, that one use a potential which was spherically symmetrical within spheres inscribed in the Wigner-Seitz cells, but which was constant in the outer corners of the cell, outside the inscribed sphere. This potential is what is now popularly called a muffin-tin potential. Then we had the possibility of expressing the wave function within each sphere in the form of Eq. 22-1, but matching it over the surface of the sphere to a solution of the constant-potential problem.

The particular constant-potential problem I used was suggested by the fact that the wave function for sodium outside the spheres was obviously very much like a plane wave, as was shown in Fig. 22-5. A plane wave of course is expressed as $\exp(i\mathbf{k}\cdot\mathbf{r})$. A plane wave satisfying the Bloch condition, with the same wave vector $\mathbf{k}$, is $\exp[i(\mathbf{k}+\mathbf{K}_j)\cdot\mathbf{r}]$, where $\mathbf{K}_j$ is any one of the vectors of the reciprocal lattice. A superposition of an infinite number of such plane waves,

$$\psi = \sum (j) A_j \exp\left[i(\mathbf{k}+\mathbf{K}_j)\cdot\mathbf{r}\right] \qquad (23\text{-}1)$$

is a three-dimensional Fourier expansion, which forms a perfectly general way of writing any function satisfying the Bloch condition. Thus the correct wave function can surely be expanded in this form.

From the nature of the wave function, as shown in Fig. 22-5, we see that between the atoms, the wave function can very well be quite free-electron like, so that we may need only a few plane waves to get an adequate expansion. Inside the atomic spheres, the wave function is like an atomic wave function, and it requires an impossibly large number of plane waves to expand such a function adequately. But I proposed to use the expansion only outside these atomic spheres, where good convergence was likely. I accordingly set up the boundary conditions for getting continuity of the function and its derivative on the surfaces of the spheres surrounding the atoms, and found that the method could be set up in a form which seemed quite adaptable to computation. This is the method now called the augmented plane wave (APW).

In recent years, the APW method has proved to be one of the very useful approximations. But before the war it involved such difficult calculations, with a desk calculator, that only a very preliminary test could be made of it. My student Marvin Chodorow undertook to test it for the copper crystal, and carried through the calculations for several $\mathbf{k}$ vectors, in 1939. He obtained reasonable results, and was convinced that the method was in principle possible, but it was not until we got large digital computers after the war that we were really able to get into operation with the method. By then Chodorow had long since lost interest in the method, had become the chairman of the applied physics department at Stanford, and is very proud and somewhat incredulous when the electronic computers, turned loose to check the few points he computed in 1939, still get the same answer he found then. His potential for determining energy bands in copper has proved as useful as Prokofjew's was for sodium. Speaking of sodium, I might mention that the surmise that the very free-electron-like energy bands of sodium are analogous to the Ramsauer effect in neon has turned out to be entirely justified.

Another possible energy-band method was also suggested in the late 1930s. It occurred to Conyers Herring, who as I have mentioned was a postdoctoral fellow at MIT during 1937–1939, that the wave functions for sodium, shown in Fig. 22-5, might be rather well approximated by adding to a plane wave a linear combination of atomic orbitals, which in the case of sodium would be atomic $1s$, $2s$, and $2p$ orbitals. If we were to represent a Bloch wave this way, it would have to be orthogonal to the LCAO combinations of $1s$, $2s$, and $2p$ atomic orbitals with the same **k** vector. By imposing these conditions of orthogonality, one could determine the coefficients of the linear combination, and for this reason the method was named orthogonalized plane waves (OPW). It was not hard to work out the techniques of handling this method, provided the inner atomic orbitals to which one was orthogonalizing the plane wave, did not appreciably overlap each other. In 1940, Herring and Albert G. Hill, who was then an instructor at MIT, collaborated on a test of the method, computing energy bands of beryllium. They came out according to expectation. Herring went on to become a leading member of the Bell Laboratories research staff, and Hill to become vice-president of MIT in charge of research. Energy-band calculations seem to have been good training for quite a number of people.

During this time, while the various methods were being worked out in the United States which have led to valuable computational schemes since the war, the main effort abroad had been to use methods which avoided computation as far as possible (aside from Hund's work with the cellular method). This was particularly true in England. The leading young man in the field of solid-state theory was Nevill Mott, whom I had first met in the early 1930s. He had acquired his training at Cambridge, and for a number of years was head of the physics department at the University of Bristol, where he established a very good school of physics. He and his associate Harry Jones published in 1936 the excellent text, *Theory of Metals and Alloys*. Mott was one of the first to point out the significance of the partially filled $3d$ bands for the energy-band theory of magnetism. He recently retired from Cambridge University, where he was Cavendish Professor for a number of years.

Jones in 1934 made some very imaginative applications of the idea of the Brillouin zone to the theory of alloys. He considered certain alloys of copper and zinc. Since copper is monovalent, zinc divalent, the addition of zinc to copper increases the number of electrons per atom. He considered the probable nature of the energy bands, and consequently the nature of the Fermi surface, the surface in the Brillouin zone which passes through all **k** values whose energy equals the Fermi energy. It is when the Fermi surface is close to the boundary of the Brillouin zone that one is concerned with the sort of gap in the energy bands shown in Fig. 22-4. Then it can well happen that the occupied energy levels are pushed down by the development of the gap, and this can stabilize the energy of those particular crystals. Jones showed that as more zinc atoms are substituted for copper in the alloy, this situation arises with the Fermi surface, and he was able to correlate this with particularly stable phases found in certain compositions of the alloys.

These gaps at the surfaces of the Brillouin zones occur very naturally for free electrons in a periodic potential of small amplitude. As a result of this, one got the impression in studying the work of Mott and Jones that they felt that the potential actually occurring in energy-band theory was a small one, which could be handled by perturbation theory. This was not justified, but it affected the thinking of the English school of physicists enough so that even now most of them are trying to get valid results relating to energy bands from simplified models, rather than through the direct types of calculations which one can make with the methods now in use.

Before we leave this section, it may pay to make a few remarks about other computational methods. The LCAO method in principle is perfectly adapted to a study of energy bands, but in practice the coulomb and exchange integrals which come in are so difficult to carry out that it can hardly compete with the APW and other methods. It has been carried through for some simple crystals, and is capable of giving useful and correct results, but for crystals made of heavy atoms it is hardly adequate. As for the OPW method, it is very good for those atoms whose inner shells are not overlapping with their neighbors. But for an atom containing, for instance, d electrons, whose wave functions are very extended, one runs into the same difficulties as with the LCAO method. At the time we are writing of, about 1940, it appeared that the APW method was probably the most promising, though it had never been really used. But as we shall see in Sec. 28, another very similar method, the KKR method (suggested by Korringa, Kohn, and Rostoker), first proposed in 1947, has many advantages.

24. General View of Energy Bands, 1933–1940

We have been looking at specific ways of finding energy bands, and at some of the results which were obtained by them during the 1930s. During this same period, however, I was trying to get a unified point of view regarding the whole subject of the electronic constitution of crystals, particularly of metals. John Tate, the editor of the *Physical Review* and of the *Reviews of Modern Physics* in those days, had asked me in the spring of 1930 if I would write a review article on the electronic structure of metals for the *Reviews of Modern Physics*. I agreed, but said that it would have to be delayed until I finished the book I was working on. This was a book which went through many changes of plan before it was finally published. While I was at Harvard, Kemble and I had thought of writing a joint book on wave mechanics. We had talked to various publishers about it, finally settling on McGraw-Hill. But then we had decided, after making a start on the manuscript, that Kemble was really interested in writing on the principles of quantum theory, and I was more interested in its application to properties of matter, so we decided, with the approval of McGraw-Hill, to split it up into two volumes. The first eventually turned into Kemble's excellent text, *Fundamental Principles of Quantum Mechanics*, which McGraw-Hill published in 1937. My book was still in its early stages when I went to MIT.

The course on introduction to theoretical physics which I started offering at MIT went through classical physics, mechanics, electromagnetism, and so on, before taking up quantum theory. I decided that all of that preparation was really needed if one was to teach quantum mechanics. Also I learned a great deal more classical physics than I had known, through Ned Frank and various colleagues at MIT. So my book on quantum theory eventually turned into a number of chapters toward the end of the book, *Introduction to Theoretical Physics*, which Frank and I published in 1933. The book on properties of matter grew into one volume after another, which have continued until the present.

It was the joint effort with Frank which I wanted to finish before writing the review article for Tate, so that it was in fact only in 1934 that the latter was written and printed. It gave me an opportunity to take a comprehensive look at the whole field of the electronic structure of matter, and I looked ahead of the thinking of the time in a number of respects. Somewhat earlier, in the summer of 1933, I had had a chance to present a similar view of the theory, and many of the ideas embodied in the 1934 review article were things I had been turning over in my mind for over a year, starting even before the appearance of the paper of Wigner and Seitz. The specific opportunity which came up in 1933 was an invitation to give a talk at a symposium on the application of quantum mechanics in chemistry, in connection with a meeting of the American Physical Society in June 1933, in Chicago. My talk was on Electron Energies in Atoms and Molecules; the second talk, by Linus Pauling, was on Quantum Mechanics of Condensed Ring Systems, Free Radicals, and Other Complex Molecules; the third, by Henry Eyring, on Quantum Mechanics and Chemical Reactions; and the fourth, by Robert Mulliken, on Band Spectra and Molecular Structure. With such a list of speakers, I felt that I had to do my best to give a good talk.

This was a rather extraordinary meeting. It was held in connection with the Chicago World's Fair entitled "A Century of Progress Exposition," celebrating the hundredth anniversary of the incorporation of the village that developed into that great city. Most of the meeting was held at the University of Chicago, but there were a few talks given at the fairgrounds, including one by Fermi on Hyperfine Structure, and one by Niels Bohr on Space and Time in Contemporary Physics. These were open to the public, and it was an astonishing sight to see poor Bohr struggling with a balky microphone in a room with people standing up and filling all the seats, mothers with babies and others of the general public, hoping to be enlightened by the great man. Fortunately the symposium I had to talk at was at the university. The main part of the Physical Society meeting dealt with high-energy particles and radioactivity. This was the moment when the transition from chemical physics to nuclear and high-energy physics was most pronounced, and we heard from Aston, Millikan, Urey, Lawrence, Cockcroft, and many others on what was going on in the nucleus. But at least I and my colleagues were there to uphold the molecules.

With Pauling on the program pushing atomic orbitals, and Mulliken advocating molecular orbitals, I felt that it was a good opportunity to point out the more general view that each of these methods was really aiming at a many-electron wave function, and the final results were identical, if each method was pushed far enough. Though I did not use many equations in my talk, I can reconstruct from my notes of the time the detailed argument that I was using. I was thinking of the potential energy of the molecular system. If the exact wave function is $\psi(1,2,\ldots,N)$, then, as in Eq. 22-2, the charge density of electron 1 is $\int \psi^*(1\cdots N)\psi(1\cdots N)dv_2\cdots dv_N$, and for all N electrons the total charge density is

$$\rho(1) = N\int \psi^*(1\cdots N)\psi(1\cdots N)dv_2\cdots dv_N \qquad (24\text{-}1)$$

In Eq. 22-2, we gave the probability that electron 1 should be found at point 1, electron 2 at point 2. From this we can get the potential energy $U(1)$ acting on the first electron, at position 1, produced by all electrons. It is

$$U(1) = \frac{(N-1)\int \psi^*(1\cdots N)\psi(1\cdots N)g_{12}dv_2\cdots dv_N}{\rho(1)} \quad (24\text{-}2)$$

where as earlier we use g_{12} for the coulomb interaction between electrons 1 and 2. As in all these equations, the integrations include summation over spin, and $U(1)$ is spin dependent. In terms of this potential, the exact electrostatic interaction energy between electrons is

$$\frac{1}{2}\int \rho(1)U(1)dv_1 = \frac{N(N-1)}{2}\int \psi^*(1\cdots N)\psi(1\cdots N)g_{12}dv_1\cdots dv_N \quad (24\text{-}3)$$

where $N(N-1)/2$ is the number of pairs of electrons. The potential energy of Eq. 24-2 is that of all the electronic charge, minus that of the exact quantity which is approximated by the Fermi hole. But the expression of Eq. 24-3 has a form identical with that met in classical electromagnetism, if we have a charge density $\rho(1)$ in a potential field $U(1)$. In addition to this, we must include the potential energy of the electrons in the field of the nuclei, which can be found from the charge density of Eq. 24-1, and also the electrostatic repulsions between the nuclei.

This argument was not new; Dirac had pointed out in 1930, in a discussion of the density matrix that the first-order density matrix, which appears in Eq. 24-1, and the second-order density matrix, which appears in Eq. 22-2, were all that one needed to find the electrostatic energy of the system. But I felt that people did not sufficiently realize that if a wave function led to correct values of the one-electron and two-electron density matrices, it was bound to give the correct electrostatic energy. And I pointed out how wave functions obtained by quite different methods, in particular from the atomic orbital and molecular orbital points of view, could lead to identical density matrices, provided one carried the approximations far enough.

I did not discuss kinetic energy, but I did ask the question, how far can one go in deriving the interatomic potential energy curve in a molecule, provided one has accurate values for the charge density, or the one-electron density matrix? I gave the following argument. If one has the charge density, one can compute by electrostatics the force on each nucleus. If one moves the nuclei, the work done against these forces will give the change of energy. In other words, one should be able to get the energy by electrostatics. This was merely a hunch at the time, but it foreshadowed the Hellmann-Feynman theorem, which was not proved for several years, and it turned out to be correct, as we mention a little later.

Almost as soon as I returned from the Chicago meeting, I realized that the virial theorem should be able to predict the relation between the kinetic and potential energies of a molecule, and thus to give information about the kinetic energy without actually computing it. If one could get the total energy from the one-electron density matrix, one could do the same for the kinetic energy. I worked out a more general proof of the virial theorem, for a wave-mechanical system, than had been available earlier, and sent this in to the newly established *Journal of Chemical Physics* before the summer was over. I shall discuss the results of the virial theorem a little later.

When I came to write up the *Reviews of Modern Physics* article in the summer of 1934, I was able to include all of these types of work: the nature of the Fermi hole, the application of the virial theorem, the cellular method and its application to the calculation of energy bands, and the use of the density matrices for finding the potential energy. It was in my mind that the potential of Eq. 24-2 would provide a very natural one to use for a self-consistent-field calculation, though I did not say so in so many words until almost 20 years later, in 1953, as mentioned in Sec. 30. But along with all of these topics, I also gave a considerable discussion of the application of the Thomas-Fermi method to crystalline problems. Dirac had shown in 1930 how to incorporate the exchange correction into the Thomas-Fermi method. Essentially, one proceeded as Bloch had done in his ferromagnetism paper of 1929, including the exchange energy as given in Eq. 16-6, as well as the kinetic energy of the electron gas as in Eq. 16-2. This theory was capable of giving a potential suitable for a crystal, as well as for an atom, and furnished a good first approximation to the self-consistent field.

There was in fact a considerable amount of activity in the direction of using the Thomas-Fermi method during the 1930s. In 1932 W. Lenz and H. Jensen, in Hamburg, had examined the Thomas-Fermi-Dirac method, as one called the method including Dirac's treatment of exchange, and had shown that various general theorems applied to it. In particular, one could derive the method from a variation principle applied to the total energy function, and one could prove the virial theorem. They had gone on to apply the method to various crystals. They could not treat the one-electron equation directly, on account of the lack of spherical symmetry, but they set up trial functions with enough flexibility to represent the charge density, and varied the parameters to minimize the energy. This gave quite good results for the alkali halide crystals. P. Gombas, in Budapest, went on with this method, applying it to metallic crystals. He continued to contribute to the method even after the war, and his book, *Die Statistische Theorie des Atoms und ihre Anwendungen*, published in 1949, is a standard work on the subject. Even now, Budapest has remained a center of work with the method, and in Sec. 30 I shall mention R. Gaspar from the group there, and contributions which he has made to the $X\alpha$-SCF method.

It occurred to me that the cellular method was adaptable to the Thomas-Fermi-Dirac (TFD) method, just as it was to the self-consistent field, and I suggested that my student Krutter investigate this scheme. Our idea was to solve the spherically symmetric problem within a sphere of the same volume as the Wigner-Seitz sphere, and to use proper boundary conditions at the surface of the sphere. As the crystal was compressed, one would have to have the same number of electrons in a smaller and smaller sphere, so that the energy would increase, and one might be able to get a good idea of the pressure-volume relations. Krutter carried the problem through, finding reasonable results, but not finding any binding in the crystal: The energy of the crystal nowhere fell below that of the separated atoms. It was generally believed that this was because we had not included any form of correlation energy in the calculations. Since the cellular calculations were going so well, we did not carry this TFD calculation further, but there was some preliminary discussion of it in the 1934 paper.

The theorem known as the Hellmann-Feynman theorem, stating that the force on a nucleus can be rigorously calculated by electrostatics from the charge distributions of all electrons and of the other nuclei, remained as far as I was concerned only a surmise for several years. Somehow I missed the fact that Hellmann, in Germany, proved it rigorously in 1936, and when a very bright undergraduate turned up in 1938–1939 wanting a topic for a bachelor's thesis, I suggested to him that he see if it could be proved. He came back very promptly with a proof. Since he was Richard Feynman, who has later gone on to be one of the leading many-body theorists, with a Nobel Prize, it is not suprising that he produced his proof without trouble. But both of us ought to have known of Hellmann's work.

As soon as we had this proof, I realized that we had a direct way of getting the total energy, potential energy, and kinetic energy for all nuclear positions, provided we knew the charge density, or first-order density matrix, for all nuclear positions. As I mentioned earlier, a knowledge of the forces on the nuclei, obtained by the Hellmann-Feynman theorem from the charge density, can be integrated to give the total energy. And the virial theorem gives a relation between total, kinetic, and potential energy. In the form which I gave in my 1933 paper, and in the 1934 *Reviews of Modern Physics* article, the relations for a diatomic molecule are the following:

$$\text{kinetic energy} = -E - R\frac{dE}{dR}$$

$$\text{potential energy} = 2E + R\frac{dE}{dR} \quad (24\text{-}4)$$

Here the total energy E is known as a function of R, the internuclear distance, from the Hellmann-Feynman theorem. For two cases, the equilibrium position where $dE/dR = 0$ and infinite separation where $R\,dE/dR = 0$, the equations reduce to those familiar for isolated atoms, kinetic energy $= -E$, potential energy $= 2E$. It is to be understood that the energy E is that found from the Born-Oppenheimer approximation, not including any kinetic energy coming from nuclear motion.

In the 1933 and 1934 papers I indicated the reasons for the behavior of the kinetic and potential energy, for a diatomic molecule. In Fig. 24-1, I give the kinetic (T), potential (V), and total energy of a diatomic molecule as a function of R, if the potential energy is given by the Morse curve. Other more accurate curves for the energy as function of R give qualitatively the same sort of behavior. The figure shows only the change in energy as R is decreased from infinity. We see that as R is decreased, the kinetic energy falls, and the potential energy rises. This, as I interpreted it, is a result of the beginning of the formation of bonding charge between the atoms. Some electronic charge moves from the neighborhood of the nucleus to the region between the atoms, thus increasing its potential energy, and correspondingly decreasing its kinetic energy.

Then as the atoms go still closer together, the kinetic energy starts to rise and the potential energy to fall. At the equilibrium distance, the kinetic energy is very rapidly rising as R decreases, and the potential energy is very rapidly falling. This is a result of the charge being squeezed into a smaller and smaller volume. The de Broglie wavelengths of the outer electrons decrease, explaining the increase of kinetic energy, while the electrons are pushed closer to the nuclei, decreasing the potential energy. This behavior is universal, holding for all sorts of binding between atoms.

One can explain the forces associated with these changes of energy as the Hellmann-Feynman theorem interprets them. For distances larger than the equilibrium distance, when electronic charge is piling up in the region between the nuclei, each nucleus is attracted by this charge distribution more than it is repelled by the other nucleus. For distances smaller than the equilibrium distance, more of the electronic charge is pushed to the far sides of the molecule, and it pulls each nucleus away from the molecule, resulting in the electrostatic explanation of the apparent repulsion between the nuclei.

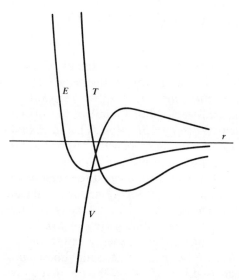

Fig. 24-1. Kinetic, potential, and total energy ($T, V,$ and E respectively) by the virial theorem. From Slater, *Rev. Mod. Phys.*, **6**, 209 (1934).

The virial theorem has been very useful in molecular theory, but the Hellmann-Feynman theorem has been less so. The reason is that most molecular work has been done with very inaccurate approximations to molecular orbitals. It turns out that the conditions under which an incorrect molecular orbital will nevertheless lead to a solution which satisfies the virial theorem are not very stringent, and many of the calculated wave functions satisfy this theorem. On the other hand, the conditions for the Hellmann-Feynman theorem to be satisfied for an incorrect wave function are much more stringent and are seldom satisfied. We bring these points out here because, as we mention in Sec. 35, the Xα-SCF approximation is one which rigorously satisfies both theorems, and we may look forward to useful results in the future coming from both sources.

The general line of development which I therefore looked forward to as the 1930s came to a close was to use a self-consistent-field calculation with a potential which was as much like that of Eq. 24-2 as possible, to find the charge density from the sum of the charge densities of the occupied orbitals, to get the forces and energies from the Hellmann-Feynman theorem, and the potential and kinetic energies from the virial theorem. This is essentially what we are achieving with the $X\alpha$-SCF method, but it has taken over 30 years to reach this point, which is discussed in Sec. 35. Before we go on to the events of those 30 years, however, let us take up one other type of advance which came in the 1930s.

25. Localized States and Magnetism, 1936–1937

In Sec. 20, we mentioned Frenkel's work on the localized excitations in crystals, based on experimental work on the exciton which had been carried out at Göttingen, in Pohl's laboratory, about 1930. In 1936, Arthur von Hippel, who had formerly been a student in that laboratory, and who was a son-in-law of James Franck, joined the MIT faculty in the electrical engineering department, after several years of travel following his departure from Germany at the time of Hitler's rise to power. von Hippel immediately founded an experimental solid-state laboratory at MIT, which he called the Laboratory for Insulation Research, patterned in the German style, and this made a great addition to the facilities of the Institute in the field of solid-state physics. One of the first things he did was to give me a preprint of a paper which he had just written, giving an explanation of the excitons observed in alkali halides which was somewhat different from that of Frenkel.

Briefly, he considered that in an alkali halide like NaCl, the most easily excited electron was a valence electron of the Cl^- ion. He considered that the smallest excitation would arise when this electron became attached to one of the nearest neighboring Na^+ ions, to form a neutral Na atom. The original Cl^- would be left as a neutral Cl atom. Such an excitation is often called a charge-transfer excitation. He worked out the probable locations of these transitions in the spectrum, making use of experimental information about the energy levels of the atoms and ions, the electrical properties, and so on, and came out with quite satisfying explanations of the observed absorption spectra.

It seemed to me that this was a quite plausible theory, but probably not so different from Frenkel's assumptions as one might think. If one estimates the probable dimensions of an excited orbital originating on a Cl^- ion, it would turn out to be large enough so that most of the charge density would lie on one of the neighboring Na^+ ions. Thus the final charge distribution in such an excited state might be quite similar to that of a mixture of von Hippel's charge-transfer states, the electron being shared by each of the nearest neighboring Na^+ ions. But either theory, Frenkel's or von Hippel's, looked very different from an energy-band theory. If one had the energy bands of the NaCl crystal, the bands arising from the $3s$ and $3p$ atomic orbitals of the chlorine would be occupied, and those arising from the sodium $3s$ and $3p$ would be empty. Thus according to the band theory the ultraviolet absorption should come from a transition across this gap, from the topmost occupied Cl $3p$ band to the lowest empty Na $3s$ band. But this should be a photoconductive state, and Hilsch and Pohl had already shown that the exciton states were nonphotoconductive. Somehow, there must be a place in the theory for the localized excitations that Frenkel and von Hippel had suggested.

This seemed to me one of the cases where we were getting different answers from a band theory and a localized electron theory, and I felt that it should be investigated. It was partly with such an investigation in mind that I had suggested to Shockley that he compute NaCl bands by the cellular method. He had already become interested in excitons, and in fact the first colloquium talk which he gave at MIT, in March 1933, his first graduate year, was on Frenkel's papers on the exciton. When Shockley was finishing up his energy-band work in 1936, preparatory to writing up his Ph.D. thesis, and when von Hippel had focused our attention on the exciton problem again, Shockley and I got busy asking how one would describe these localized excited states in wave mechanics.

The first thing we did was to get enough information together, both from experiments such as von Hippel had used and from Shockley's energy bands, to make the best estimates we could of the one-electron energies of the NaCl crystal, as a function of internuclear distance. In Fig. 25-1 I give the results we came to. As I have mentioned before, the top of the $3p$ chlorine band is the topmost occupied level, and the bottom of the $3s$ sodium band is the lowest empty level. The lowest photoconductive transition should be across the gap between these bands, and the calculations indicated that it should be at about the place where it was found experimentally. But this was analogous to the transition to the bottom of the continuum of excited or ionized levels in an isolated atom, and we still had to explain the discrete states that experimentally occurred below this continuum, to form the Frenkel or von Hippel excitons.

Only energy-band states, with wave functions satisfying the Bloch condition, occur in a perfectly periodic potential. But we realized that in a self-consistent treatment of a localized excited state, the Cl neutral atom from which the electron had been removed would have a decidedly different potential around it from what would be found in the unexcited crystal, as would the Na atom on which, according to von Hippel, the electron would lodge. The self-consistent problem of the excited state, then, would be like that of a perfect crystal, with a superposed localized perturbative potential. What would this localized potential do to the wave functions and energy levels?

We looked into this problem, using a very simplified one-dimensional model, in which the potential on one atom was different from that on the rest of the atoms. It was simple to show that one energy level would

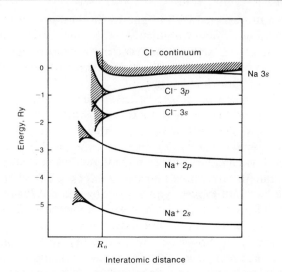

Fig. 25-1. Boundaries of energy bands of NaCl, as functions of interatomic distance, as estimated by Slater and Shockley. From *Quantum Theory of Molecules and Solids*, Volume 4, by J. C. Slater. Copyright 1974 by McGraw-Hill, Inc. Used with permission of McGraw-Hill Book Company.

become detached from the band, acting like an exciton level below the continuum of states. If an electron were located in this localized state, it would set up such a charge distribution that the situation could be self-consistent. Furthermore, we showed that the oscillator strength associated with a transition to such a localized state was very much larger than the oscillator strength of a transition to the continuum. In fact, by the use of the oscillator sum rule of Kuhn and Thomas, which we mentioned in Sec. 3, we were able to show that if the perturbation of potential on the excited atom was sufficiently great, almost the whole oscillator strength would go into the transition to this excited state. This seemed to give a framework around which a more complete theory of the exciton could be built. Unfortunately, however, we had no time to go further with the theory at the time. Shockley obtained his degree and left MIT, and I was very anxious to get into the theory of ferromagnetism, using the energy bands which Krutter had computed for the copper atom in order to estimate the energy bands for nickel.

In Sec. 23, I mentioned that one reason I was glad to have Krutter calculate the energy bands of copper in 1935 was to be able to extrapolate back to the next preceding element, nickel, and find something about its $3d$ bands. In 1936 I carried this through, and used something like the spin-dependent exchange which Bloch had suggested in 1929 to find whether we should expect nickel to be ferromagnetic. I did not actually use just this method, but rather took advantage of the fact that the $3d$ atomic orbitals coming into the energy-band wave functions were not very far from orbitals in the isolated nickel atom. Consequently I could calculate the exchange terms from the known atomic spectra. Later calculation has shown that the exchange effect which I calculated in this way was approximately what I would have found from Bloch's method.

In the first of several papers on ferromagnetism, I carried through a calculation of this type for the elements in the neighborhood of iron, cobalt, and nickel. I realized, of course, as is pointed out in Sec. 16, that I could not prove that a substance had to be ferromagnetic if its partly filled d band was narrow enough. I could, however, prove the opposite: If the bands were wide enough so that the spin-down band overlapped the spin-up band, electrons would fall from the bottom of the spin-down band to the top of the spin-up band, decreasing the magnetization. This in turn would bring the bands closer together, increase the tendency toward the disappearance of the magnetization, and thus destroy the possibility of ferromagnetism. In the calculations I made, it appeared that only the three elements iron, cobalt, and nickel should be ferromagnetic. Chromium and manganese should be on the edge; they are now known to be slightly antiferromagnetic, and my arguments could not distinguish between the two possibilities of ferromagnetism and antiferromagnetism. None of the other elements in the $3d$ group of transition elements would be expected to show ferromagnetism. To this extent, then, the results were satisfactory. I might say here that calculations in the 1960s have verified the essential correctness of these conclusions from the 1930s.

This calculation, however, did not do what I thought desirable: to make connection between the energy-band and the Heisenberg or atomic pictures of ferromagnetism. I was at this point when I went to the Institute for Advanced Study in the spring of 1937. I had realized that there must be a close relationship between the theory of excitons, which I had been working on with Shockley in 1936, and the theory of spin waves. In a ferromagnetic metal, all magnetic electrons will have spin up in the ground state. However, if just one electron reverses its spin, we have the equivalent of a localized exciton. We must realize that since there are quite different potential energies for a spin-up and a spin-down electron, on account of the spin-dependent exchange term of Eq. 16-6, there will be two different energy bands, one for spin up, one for spin down, of which the spin-up band will lie lower. Thus to transfer an electron from the spin-up to the spin-down band will be an excitation of the electron, and will result in a very considerable increase of energy.

This energy change, as one can demonstrate, is of the order of magnitude of the energy difference between the multiplets of highest multiplicity in the magnetic atom and those of next highest multiplicity. It was from these multiplets in nickel that I had estimated the exchange energy, in my 1936 paper. But as a matter of fact, to rotate the spin of one magnetic electron is known to require only a very much smaller amount of energy, measured by the Heisenberg exchange integral. The energy-band theory, in other words, can give quite a wrong idea of the excitation energy. It is this same error we commented on in Sec. 16, where we discussed the way in which the diagonal energy of a magnetic state fell way below that of a nonmagnetic state, if the energy bands were narrow enough, and referred back to Fig. 14-1.

In my 1937 work on ferromagnetism, I showed that the spin-wave states lay far below the energy-band states, by arguments which were essentially similar to those which had been used in the discussion of excitons. On the other hand, they lay only slightly above the completely ferromagnetic energy levels: It required only a slight energy to excite a spin wave, and from this calculation one could estimate the Heisenberg exchange integral. I then asked how the Heisenberg exchange integral would be expected to change sign, from positive to negative, as the energy bands became broader, the case in which we knew that ferromagnetism would have to disappear. I was able to show that this change of sign was produced by the effect of ionic states of the type which one had to use in the theory of the hydrogen molecule to get a wave function which reduced to the proper behavior at infinite internuclear separation. These ionic states became more and more important as the energy bands got broader, and one could see very clearly in a qualitative way how the nonferromagnetic nature of metals with broad energy bands came about. It was not possible at that time to compute Heisenberg exchange integrals quantitatively, but the work done in this 1937 paper pointed the way to calculations which are now becoming practicable with recent advances in the $X\alpha$-SCF method.

There was another advance in the theory of localized excitations during my period in Princeton in 1937. A young postdoctoral worker from Switzerland was there, Gregory Wannier, who had just figured out a very ingenious technique for handling the excitations. He started by building up a set of convenient atomic excited orbitals, of a type now called Wannier functions, one of whose useful properties was that those on different atoms were orthogonal to each other. When one set up the exciton theory in terms of them, one then avoided the overlap integrals which had been such a nuisance in the Heitler-London method. But quite beyond this convenient feature, he showed how to build up the localized functions representing an exciton.

One made a linear combination of Wannier functions on the various atomic sites of a lattice. There was a coefficient for each of these atoms. Wannier then made a very ingenious mathematical transformation in which he could replace these discrete coefficients by a continuous function of position throughout the lattice, such that the value of this continuous function at a given lattice site gave the coefficient by which the Wannier function on that site had to be multiplied to get the wave function. He then was able to write a differential equation for determining this continuous function, a differential equation which was like a Schrödinger equation for a particle of a given effective mass, moving in a potential which represented only the deviation of the local potential from the value it would have in the unperturbed crystal. In ordinary cases, this differential equation was like that of a hydrogen atom, only with a much larger effective mass and a much smaller coulomb attraction. The resulting function and energy level were similar to those of hydrogen, but the excitons were quite spread out in space, and their energies were only slightly below the continuum. As time went on, these Wannier excitons, or large excitons, were distinguished from the much more localized Frenkel excitons, or small excitons. It was only in work which I and my student George Koster carried out in 1954 that the theory which went continuously from the one limit to the other was worked out, and we could see that both types of excitons were really examples of a broader theory.

The work I carried out in Princeton represented nearly the last work along these lines which I undertook before war came in the 1940s. The reason was that I was teaching courses on thermodynamics and statistical mechanics, was learning a great deal about those topics, and was working for several years on a textbook on the subject, which I published in 1939 as *Introduction to Chemical Physics*. I carried out a few other minor pieces of research during the period, but the work I have described represents pretty much the state of the art, as far as I was concerned with it, at the time the war came on. There was then an interruption of ten years or more, which I describe in Book III. It was only in 1951 that I returned to these topics of the electronic structure of molecules and solids, which were always my principal interest.

BOOK

III

War and Postwar Years,

1941–1951

26. Radar, 1940–1945

The war came to the MIT physics department in October 1940, more than a year before Pearl Harbor. Evans had planned a conference at the institute on applied nuclear physics toward the end of that month. It was a very successful conference: 90 invited papers were presented, and 590 persons registered from 100 separate institutions. Most of the well-known nuclear physicists of the country were present, but they did not all appear at the open meetings. Instead, they assembled behind closed doors, and talked over the plans for the new microwave laboratory, which the NDRC, the National Defense Research Committee, had just decided to establish at MIT, and which later turned into the MIT Radiation Laboratory. Things happened rapidly, and ten days after the conference was over, Lee DuBridge had been appointed director of the new laboratory, staff was beginning to arrive from all over the country, Alfred Loomis was telling them in rather breathless excitement of the possibilities of microwave radar (though that name was not yet in use) as it was being developed in England, and the new laboratory was set up in space in MIT Building 4, which years before had been the heat laboratory of the physics department, though it had been converted to other uses long before.

This was the beginning of the invasion of the Institute by physicists from all over, working mainly on radar but also on other projects, an invasion which lasted during the war, and had profound effects on physics at MIT. It was also the beginning of the diversion of MIT physicists to war projects. The physics department was not consulted or even informed by the administration about the plans for the Radiation Laboratory, and did not take part in its organization. This was part of Compton's plan, by which, though the Institute was host to the laboratory, it would make no effort to influence its policies in a scientific way. Nevertheless individual members of the department began to be consulted and drawn into the work of the laboratory from the start.

One of the first moves of I. I. Rabi, of Columbia, one of the moving spirits of the Radiation Laboratory from the beginning, was to walk into my office and ask my help in understanding the theory of the magnetron, the power oscillator whose great development by the British had made microwave radar practical, but whose workings were very poorly understood. When Rabi first asked me how it worked, I knew nothing of what a magnetron was, except for the knowledge that molecular spectroscopists had made very primitive versions for far infrared or millimeter spectroscopy, so I asked Rabi what he was talking about. He drew me a picture of a cavity magnetron. I insisted on knowing at least where the electrodes were located, what sort of electric and magnetic fields were applied to the thing, what its dimensions were, and what came out of it. Enormous microwave power came out, said Rabi, as the English had found. But its operation was an art, not a science, and they had only the remotest idea how it worked. I told Rabi I also didn't know, but would try to find out.

This started me on several weeks of very concentrated calculation. I knew nothing of vacuum tubes, except what I had learned as a graduate student, but it is an interesting example of the way one problem of physics can throw light on another that a couple of years earlier I had been working on a problem which was mathematically quite similar. Though I have not mentioned it in earlier sections, I had been trying to understand the phenomenon of superconductivity, and one of its interesting features is the Meissner effect and related phenomena which come up when a magnetic field is superposed on the flow of electrons. I had thought it would be informing to consider a space charge of electrons in a cylindrical container, with a magnetic field applied along the axis of the cylinder. This had not led to anything very remarkable concerning superconductivity, but it did show me just how to set up such a problem mathematically, how to set up the Hamiltonian function, and to proceed with the solution of Hamilton's equations to find the paths of the electrons. I saw at once that mathematically the magnetron problem was very similar. Let the outer wall of the cylinder be the anode surface of the magnetron (the resonant cavities being outside), put in a smaller concentric cathode, add a radial electric field between anode and cathode, and superpose an oscillating electromagnetic field, such as would obviously be present if the magnetron were putting out power, and then find how the electrons moved.

It was immediately obvious to me how to go further, on account of the resemblance of the problem to that of the self-consistent field in an atom. That problem, of course, is one in which we study the motion of an electron in the presence of the space charge produced by all other electrons and of the nuclei (which take the place of the electrodes). As we know, this problem is too hard to solve all in one step. One proceeds instead by iteration, assuming a form for the space charge, solving for the motion of an electron in that field, studying the space charge produced by all the electrons moving in the way just described, and demanding that the final space charge be identical with that originally assumed. By successive iterations, we know that we can achieve self-consistency. I resolved to try to carry through such a self-consistent calculation for the magnetron, and see if the result would include not only a time-independent space charge, but more interesting, an oscillating, or rather rotating, space charge which would produce the oscillating or rotating electromagnetic field which corresponds to the radio-frequency output of the magnetron.

This was quite a piece of arithmetical calculation, but I had done similar things for the atomic problems, so I set to work, and after very few weeks I had a satisfactory answer. The rf field allowed streams of electrons to travel over from cathode to anode in such a pattern as to produce the rf field, and at the same time the power fed from the dc field to the electrons, as they streamed from cathode to anode, very largely reappeared as rf output, explaining the very high efficiency of the magnetron. Furthermore, the theory was sufficiently straightforward so that one could put it in dimensionless form, which made it possible to study the scaling, the way in which the voltages, magnetic fields, and dimensions appropriate for operation at any arbitrary wavelength could be found from the single calculation I had carried through. This was the beginning of the understanding of magnetrons of different wavelengths from the original ones which were being studied, which were in the 10-centimeter range.

Naturally there were interesting features connected with this calculation. My colleague Manuel Vallarta had been carrying through similar numerical integrations as a check. Vallarta had been interested in the deflection of cosmic ray particles by the magnetic field of the earth, and this problem was mathematically similar enough to the magnetron one so that, just as in my case, he was familiar with all the required mathematics, though he had used it for quite a different purpose. He carried through calculations like mine, and checked the results very satisfactorily. This seemed to make the situation clear enough so that I wrote a preliminary report on the magnetron operation. Of course, all these things were highly classified, but the report was duplicated and sent to England, and then a very amusing thing appeared.

It was reported back to me that Hartree was also working on the same problem, again with the self-consistent field as a background, and he had come to the same conclusions I had, but his results differed from mine somewhere by a factor of 2. I knew Hartree well enough to know he would not make a mistake, but I was mystified as to how I and Vallarta had checked each other. We looked into this pretty carefully, and found I had made a mistake of a factor of 2 in one place, and Vallarta had done it in an entirely different place. The result was that we checked each other, both of us were wrong, Hartree was right, but fortunately the general results did not depend on the numerical error, and we all agreed that we pretty well understood how the magnetron worked. This fundamental theory was slightly refined later, but in its broad outlines it still gives a proper understanding of the working of the magnetron, and we had worked it out, as I said, within a period of a very few weeks in the winter of 1940–1941.

Let me now say something about what was going on in the Radiation Laboratory. That laboratory was an extraordinary assemblage of very able and intelligent people who had almost no experience with the problems they were working on. Many of them were very young, pulled out of graduate school. The fact that they could get so far was described later in the war as the "miracle of the children." But there was almost no one in the laboratory with any particular feeling for microwave problems. The people, such as Stratton, Barrow, and Chu and others at MIT, who had long experience with such questions, were not very closely incorporated into the laboratory.

And yet the laboratory members were doing very effective things, treating the microwaves by the methods we know as "equivalent circuits." They talked impedances, impedance matches, and so on, and it was clear to me that this was a very fruitful way to handle the problems, but that it was based entirely on the analogies between oscillating cavities and ordinary electric circuits. Once the general principles of magnetron operations were clear, it was necessary to go further: The behavior of a magnetron depended very profoundly on the impedance of the output circuit, and yet this was not an ordinary lumped circuit, but a set of wave guides or coaxial lines, with the type of connections which later became faimiliar as we understood microwaves better. I had been treating the microwave cavity as a real cavity, in which I was dealing with solutions of Maxwell's equations, rather than as a circuit element, and it was clear that the same thing had to be done for the rest of the microwave circuit.

Fortunately Edward Condon had taken a look at this same problem, and had written a very informing paper on the relations between microwave cavities and resonant circuits. I studied this paper, which had a very close resemblance to the fundamental Hamiltonian theory of radiation fields which we met in Sec. 17, and found how it could be applied to the type of circuit problems one met in cases of wave guides, coaxial lines, antennas, and so on. It seemed to me that these were approaches that should be made available to the members of the laboratory, and consequently I took time off to write a book, *Microwave Transmission*, which was sufficiently fundamental so that it could be published as an unclassified book, and yet which I felt would help along the research effort.

By this time, the United States was approaching involvement in the war. In the fall of 1941, I became aware that Jim Fisk at the Bell Laboratories was working on magnetrons in a very fundamental way. It seemed to me that I might be able to make more progress myself by going to the Laboratories to work in Fisk's group than by staying at MIT. I was anxious to do some experimental work on magnetron cavities, and to test out my theories of the resonant modes of these cavities. In particular, I wished to look into the problem of avoiding the excitation of unwanted modes of oscillation, which I felt convinced was the main difficulty in the way of efficient operation of some magnetrons. Consequently I arranged, through the authorities of MIT and BTL, to transfer my personal activities to the Laboratories, and to carry on enough personal experimental work on magnetrons so that I was sure I understood their properties as circuit elements. I went to New York in December 1941, the arrangements having been made just before Pearl Harbor, and naturally there was a particular urgency since by the time I arrived we were at war.

Work with Fisk and the others in his group and related fields at the Laboratories was a stimulating experience. This was before the days of Murray Hill, and the Laboratories were still centered in downtown Manhattan, at 463 West Street. The microwave work was across the street, in a remodeled soda-cracker factory. Though the surroundings were not what one now expects of a laboratory, the companions were superb: not only Fisk, but John Pierce, A. L. Samuel, Homer Hagstrum, Paul Hartman, Larry Walker, Arnold Nordsieck, and others, a group all of whose members later went on to distinguished work. Fisk already showed the qualities of light-hearted and yet efficient leadership which carried him later to the presidency of the Laboratories. I well remember the lunch hours. We were on the edge of Greenwich Village, surrounded by interesting restaurants, and each lunchtime Jim Fisk would lead us out to a different restaurant. For the first month we had lunch in a different restaurant every day, always a good meal, always good talk, and yet always working very hard on the problems that faced us.

I shall not go further into the details of the work with which we were concerned. The main problem was to produce an efficient 3-centimeter magnetron, and if possible to make it tunable. This demanded thorough knowledge of both the electronic and the circuit properties, but we were well set up to carry this through. Most of the rest of the war I was at the Bell Laboratories, and I was able to see all the stages of the development of a piece of equipment, from the original theory through the design, the practical features of carrying out the construction in a way that could be manufactured, and the detailed production schedule. Part of the strength of the Bell Laboratories is the way in which all of these features are given appropriate emphasis. Jim Fisk and I made visits to the manufacturing works, and we looked into ways in which magnetrons could fail. I thoroughly learned the lesson that if one element in the construction is successful only 90 percent of the time, and another and still another element are similarly only partially successful, the fraction of satisfactory units is 0.90 raised to a very high power, which is very close to zero. This is a lesson that anyone wanting to translate mathematics into production equipment must learn. By the latter part of the war, the equipment we were working on was in production, worked reliably, and proved to be useful for practical radar.

In the winter of 1942–1943 the English requested that I come over during the following spring, to tell the British scientists about the work we had been doing. This trip to England in wartime was one of the entertaining features of my wartime experiences. I took off on an RAF plane from Montreal, in March 1943, where deep snow still covered the ground. This plane was a converted bomber. The passengers—an assorted set of important characters—sat on the floor of the plane (it had no seats) on mattresses, dressed in flying suits, with heavy blankets to keep warm, and oxygen masks to be worn the whole time, since we were unpressurized, unheated, and were flying at 20,000 feet. We wore parachutes, and told each other that when we looked back at our experiences after the war, we would feel that we had been real pioneers.

It was a striking and anomalous experience to land the following morning at Prestwick airport, and get taken to the officers' club, in a beautiful garden with the daffodils coming out. Going from the American continent to England, far from seeming like going from peacetime to wartime, seemed almost like the opposite. This feeling continued when I put through a long-distance call from Prestwick to the American liaison office in London regarding my train trip down to London, found that the reservation had already been made, had a very comfortable sleeping-car ride to London, and the first evening, when I opened my hotel window, was greeted not by the sound of air raids, but by a Danny Kaye record being played in the room across the courtyard. The person with whom I had closest contact in the American liaison office was Guy Stever, who later was professor of aeronautical engineering at MIT, and is now director of the National Science Foundation and science advisor to the President. He, trained at Cal. Tech. as a cosmic ray physicist, proved to be equally successful in all branches of science and engineering, as well as a very genial friend.

During my three months in England in 1943, I had plenty of opportunity to observe England at war. I visited laboratories, including universities, government laboratories, and also industrial companies. I talked with many physicists. In Manchester I saw Hartree, and in Malvern, where the main British radar research laboratory was located, I saw many friends. Herbert Skinner, whom I had known at MIT in 1932, was there, and I not only had a visit to his laboratory, but a pleasant Sunday dinner with him and his talented wife. She was such an expert in comparative philology that her first request, when they had arrived at MIT, was to have an arrangement set up by which she could attend the all-masculine Harvard seminar in Icelandic philology! A lady of eastern European origin, who required 18 pieces of luggage when she traveled to enable her to dress like a fashion model, she made something of a sensation in Cambridge.

At Malvern I also met R. A. Smith, an expert in electronics, whom I came to know very well after the war, and who served for a number of years in the 1960s as the director of the MIT Center for Materials Science and Engineering. England in wartime was an interesting experience, though fortunately I was not there at a time of very heavy air raids. But in spite of that, I remember one conference with industrialists, held in the morning in a small industrial town, when the opening session had to be postponed for a while because the presiding officer was a few minutes late. He explained when he arrived, that during the night he "had had a bomb in the bottom of his garden." People took such things in their stride.

The trip home from England was very different from the trip over. We were sent back on Pan-Am, in one of the great flying boats which they used in those days. Karl Compton had been in England during the last few days I was there, and we traveled back together, along with a passenger list of VIPs. We had to lay over a day at Shannon Airport, and Pan-Am had us taken to Adare, a small town 20 or 30 miles away, to spend the night. We stayed at the Dunraven Arms, a delightful Irish country inn. Adare is the seat of the Earl and Countess of Dunraven, and at dinner in the inn, the waiter pointed out the Earl and Countess having their dinner at the table in the corner. And during the day, Compton and I walked around the country town, with its thatched houses, and the pigs and chickens in the yards. Ireland seemed like the only really peaceful place in the world at the time, with even real bacon and eggs for breakfast, in place of the synthetic substitutes that we had been used to in England.

After the overseas trip, I spent another year at Bell. By the time the European phase of the war was starting to wind down, in the summer of 1944, I was beginning to feel that my main contribution there was over. A number of us at MIT were becoming conscious of the close interplay between science and technology in the field of microwaves, and we began to lay plans to try to prolong this interplay into our postwar planning. There was a very real danger that at the moment when the war was finished, the Radiation Laboratory would close down instantly, and this facility, which was so useful to the country as well as to MIT, would be lost.

In August 1944, I had a visit in New York from Julius Stratton. He had received a flattering offer from the Bell Laboratories, to head up a considerable segment of their research. He preferred to return to MIT, but the Bell offer was too good to give up without substantial prospects at MIT. We had already talked with Compton about the hope of establishing an electronics laboratory, a cooperative venture between physics and electrical engineering, after the war. There had, in fact, in prewar days been definite suggestions both from Compton and from Bush, who was then dean of engineering, that various interdisciplinary laboratories should be set up to encourage cooperation between departments. As Stratton and I talked in New York, we evolved the idea of suggesting to Compton that such a laboratory definitely be planned, with Stratton heading it. In a long letter the next day, dated August 23, 1944, I proposed this to Compton, and outlined a possible research program which I might like to undertake in such a laboratory. I suggested to him that I would be glad to return to Cambridge during the fall to help start this program.

Several days later I was in Cambridge, talking over these plans with Compton and Stratton. My own suggestion was that I should return to Cambridge in October 1944, spending one day a week in New York at BTL, which would be reduced to a monthly consulting visit starting in January 1945. As for the broader plans for an electronics laboratory, Compton approved them, and within a few days, in September 1944, an MIT appropriation was made to establish a new interdepartmental electronics laboratory to conduct research in the field of electronics in association with the departments of physics and electrical engineering, and approving the appointment of Stratton as director, beginning at a date to be determined by him. Stratton was acting as a consultant for the War Department, dealing mostly with radar for air defense, and I had been on a committee of which he was chairman, which had made a number of visits to air defense installations, including one at Orlando, Florida, to inspect their progress. He was really needed in Washington, and agreed to return to MIT at the close of the European war to assume his duties as director of the new laboratory.

The people in charge of the Radiation Laboratory were as anxious to see some of its work perpetuated as we were, and they set up a Basic Research Division within that laboratory in the winter of 1944–1945. In it we could start some work more fundamental than the applied radar research, and it could form a mechanism for transferring some of the equipment of the Radiation Laboratory to MIT when the opportunity presented. During the year 1945 the Research Laboratory of Electronics took form, as an outgrowth of this planning, with Stratton as director. As soon as the war in Japan was at an end, with the explosion of the atomic bombs in August 1945, we were prepared to start this laboratory in earnest, and to begin to recruit an enlarged staff to man it.

27. Postwar Development at MIT, 1945–1951

During the war practically every member of the MIT physics department had been associated with war work in one way or another. As the end of the war approached, many of them turned back to Cambridge, as I did, full of new plans for the development of physics at the Institute. The war had brought physics to the attention of the public, of industry and government, as had never happened before. It was obvious that this would result in a greatly expanded interest in physics after the war, just as the first world war focused attention on chemistry. In particular, the two fields of electronics, as exemplified in radar, and of nuclear structure, as applied in the atomic bomb, were bound to lead to greatly accelerated research and application, and greatly increased numbers of students and opportunities for their employment. The institute had been a leader in both these fields before the war, and everything pointed toward our taking an even more leading position after the war. The department plans, we felt, should be such as to assure this leading position.

We were bound to grow; but perhaps the most crucial question was how much we should grow. A rather horrible possibility could be seen by contemplating the Radiation Laboratory. This organization, with its almost exponential rate of growth during the war, had been a very effective way to get its emergency work done. It gave the maximum chance for each individual to use his own initiative, with very little direction or supervision from anyone else. So many young and untrained men had been taken in, in proportion to the senior members, that the young ones had small chance to learn from their older colleagues, and we had the "miracle of the children" which I have already mentioned. But this is no way to run a university in peacetime. The essence of a university is the training of the students by their older associates and professors, and to do this properly demands enough professors, or few enough students, so that they can effectively keep contact with each other. Early in our thinking, we began to see that the overall size of our postwar department had to be set by the size of the senior staff, with few enough younger men so that they could really get maximum benefit from their association with men of experience.

The next question was, how big should the senior staff be? First, should there be research appointments, men not taking part in the teaching, or should we be limited by the size of the teaching staff? Here our answer was clear. Any division of the staff into separate research and teaching appointments would bring about an invidious distinction. Research always seems at first sight to be a more attractive career to the physicist than teaching, though on further acquaintance he comes to realize that the satisfactions of teaching are as great as any he meets. The teachers in such a scheme are bound to be envious of research men who have no teaching duties. This principle made us from the beginning dismiss the idea of a separate research institute or laboratory—a continuation, for instance, of the Radiation Laboratory, existing in parallel with the physics department, but just carrying on research. Any postwar developments should be carried out by senior staff who were members of the ordinary departments and took part in teaching in the normal way.

Thus we met the question, how much, and by what type of appointments, should the existing staff be supplemented? And here the criterion was twofold. Financial caution demanded as small additions to the department as possible. Even though we might find funds plentiful for a few years after the war, this might not last forever. A professor implies a continuing commitment; and we felt that the Institute should make no more such appointments than it could carry at some future time, even if funds were not plentiful. The size of the staff, therefore, should not grow larger than could be justified by the size of the department teaching load. Even this, however, allowed a considerable expansion, for we expected, and actually had, enough increase in students of physics so that a good many more staff members were needed to maintain a reasonable ratio of staff to students.

The other side of the argument on increasing staff was the matter of having different fields of physics represented in the department, so that we could give adequate instruction in them. Our account of the department before the war, in Sec 21, must have shown the reader that we had a wide diversity of interests, wider perhaps than almost any other physics department in the country. We had never believed in the extreme concentration in one or two specialties which some departments have chosen. We could afford to diversify, partly on account of the size of the department, made necessary by the large teaching load particularly in the elementary courses, and partly because we felt it almost a duty in a technical institution to carry on work in various applications of physics which do not usually attract interest in an arts college. Thus our work in electronics, x-rays, optics, and acoustics, was in each case in a field pursued in only a very few institutions in the country, and consequently not well known by physicists in general. Our department was often looked down on, to some extent, by those who felt that no physicist of any imagination would be in any field except nuclear and high-energy physics. And yet in each of these less popular fields, our department was looked up to by the industrial leaders as the best department in the country, and we were constantly urged to turn out more students in each of these fields. We felt firmly, as we faced the situation after the war, that this diversity was a good thing, and that we should not alter it, or reduce our emphasis on any of the fields in which we were already working.

As I have already mentioned, the two fields where we felt we should expand were the two made most famous by the war work, electronics and nuclear physics. The first of these represented the most striking area in which the well-known and classical principles of physics were finding quite new and fresh applications; the second was the most important new field of physics, in which even the fundamental principles were still in doubt, and where the most important discoveries of the next decades were bound to be made. In the first field, microwave electronics, we had a great headstart as compared with the rest of the country: We had the prewar experience of Stratton and others, and we had the presence of the Radiation Laboratory and the experience of a number of staff members in that laboratory. In the second field, nuclear physics, though we had a number of experts in the department, most had not taken part in the Manhattan Project, and we lacked the knowledge of the newer developments. I had, as a matter of fact, been a member of a very early committee appointed by the National Academy of Sciences, to advise as to whether the atomic bomb project was sufficiently promising to warrant the big government effort later taking form as the Manhattan Project. Thus I had followed the progress of the enterprise with interest, as far as it was public knowledge, but I had not had detailed information about it since those early days.

Our first and immediate problem toward the end of the war was how to consolidate our position in the field of microwaves, and I have already described the first steps in that direction in Sec. 26. The first requirement for the new Research Laboratory of Electronics was an adequate staff, and the most important part of this was the senior staff. We were better off here than we might have been, on account of the various staff members who had been in the Radiation Laboratory. Stratton, Frank, and I were well acquainted with the theoretical side of the field, and knew something of the laboratory aspect; George G. Harvey had been in the laboratory side of microwaves. Still, this was a rather small nucleus to start with. The Radiation Laboratory was a tremendous organization, and while we expected our continuation of it to be on an almost infinitesimal scale compared with it, still we needed a number of people who were well trained in its experimental techniques. With this requirement in mind, the department was authorized by the administration to make a very few appointments in the field of microwaves, enough to ensure continuity between the Radiation Laboratory and the new Research Laboratory of Electronics. We made several unsuccessful efforts to induce physicists to join the department. Finally our efforts were successful, and we made two invaluable appointments in physics from the Radiation Laboratory staff: those of Jerrold R. Zacharias, a former associate of Rabi, who was head of the physics department at Hunter College when he came to the Radiation Laboratory, and of Albert G. Hill. Hill had been with the department before the war, but had left it to go to the University of Rochester. Lee DuBridge, the director of the Radiation Laboratory, had come to MIT from Rochester, and he brought Hill with him.

Both Zacharias and Hill were appointed with complete freedom as to what type of research they wished to take up. Zacharias decided not to follow microwaves, but to return to his earlier field of molecular beams; Hill decided to stay with microwaves. In spite of Zacharias's change of field, he nevertheless was of great service in the organization of the Research Laboratory of Electronics, in which his research on molecular beams was located, as well as of the Laboratory for Nuclear Science and Engineering, organized later, which he headed for a number of years. Hill became an associate director of the Electronics Laboratory, as did Jerome B. Wiesner, who joined the electrical engineering department from the Radiation Laboratory. Out of the three leading senior members of the Research Laboratory of Electronics in the early days, Stratton, Hill, and Wiesner, two of them, Stratton and Wiesner, have since become presidents of MIT, and the third, Hill, as I mentioned earlier, is vice-president for research. We had a strong team to organize the laboratory.

With these senior members assured, the next problem was a junior staff. Here we remembered that MIT is first of all an educational institution, and we took advantage of a situation that existed in the Radiation Laboratory as a result of the war. Many young men from all over the country had been drawn into that laboratory after taking a bachelor's degree in physics, but before completing their graduate work. They were thus a number of years older than ordinary graduate students, had become accustomed to good salaries, many of them had married and acquired children, and we knew that all the industries and government laboratories in the country would be competing for their services at the end of the war.

Gloomy predictions were being made about the deficiency of trained physicists that would result from the unwise draft policies of the country during the war. These policies, as a matter of fact, had resulted not so much in potential physics students, at least on the graduate level, being drafted into the armed services, as in their joining the Radiation Laboratory and similar organizations, and devoting their time to physical research. They, in other words, formed this generation of physicists who were, or threatened to become, a lost generation; for if they stepped immediately into well-paying positions in industry, the only future which seemed to promise them enough financial rewards to live on at the scale they were used to, they would never finish their graduate work, and while they might do well for a few years on the momentum they had acquired in the Radiation Laboratory, they would lose out in the long run on account of their lack of breadth of training.

We resolved to try by all means to get some of the most promising of these young men to stay on in the Research Laboratory of Electronics. There they could continue their graduate work, if we could establish some sort of part-time positions. But further, they could be of invaluable help to the department in two ways. First, they had an extraordinary knowledge of the microwave methods that had grown up by long experience in the Radiation Laboratory. They could apply the practical touch that the senior members, in their more executive or theoretical capacities, had not gained. Second, we looked forward to a great overcrowding of the undergraduate part of the Institute in the years after the war, as veteran students returned, and we knew that we needed a teaching staff of unprecedented flexibility to meet this teaching load. The same situation had been met after the first world war, but with less success: Then the high salaries in industry had attracted many of the best young teachers in the department, and the staff was left in a depleted state to handle the extremely large classes. We determined that this should not happen again. And since many of these young men from the Radiation Laboratory wanted eventually to become teachers, and were anxious to get teaching experience, we felt that they would form a reservoir of teaching talent which would give the department the flexibility it needed.

By making a temporary modification in the nature of the rank of research associate, which was ordinarily a postdoctoral position, we were able to set up those appointments so as to take care of the situation we had in mind. We resolved to appoint a number of predoctoral men from the Radiation Laboratory as research associates, a type of appointment of considerable standing and dignity. We hoped to offer them salaries of the order of three-quarters of what they had had in the Radiation Laboratory, feeling that they should be willing to take this much reduction for the privilege of being graduate students. Their appointments, as full-time staff members, would in accordance with the usual Institute policy allow them to take about a third of a normal load of graduate work. And their duties would be carrying on research in the laboratory, research that would contribute to the program of the laboratory but might in some cases be used for theses toward their degrees, accompanied in many cases by a small teaching load, not to exceed 6 hours per week.

The decision of the department to make appointments of this nature was all very well, but nothing could be done without money to support it. We all felt, during the last few weeks of the war, that somehow funds would be found to continue research on a large scale, but it was not obvious where the money would come from, whether from the government, industry, or other sources. Time was pressing, however. We felt that unless we were ready to offer research associateships to promising candidates as soon as the war was over, they would all take other positions almost immediately. Accordingly we put the situation up to the administration of the Institute, and they appreciated the problem, and very generously appropriated a considerable sum as a revolving fund, to allow us to offer appointments for which we saw no outside sources of funds, reimbursing the fund as soon as other financing should appear. This revolving fund made all the difference between success and failure in our program. We made a great many appointments, which were eventually covered by government financing, but which we certainly should never have been able to make if we had had to wait for the government contracts to be set up first. In fact, if the appointments had not been made in advance, we might not have had a sufficiently good staff to justify the government contract at all.

All of this negotiating had been concluded before the end of the war, and we were ready on August 15, 1945, immediately after VJ day, to send out printed notices to all parts of the country describing the new appointments. These appointments attracted much interest, not only in the Radiation Laboratory, but in Los Alamos and other centers as well. Our planning had been done sufficiently early so that we were ready with a definite plan at the crucial moment, which no other universities were. The result was that we were able to make appointments of a group of young men of unequaled quality, many from the Radiation Laboratory but many from elsewhere. We had already decided to make appointments in nuclear physics and other areas, even though our plans for research in the nuclear field had not progressed nearly as far as those in electronics. By the beginning of the first postwar term, in November 1945, we had appointed about 25 research associates, and had admitted enough other graduate students to bring the graduate enrollment up to about 90, far ahead of any prewar year. By the March 1946 term we had many more research associates in residence and a total graduate enrollment of about 125. By then, the organization of the research had progressed much further.

When we appointed our first research associates, we were operating only on the basis of the revolving fund set up by the Institute. Immediately after VJ day, however, the situation of the Electronics Laboratory was clarified. President Truman noted with alarm the intention of the Office of Scientific Research and Development (OSRD) to terminate its activities completely and suddenly. Thus the initial intention of the Radiation Laboratory had been to stop all research and development at once, and to retain only enough staff to liquidate its activity, except for a group of leaders who would write a series of books describing the activities of the laboratory. President Truman felt that by such a termination a great deal of scientific knowledge would be lost, echoing our own fears which had led to our decision to establish the Research Laboratory of Electronics. He hoped that in the course of time the armed services would be able to set up contracts to keep the sort of work which the OSRD had carried on going in peacetime. In order not to have all scientific work lapse while these contracts were being set up, he gave directions to the OSRD to have certain projects continue until June 30, 1946, and to have their attention devoted to basic research of the type more suitable for peacetime applications. One of these projects was the Radiation Laboratory; and it became at once obvious that this directive on his part, and our plans for the Research Laboratory of Electronics, complemented each other perfectly.

I shall not go into all the planning that went into practical matters, the administration of the new laboratory, providing space for it, arranging with the government to transfer a great deal of equipment to it from the Radiation Laboratory, but these matters were all taken care of. The main activities of the Radiation Laboratory ceased on December 31, 1945, and by then the Research Laboratory of Electronics was in very satisfactory and smooth operation. The army and the navy indicated interest in its program. Professor Stratton's close association with the army during the war helped in the negotiations to set up continuing government support. The net result of these negotiations was a joint contract between the Institute and the Signal Corps, with additional funds from the air force and the navy, to take effect at the expiration of the OSRD contract, in July 1946. This contract has been in effect ever since, much additional financial support has come to the Electronics Laboratory, and it has been an important part of the Institute research program ever since.

At the same time that all these negotiations were going on leading to the establishment of the Electronics Laboratory, the administrators of the department and of the Institute were very active in establishing the nuclear program on a similar basis. Here, as we have intimated earlier, the problem was in a way much more difficult, since department staff members had had much less contact with the atomic bomb project than with the microwave work. Professor Zacharias's decision to return to his field of nuclear physics proved a very great help in building up the nuclear program. His energy and initiative led first to building up a satisfactory senior and junior staff in the field, and eventually to the establishment of a Laboratory for Nuclear Science and Engineering, which he directed for a number of years.

Shortly before VJ day, Zacharias was released from the Radiation

Laboratory to go to Los Alamos to head an important division. He arrived there just after the trial explosion in New Mexico, and found life there, at that particular moment, to consist of a series of parties: celebration of the trial explosion, celebration of Hiroshima, celebration of Nagasaki, celebration of VJ day. And then that laboratory had much of the same letdown that the Radiation Laboratory did at the close of hostilities. Under the circumstances, it was natural for Zacharias's six-month stay there to turn into a good deal of a recruiting mission. He soon convinced himself that, of the senior staff there, the two men whom we most wanted for the Institute were Victor F. Weisskopf, in theoretical physics, and Bruno B. Rossi, in cosmic rays. Weisskopf, a student of Wigner's during his days in Germany, had come originally from the continent and more recently from the University of Rochester. He had been, next to Bethe, the principal theoretical physicist at Los Alamos. Rossi, originally from Italy, most recently from Cornell, had been a leader in the field of cosmic rays from the early days. The administration of the institute agreed to make offers to both of them; and after visits to Cambridge, and with the urging of department members there and of Zacharias in Los Alamos, they both accepted.

The trio of Zacharias, Weisskopf, and Rossi then began looking for junior staff, just as we had done in electronics. Our announcement of the positions of research associate was timed just right at Los Alamos. The young men there felt that at least one university understood their problems, and there, as in the Radiation Laboratory, many of the ablest of them accepted positions with us. I shall not go into details regarding how we obtained government funding, space, and equipment, but all of these things were handled promptly. Both electronics and nuclear laboratories were in good operation by the close of the academic year 1945–1946, or in less than a year after VJ day.

In our planning of the nuclear laboratory, there were two sorts of programs which we omitted: nuclear reactors and extremely big accelerators. We felt that such projects were too big for any one university. There was at that time a great movement among a number of the eastern universities to establish a joint laboratory for such work, and members of the department and the Institute, of whom I was one, took a leading part during the year 1945–1946 in the negotiations leading to the establishment of the Brookhaven National Laboratory. Professor Morse from our department was the first director, and Professor Livingston spent much time, including a year in residence there, leading to the design of the very high-energy accelerator which they built. Relations between MIT and Brookhaven have always been very close, and as I shall describe in Sec. 29, I spent a year, 1951–1952, in residence there.

We have now seen the history of the development of the interdepartmental laboratories in which members of the physics department took a leading part. It has seemed to those in charge of the department that their mode of organization solves in a very satisfactory manner the problem of setting up research laboratories in a university. They have their own independent organizations, their own administrative staffs, budgets, and accounting systems, their own technicians and other employees, so that the large administrative problems associated with carrying on research on such a scale are not faced by the department itself. Yet their scientific staffs, both on the senior and junior levels, are drawn from the departments, maintain membership in the departments, and are responsible to the department chairmen for their appointments, teaching duties, and such things. It is a system that has been worked out thoroughly during the past 25 years, and that seems to avoid difficulties which were met in postwar years by a number of other institutions, which endeavored to set up research laboratories or institutes rather too separated from the other activities of their universities. The experience I accumulated in the establishment of these two laboratories was very helpful in later years, when we came to build up the Solid-State and Molecular Theory Group, and after that the Center for Materials Science and Engineering.

28. Postwar Physics, 1945–1951

The establishment of the Research Laboratory of Electronics and of the Laboratory for Nuclear Science and Engineering, with the new staff appointments I mentioned in the preceding section, carried the MIT physics department through the next few years. Interest in physics grew greatly, as we had anticipated. My own interest during these years was gruadually changing. For several years I continued to be associated with the Research Laboratory of Electronics, since I wished to write a book on microwaves which would embody some of the background in that field which I had picked up during the war. I pushed the application of low-temperature physics to microwave electronics, a practical thing to do since Samuel C. Collins, of the mechanical engineering department, had worked out a much improved technique for making liquid helium, which we wished to exploit. Also I realized that electron and proton accelerators operated on much the same basic principles as microwave devices, and I initiated the construction of a microwave linear electron accelerator to demonstrate the principles involved.

One of my young associates in that venture, Peter Demos, became so useful that in 1961 he stepped into the position of director of the Laboratory for Nuclear Science, after serving for some years as associate director. The name "engineering" in the title of that laboratory was dropped after the chemical engineering department started teaching nuclear engineering in 1951, followed by the formal inauguration of a Course in Nuclear Engineering in 1958.

My group at the Electronics Laboratory also did some of the early work in microwave magnetic resonance, a field which developed a few years later into a major concern of solid-state physicists. There were times during those years from 1945 to 1951 when we had in the group many physicists who have since gone on to leading positions: Charles Kittel, now at the University of California at Berkeley; Francis Bitter, magnetic pioneer (now deceased); Benjamin Lax, now director of the National Magnet Laboratory named for Bitter, and professor at MIT; Arthur Kip, now at Berkeley; J. M. Luttinger, now at Columbia; Emmanuel Maxwell, who later joined the magnet laboratory at MIT, and who was one of the discoverers of the isotope effect in superconductivity, which led to the famous theory of Bardeen, Cooper, and Schrieffer; Julius Halpern, long at the University of Pennsylvania (now deceased); Charles F. Squire, now of Texas A. and M.; Michael Tinkham, now professor at Harvard; and numerous others.

During these same years, 1945–1951, it became very obvious that solid-state physics was going to turn into a major field of endeavor. The invention of the transistor, by Shockley, Bardeen, and Brattain at the Bell Laboratories, revolutionized the whole electronics industry. It was clear that solid-state theory, my first interest, was going to form the theoretical basis for a great deal of future work in pure and applied physics. I was consulting at that period for the Bell Laboratories, for a Cambridge outfit called the Photoswitch Corporation, which later changed its name to the Electronics Corporation of America, and also, somewhat later, for Raytheon. My relations to Photoswitch were particularly close, for not only was it located in Cambridge, but my colleague Wayne Nottingham was also a consultant, as was R. A. Smith, whom I have earlier mentioned as having been at Malvern in England, and in addition Karl Lark-Horovitz, of Purdue. Lark-Horovitz was an older man, who had been interested in semiconductors for a long time, and who had built up a very strong group in semiconductor experiment and theory at Purdue. I learned a great deal from these consulting connections.

During the war, the method for detecting the radar signals was the use of a crystal of silicon. This acted as a rectifier, in much the same way that the old galena, or PbS, crystals had acted in very early crystal radios. Naturally this set people thinking about the rectifying properties of a semiconductor. Bethe, who had not forgotten his early training with Sommerfeld on energy-band theory, wrote a classified report on semiconductors, which proved valuable at the Radiation Laboratory. Toward the end of the war, it was found that germanium, the element similar to silicon in the next row of the periodic table, showed even more remarkable properties than silicon. Naturally there was a great deal of study of the properties of these materials.

It was clear that their properties were strongly dependent on their impurity content, and the industrial laboratories, particularly Bell, worked very hard to develop methods of obtaining crystals of great purity, or of known amounts of impurities. The group at Bell and that at Purdue were the leading ones in this study of semiconductors, and the invention of the transistor at Bell was a direct outcome of the study of the properties of the energy bands which went on there, principally under Bardeen and Shockley. Through my connections with Bell and Photoswitch, and thus indirectly with Purdue, I was following these developments closely. The Photoswitch work was with infrared photodetectors, principally PbS, and that involved the same sort of study.

The way in which my interests were changing was shown very clearly by several conferences which I attended, or helped organize, in the late spring and summer of 1948. In the first place, that was when I was starting with

Photoswitch, and in May 1948, Nottingham and I went to visit Purdue to get a thorough view of the research going on there under Lark-Horovitz. Then in early June, I helped with the organization of a conference on low-temperature physics, to be held at Shelter Island, on Long Island, organized by D. A. MacInnes of the Rockefeller Institute. We had talks by F. London, Van Vleck, and others, and much informal discussion on superconductivity, adiabatic demagnetization, liquid helium, and related problems. Some of our work in the group at the Electronics Laboratory had dealt with superconductivity at microwave frequencies, a subject which was soon taken up by Pippard at Cambridge University, and proved to throw a good deal of light on the problem of superconductivity.

But then the following week, in Cambridge, we had an MIT conference on high-energy accelerators, which I had organized. We had discussions of proton synchrotrons, electrostatic accelerators, linear accelerators, and also of the problems in the high-energy range which could be investigated by accelerators. Among the participants were Feynman, then at Cornell, Breit, Panofsky, Haworth, Hansen, Williams, Kerst, and many others.

All this was preliminary to a trip to Europe, the first I had made since the war, which came in late June and July, and included several more conferences, one after the other. The first of these was an international conference on low-temperature and high-energy physics at Zurich. Immediately following that came an Amsterdam conference on low temperature, followed by a meeting of the general assembly of the IUPAP, the International Union of Pure and Applied Physics, also in Amsterdam. I was asked to be one of the American delegates to this IUPAP general assembly, which is what decided me to go to the various meetings. The Amsterdam meeting was to be followed by a conference at Oxford, on microwave spectroscopy. All of this seemed very interesting, and I made plans not only to go to the meetings, but to visit various laboratories as well. Al Hill decided to go along, and we traveled together, my only Atlantic crossings on the Queen Mary and the Queen Elizabeth.

After landing at Southampton, I went to Malvern, where I had a chance to talk to Smith about Photoswitch business and about solid-state physics in general. Then I went to the Zurich meeting. This was the first chance I had had since the war to see many of the leading European figures: Heisenberg, Peierls, Mott, and numerous others, and the 80-year-old Sommerfeld, who gave a genial speech after the dinner. Pauli at the time was living in Zurich, and he had an evening party at his apartment near the Zurichsee. It was a very relaxed meeting, at which the stresses of the recent war were largely forgotten. Next was the Amsterdam meeting, not merely about low-temperature, but about solid-state physics as well. Before it was well under way, I worked in a one-day visit to Delft, to visit Kronig, whose laboratory was there. At the first Amsterdam session, he gave a general survey of the band theory of metals. The Kronig and Penney paper of 1931, relating to a simplified model of energy bands, is known to every physicist. Other talks were by Mott, Jones, Bragg, Pauling, Gorter, Borelius, Van Vleck, and many others. They provided a good survey of the current state of solid-state physics.

At the general assembly of the IUPAP, to my amazement, I was elected one of the vice-presidents of the organization, and this involved almost countless trips across the ocean during the following six years while I was vice-president. Just after this, however, while I was still in Amsterdam, I found myself feeling decidedly unwell at an evening reception. I happened to look into a mirror, and found unmistakably that I was coming down with spots which I felt must be measles, a disease I had never had. I was right in my diagnosis. I thought back to see when I could have caught it: Just about the proper number of days before, during the Zurich meeting, I had taken a ride on the steamboat on the Zurichsee, and many young Swiss children were on the boat with their elders. I must have caught it from the crowds on the boat.

Anyway, my kind Dutch hosts nursed me through the disease, and in a week I was ready to travel again, but I had missed the whole schedule of the Oxford conference and the visits to laboratories which I had planned in connection with it. However, I went to Oxford anyway, a few days after their conference was over, had a very pleasant visit there with Simon, the low-temperature man, and Lord Cherwell, who was Churchill's scientific adviser during the war, and managed to return with Al Hill according to schedule. I also worked in a two-day visit to Cambridge, where Sir Lawrence Bragg had asked me to stay at his house; he was then the Cavendish Professor. It was a pleasure to see the great man in his home surroundings. When I came downstairs in the morning, I found him and Lady Bragg in the kitchen, she cooking the eggs, he preparing the bacon. They were always the informal type, easy to get on with.

By the time these conferences in the summer of 1948 were over, I had rather definitely made up my mind to return to solid-state theory. There had been several advances made in 1947, quite independent of each other, which showed that the switch of interest of physicists to nuclear problems had not entirely interrupted the older lines of thought. One of them, which I did not know about until 1949, was a calculation by Per-Olov Löwdin, of Uppsala, of the cohesive energy of the alkali halide crystals.

I first heard about Löwdin and his work from Ivar Waller, of Uppsala, whom I had known many years before when he had visited Harvard for a few weeks in 1933. One day in 1949 Waller appeared in my office at MIT, full of enthusiasm for his student Löwdin, who had taken his degree not long before. What Löwdin had done was to make a complete Hartree-Fock calculation of the energy of an alkali halide crystal as a function of lattice spacing. He had calculated all the difficult exchange integrals, and his result was to find the energy as a function of the volume, and hence the lattice spacing, dissociation energy, compressibility, and in fact a complete calculation of the pressure-volume relations up to a high pressure. He was intercomparing with my old 1923 experiments, and with more recent ones which had been made in Bridgman's laboratory and elsewhere, and the agreement was remarkable. Needless to say, I was enormously pleased. I immediately gave a colloquium talk on the subject, and looked forward to meeting this remarkable young man whom Waller had praised so highly. He will appear many times in the rest of this volume.

A second very interesting development was the paper by J. Korringa, a young Dutchman who had studied with Kronig in Delft, had then gone to Leiden, and is now at Ohio State University. He wished to consider the solution of the Schrödinger equation in the muffin-tin potential, but did not wish to use the APW method which I had suggested in 1937. He preferred to go back to methods which Ewald and other x-ray pioneers had used to study x-ray scattering in the period around 1916. They had used a multiple-scattering method, in which an incident wave strikes an atom and sends out a scattered wave which can be expanded in spherical harmonics of the angles and radial functions of the form of spherical Bessel functions. Each of these scattered waves in turn would be scattered by each atom, and this multiple scattering was treated as a problem in wave theory. Instead of thinking of the scattering by atom after atom, the way one handles this problem mathematically is the following. One can expand the sum of all the incoming waves approaching a given atom from all other atoms in terms of spherical waves centered on the atom in question, by use of powerful theorems relating to the spherical Bessel functions and spherical harmonics. Then one demands that the solution of the wave equation outside the sphere representing the atom in question, set up in this way, should join smoothly onto the wave inside the sphere.

This scheme could be easily adapted to the Schrödinger equation problem which one met with Bloch waves, and Korringa set up the necessary formalism for doing so. Much of it resembled the Morse-Allis scattering technique of 1930, which we described in Sec. 23. I was very much pleased with this scheme. Its relation to my APW method was straightforward. Inside the spherical atoms, one used exactly the same expansion of the wave function in terms of products of radial functions and spherical harmonics which I had used, but outside, one superposed these spherical wavelets emerging from each atom of the crystal, rather than using a linear combination of plane waves as I had done.

The advantage of Korringa's scheme was obvious: The superposition of spherical wavelets was an exact solution of Schrödinger's equation in the region between the atoms, which my superposition of plane waves was not. Either expansion could give a representation of the same function, but since Korringa's expansion automatically satisfied Schrödinger's equation, and was not encumbered by having components which did not satisfy this equation, it was more rapidly convergent, and thus would be expected to be easier to compute. I at once gave a colloquium talk about this method, but unfortunately the time was not ripe to make use of it, either at MIT or elsewhere. Korringa's paper was for some years practically forgotten, and he did not make any practical applications of it. Kohn and Rostoker, in 1954, unaware of his work, rediscovered the same method, and it is now known as the KKR (Korringa-Kohn-Rostoker) method, and has come into general use. We shall hear much of it in later parts of this volume, for it forms the basis for the so-called multiple-scattering method which we are now finding to be very practical for molecules, as well as for crystals.

The third of the contributions made in 1947 was by Bethe and his student von der Lage. Bethe by then had gone to Cornell, and was interested almost entirely in nuclear and high-energy problems, but he and von der Lage made this interesting contribution to the use of the cellular method. The main part of their paper is devoted to setting up linear combinations of plane waves which have the correct behavior to form symmetry functions for the type of symmetry one finds in a crystal. This is a step which every worker with energy-band theory has had to take in one way or another. But another part of the paper was devoted to the best method of satisfying the boundary conditions for the cellular method, to get continuity of the function and its derivative over the faces of the Wigner-Seitz cell.

The reader will remember that Wigner and Seitz and I had satisfied boundary conditions at the centers of the faces of the polyhedron. What von der Lage and Bethe did was to satisfy the conditions instead around a circle that was the intersection of the cell and of a sphere of the same volume, as shown in Fig. 28-1. They tested their method by applying the empty-lattice test. They found that it was satisfied to a high accuracy, whereas the method Wigner and Seitz and I had used proved to be just about the worst possible approximation. This idea of von der Lage and Bethe's seems to be coming up again in the most recent applications of the $X\alpha$ method, as is mentioned in Sec. 37.

These papers were among the factors which inclined me in the direction of returning to solid-state theory. Also, in connection with my work at Bell and Photoswitch, I felt that the experimentalists in the field of semiconductors did not understand the theory as well as they should. Bethe's classified memorandum was not generally available, and one could go much further than it did in any case. Consequently I put a general discussion of the theory of impurity states in crystals into shape, and made it the basis of an invited talk which I gave at one of Nottingham's electronics conferences, in April 1949, and which I published in the *Physical Review* later in the year. While there was not very much that was really new in this paper, it brought together many points relating to localized states in crystals, which by that time we could understand a good deal better than we had been able to do in the 1937 period, when I had last worked along those lines.

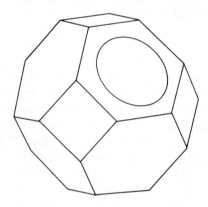

Fig. 28-1. Wigner-Seitz cell for body-centered cubic metal, showing circle of intersection of hexagonal face of cell with the sphere having the same volume as the cell. After von der Lage and Bethe, *Phys. Rev.* **71**, 612 (1947).

Another interaction which I had with the applications of solid-state physics arose in the summer of 1949. The Atomic Energy Commission asked me to undertake a study of radiation damage in solids, with particular reference to reactors. It had been known from the beginning of the work with reactors that the heavy ionizing radiation going through the materials of the reactor damaged them seriously. Wigner had been a leader in observing and understanding these troubles. In 1948, Bohr had written a fundamental paper studying the effects of a fast particle passing through matter, in the way of ionizing it, displacing atoms, and so on. Seitz, James and others had written valuable classified reports on the subject.

But the AEC felt that it would be worthwhile for me, as one who had not been concerned with the program at all, to organize a group to visit the laboratories and reactor sites, and come up with an independent report, part of which could be declassified, to give the type of information which could well be made public. I agreed to do this. We set up a group of a number of scientists from industry, government laboratories, and universities, and we spent a number of weeks during the summer of 1949 on the project. By 1951, after the study had been made and I had written up a general report of our findings, such parts of this report were declassified as were of scientific interest rather than detailed information about what went on at the various reactor sites, and it was published in the *Journal of Applied Physics.* While I had a fairly good view of what was going on in this project, I did not find myself sufficiently interested to want to carry it further.

My connection with the IUPAP started a perfect deluge of going to conferences and organizing them. There was a fascinating conference on thermodynamics and statistical mechanics in Florence in 1949. My old friend Kramers was at that time president of the IUPAP, and he and his animated daughter took it on themselves to show me the sights of Florence. The meetings were held in the Museum of the History of Science, where apparatus of Galileo and Torricelli was to be seen, apparatus so beautifully made that it would be the envy of a modern instrument maker, and where we could talk during the intermission to Signorina Bonelli, the lady who was the curator of the museum. It was interesting to read, many years later, when the disastrous floods in Florence swept through the street where that museum was located, how this same Signorina Bonelli had personally carried this apparatus over the rooftops to safety. The next September I organized a low-temperature conference at MIT. All of this was happening the same year I was organizing the AEC visit to the reactors, and when Compton resigned as head of MIT, Killian took his place, and Stratton was appointed provost, Hill becoming director and Wiesner associate director of the Electronics Laboratory.

Things kept going at the same breathless pace. I was at New York meetings, at Washington meetings, at Oak Ridge meetings. Scherrer and Amaldi were giving summer lectures in 1950 at MIT, while I was lecturing in Ann Arbor. There was a conference in October 1950 at Madison on the application of x-ray spectroscopy to solid-state problems. Mott had arranged this and presided, and it was an excellent conference. I had to give a talk at Brookhaven. There were visits to Chicago and Northwestern. I lectured at the University of Missouri. Plans were going on for meetings of the general assembly of IUPAP in Copenhagen in September 1951, with the general assembly of the International Union of Crystallography preceding it in Stockholm. Plans were also underway for a conference in June 1951, at Shelter Island, on quantum chemistry, arranged by Mulliken. And while all this was going on, I was writing a paper on the ferroelectricity of barium titanate, which von Hippel was interested in, and was finishing a book entitled *Quantum Theory of Matter,* having already written two joint books with Ned Frank, *Mechanics* and *Electromagnetism,* some two years earlier, and having finished the book *Microwave Electronics*. It was a very hectic time. But it led up to a really new and different period, starting in 1951.

I felt that the time had come to set up a research group in this field of solid-state and molecular theory, which I had decided to go into again. My first thought was to carry this through as a branch of the Research Laboratory of Electronics, but it soon became evident that that laboratory had already reached about the limit of its useful size. Consequently I decided to start an independent theoretical group, which might later form a nucleus for drawing together at MIT the whole field of the properties of materials, both theoretical and experimental. This formed an area in which there had been a good deal of interest before the war, and Bush had appointed me in the 1930s to head a committee to investigate the possibility of such an interdepartmental effort. It had come to nothing then, but by 1950 or 1951 I felt the time had come to revive the idea.

The period of 1950 was the time of the Korean War, and of increased interest in applications of science. The Institute was being pushed into greatly enlarging its practical effort in microwave electronics, the province of the Research Laboratory of Electronics. The decision was made not to incorporate this enlarged effort into the Electronics Laboratory, but instead to use the technical skill of that laboratory to start a separate and more practical laboratory, largely devoted to military research. This was the origin of the Lincoln Laboratory. That laboratory was also much interested in solid-state work, and it occurred to me to have some tie between my proposed theoretical research group in solid-state and molecular theory and the Lincoln Laboratory. I shall go further into these developments in the next several sections.

BOOK IV

The MIT Solid-State and Molecular Theory Group, 1951–1964

29. The SSMTG and Brookhaven, 1951–1952

In the preceding section, I indicated how my interests in the years before 1951 were turning back to solid-state and molecular theory, and how I felt that it would be wise to start a small research group in this area, independent of the other research laboratories which had already been set up at MIT in the immediate postwar years. This is what led to the organization of the Solid-State and Molecular Theory Group (SSMTG). This research group started in a small way in 1950, and by the beginning of 1951 it had reached the point where it was ready to break away from the Research Laboratory of Electronics, which had provided a home for it for a few months, and to start on its own. Through a grant from the Office of Naval Research we were able to start independent operation on February 1, 1951, and the first Quarterly Progress Report was issued July 1, 1951. In that report I indicated that not only would we conduct research on solid-state and molecular theory, using the facilities of quantum mechanics, but also we hoped to be a unifying influence in the whole field of the science of materials, which was widely spread through many departments at MIT. It was a number of years before this goal was really reached, but it was one of the first steps toward the establishment of the Center for Materials Science and Engineering about ten years later.

The start of the SSMTG in the spring of 1951 was complicated by a change in my own status. Dean Harrison, whom I had induced to come to the Institute as a professor of physics as my first item of business after becoming head of the physics department in 1930, had been trying to secure funds for the establishment of Institute Professorships, research professorships of a type which had become fairly common in leading universities in the postwar period. Harvard's University Professorships formed a precedent. By the spring of 1951 he had secured funds for one such endowed professorship in physics, and was looking for someone to occupy it.

At the same time, quite independently, I was invited to spend the academic year 1951–1952 at the Brookhaven National Laboratory. I have already said something of its organization, with which I had been connected several years before. Though my field was far from nuclear physics, still the phenomenon of neutron diffraction, which had been discovered as soon as neutron beams were available from reactors, forms a very valuable tool for investigating some properties of the solid state, particularly the magnetic properties which greatly interested me. I felt, and the authorities at Brookhaven also felt, that I could be of help there.

For these reasons it seemed to me that it would be a good idea for me to spend a year at Brookhaven. Another point was that I had worked very constantly, with no break, at the reorganization of the department from the time I returned from Bell Laboratory in 1944, until 1951. MIT did not have a system of sabbatical years at the time, but I realized that it was seven years since I had had a break. I therefore suggested to Harrison that he give me a year's leave from the duties of department head and professor, and let me spend the time at Brookhaven. He at once countered with the suggestion that since I had been head of the department for 21 years, it was a reasonable thing for me to give up the administrative post permanently, and step into the newly created Institute Professorship. This would give me added time for the establishment of my research group and my own research, and would give me freedom to work in an interdisciplinary way with members of other departments in building up work in the field of materials.

This seemed like a good idea all around, and I decided to go ahead with it. My colleague N. H. Frank, who had been executive officer of the physics department for some years, took over as acting department head, and after that as the department chairman, a position which he held for a number of years. I went to Brookhaven. The only problem was how to keep up the momentum I had already developed in setting up the SSMTG. With the cooperation of the Brookhaven administration, we worked out a very practical scheme: Several of the young colleagues, students and postdoctoral workers, whom I had already recruited for the group, were given temporary appointments at Brookhaven for the year, and went down there with me. We conducted a small group there, with another part of it remaining in Cambridge, and I made trips approximately every two weeks to keep liaison between the two groups. This worked much better than one might have expected, and when those of us at Brookhaven returned to Cambridge at the end of the academic year 1951–1952, the two parts of the group amalgamated very effectively. In the meantime we had been given new quarters in the physics laboratory, using space which had been vacated when the old physics-chemistry-mathematics departmental library moved into the new Hayden library building. We managed to make connection with the Lincoln Laboratory, which had started in the meantime, and several of the members of the SSMTG had the status of staff members of the Lincoln Laboratory, stationed on the campus with my group.

The division of time between Brookhaven and Cambridge made a complicated year; but even more complications arose because I was just as busy during the year with conferences and the business of the IUPAP as I had been at any time. There tended to be a three-year periodicity in this business, on account of the fact that the general assemblies of IUPAP came every three years. The same thing was true of the Union of Crystallography, which was rather closely allied with the Union of Physics. Thus 1951 was as busy a year as 1948 had been, and as 1954 was to be. But just by chance, many other conferences bunched together at about the same time. Let me list some of the meetings that came up during the year 1951, to complicate the start of the Brookhaven year.

First, on January 31–February 3, 1951, was a meeting of the American Physical Society in New York, at which I had to give an invited lecture, the Richtmyer lecture, on the electron theory of solids. Next, on February 12, was a meeting at MIT of the Solid-State Panel of the Office of Naval Research, a panel of which I was a member for many years. This panel visited laboratories doing solid-state work, and tried to carry news of what was going on back and forth between different centers. I had to talk, together with von Hippel, Warren, Egon Orowan, an expert in dislocations and similar problems who had joined the MIT staff some years earlier, and Morse. Then February 22–24 I visited Columbia, Missouri, to give a series of lectures on solid-state theory. Newell S. Gingrich, who had been with our department before the war, was at the University of Missouri, and I had promised to visit him. On March 5–10, there was a meeting of the American Physical Society at Pittsburgh, with a joint symposium of the divisions of solid-state and chemical physics on Electronic States of Molecules and Solids, with talks by Mulliken, Löwdin, and myself. Löwdin was in the United States at the time, at Duke University, and this gave me my first chance to meet him. We were already trying to work out arrangements for him to spend some time during the year with the SSMTG. Next, on March 24–27, I had agreed to talk at the Rochester meeting of the Optical Society of America. Since Rochester had been my boyhood home, and I had graduated from the university there, I felt I should accept their invitation to talk.

Next, on April 7–14, was another meeting of the same ONR Solid-State Panel in California, this time at Inyokern, San Diego, and Pasadena. On April 23–28 came the Washington meeting of the American Physical Society, with a meeting of the IUPAP United States National Committee, preparatory to the general assembly which was to take place in Copenhagen during the summer. The European trip to attend this general assembly was from June 14 to August 23. I spent the period June 23–July 5 in Stockholm, where the International Union of Crystallography was having its general assembly, to which I was given a courtesy invitation, on account of my vice-presidency of the IUPAP. It was my first visit to Sweden, including a trip to Uppsala, though since Löwdin was in the United States I did not have a chance to see him there. However, I had many friends among the crystallographers—Bragg, Ewald, Clifford G. Shull, the neutron diffraction man from Oak Ridge, Borelius, the elderly Swedish professor of solid-state physics, whom I knew well as another of the vice-presidents of the IUPAP. I had a chance to visit Erik Rudberg, whom I had known many years before at MIT, at his home in the country near Stockholm.

Then July 5–14 I was in Copenhagen. First came a conference at Bohr's laboratory on Problems of Quantum Physics. This was the only time I had been back in Copenhagen since 1924. Many of my friends were there. These included Amaldi, whom I knew through IUPAP, Bethe, de Boer, Brillouin, Darwin, Dirac, Gorter, Heisenberg, Heitler, Kotani, from Tokyo, whom I knew through IUPAP, Pauli, Peierls, Rossi and Weisskopf from MIT, and many others, as well as Bohr and Kramers. They were very much the genial hosts. Kramers at the time was president of IUPAP. Bohr was the Great Man of Denmark. Carlsberg, proprietor of one of the greatest breweries in Denmark, had left an endowment to allow the leading man of the country to live in the magnificent house which he had built for himself in the midst of his brewery, and Bohr was installed there in grand style, complete with an impressive garden and conservatory. I was invited there for a small party, as well as for the great reception, and the only way you discovered you were inside a brewery, once you had entered the house, was that on the table at meals one had not only wine, but also Carlsberg beer.

The conference itself was interesting. If it had not been for one thing, I would have thought that Bohr had mellowed completely in his old age, and I would have forgotten about my feelings concerning him and Copenhagen in 1924, more than 25 years earlier. But this one thing was an encounter following a talk which Brillouin gave at one of the sessions of the conference. Brillouin at the time lived in the United States, working for IBM, and he had gone into the general formulation of information theory and statistics. There are close relationships between the thermodynamic probability which one meets in Boltzmann's relation $S = k \ln W$, which we saw in Eq. 11-10, and information theory, and Brillouin gave an interesting exposition of these relations. When he finished, Bohr got up, and attacked him with a ferocity that was positively inhuman. I have never heard one grown person castigate another in public, emotionally and without any reason whatever, as far as I could judge, the way Bohr mistreated Brillouin. After that exhibition, I decided that my distrust of Bohr, dating from 1924, was well justified.

Following this scientific conference was the business meeting of the general assembly of IUPAP, at which I was elected vice-president for another three-year term. Then after an interval for vacationing on the continent and in England, there was a conference in Liverpool, where

Fröhlich was located, on dielectrics. I made short visits to Oxford and Cambridge, and returned home in August. It was only then that I was able to get started at Brookhaven. But almost as soon as I arrived, on September 8–10, came the Shelter Island Conference on Quantum-Mechanical Methods in Valence Theory, which Mulliken had arranged. I gave two talks, one on Determinantal Wave Functions and Magnetic Problems, and one on Self-Consistent Molecular Orbitals, which I shall mention later. The participants included most of those who were going into quantum chemistry after the war: Charles A. Coulson, a brilliant young Englishman who had entered the field during the war; Lennard-Jones, Löwdin, Raymond Daudel from Paris, Harrison Shull, Kotani from Japan, Robert G. Parr, Michael Barnett, a young associate of Coulson, Clemens Roothaan, a young associate of Mulliken from Chicago, Van Vleck, Joseph Hirschfelder, Henry Eyring, and many others. This conference really set the tone for the work going on during the next few years in the application of quantum theory to molecular structure.

That still was not the end of the year, however. On October 23–27, there was a twentieth anniversary meeting of the American Institute of Physics in Chicago, with a joint symposium in an immense auditorium, at which Fermi, Condon, and I talked. My talk was on the solid state. Also following this was a conference at Urbana, where Seitz was then living, on electrical and optical properties of ionic crystals. Then in Brookhaven, November 10–11, a meeting on neutron physics, with Clifford Shull, Bethe, Harvey Brooks of Harvard, and so on, at which I talked on radiation damage. The following week, in Cambridge, I gave the first of several talks at Raytheon; other talks came at two-week intervals, during my periodic visits to MIT. The last one came on December 14. But in between these, on December 4–7, there was a visit of the members of the SSMTG, both those from Brookhaven and from Cambridge, to Bell Laboratories, the IBM laboratory at Poughkeepsie, and the IBM Watson Laboratory in New York, where Brillouin and L. H. Thomas were located, as well as to Brookhaven. I felt that in this way the students would get some idea of what was going on in industrial and government laboratories, and at the same time would let the people in charge of those laboratories know of our plans for the SSMTG. It was a very successful visit.

Then on December 18–20, there was a conference at Brookhaven, which I had arranged, on X-Ray and Neutron Diffraction. I felt that the x-ray and neutron diffraction people did not get together enough, and did not know enough of what the solid-state theorists were doing, so I brought people from these various groups together. This small conference was rather instrumental in getting neutron diffraction people, not only at Brookhaven, but also at Oak Ridge, Chalk River, and various other installations, interested in the use of neutron scattering for investigating the thermal vibration spectra of crystals, an area of research which in the next few years produced a great deal of useful information. It also was the time when antiferromagnetism was beginning to be investigated by neutron diffraction, and the conference presented a good opportunity to talk over those problems.

The following spring of 1952 was not quite so hectic as 1951. I had to give an invited paper at the March meeting of the American Physical Society, at Columbus, and shortly after that I visited Oak Ridge, where Shull had invited me to give a talk. During this period a neutron diffraction program was getting established at Brookhaven; until that time, Shull's work with his collaborators at Oak Ridge had been the main neutron diffraction program in the country. Lester Corliss and Julius Hastings had a good program started at Brookhaven. Ray Pepinsky, from Penn State, was very interested, and frequently spent periods at Brookhaven in the following years. To finish off the Brookhaven year, I made a quick trip to Paris, on May 29 to June 9, to attend the executive committee of IUPAP, and a conference on Change of Phase. I was sorry at this IUPAP meeting not to have a final chance to see my friend Kramers. He had died in the interval since the Copenhagen meeting, and his place as president of IUPAP was taken by Mott. Almost immediately after returning from Paris, the Brookhaven year was over, and I and the group moved back to Cambridge.

This catalog of doings is meant to illustrate the fact that I was having a busy life. But nevertheless I was able to get the young members of the SSMTG interested in many aspects of their problem. In the next section I shall take up the general picture of the electronic theory of molecules and solids, which I outlined in some of the talks I have mentioned, as well as in several papers I wrote in 1951. These papers really formulate the method which we now call the $X\alpha$ method, giving it the form it maintained through the 1950s and through the larger part of the 1960s, up to the last few years.

30. The Statistical Exchange, 1951-1955

As soon as I decided to organize the Solid-State and Molecular Theory Group, I realized that I possessed a great deal of background concerning the development of quantum theory in the 1920s and 1930s, which the new students coming along did not have. I felt it desirable to pick out the features which I most wished to emphasize in the research of the group, and to write papers outlining these features in a convenient form, so that the students could get started on significant problems at once, without a long period of learning. I had no question as to what the first point to emphasize would be: it would be the statistical exchange, which was first discussed in Sec. 16 in connection with Bloch's treatment of ferromagnetism, and later in Sec. 22 in the treatment of the Wigner-Seitz method and the Fermi hole. I wrote two papers, going into these methods in detail, the first handling the nonmagnetic case, the second the magnetic case, which were sent to the *Physical Review* before the end of 1950, and which both came out in 1951. These papers in a sense mark the real beginning of the $X\alpha$ method.

I decided to base my discussion in these two papers on the Hartree-Fock method; the first paper was called "A Simplification of the Hartree-Fock Method." Thus I started with the Hartree-Fock equations 10-4, which are

$$f_1 u_i(1) + \sum (j \neq i) \int u_j^*(2) u_j(2) g_{12} dv_2 u_i(1)$$

$$- \sum (j \neq i; \text{spin } j = \text{spin } i) \int u_j^*(2) u_i(2) g_{12} dv_2 u_j(1) = \epsilon_i u_i(1) \quad (30\text{-}1)$$

These equations for an n-electron problem suffice to determine n spin orbitals u_i, where i goes from 1 to n. To derive the equations, one starts with a determinantal function formed from these n spin orbitals, computes the average energy, and varies the spin orbitals to minimize this energy. I noted that the terms $j = i$ omitted in the summations for coulomb and exchange would automatically cancel, so that one could include them without error, and I rewrote Eq. 30-1 in the form

$$\left[f_1(1) + \sum_{j=1}^{n} \int u_j^*(2) u_j(2) g_{12} dv_2 \right.$$

$$\left. - \sum_{j=1}^{n} (\text{spin } j = \text{spin } i) \frac{\int u_i^*(1) u_j^*(2) u_j(1) u_i(2) g_{12} dv_2}{u_i^*(1) u_i(1)} \right] u_i(1) = \epsilon_i u_i(1)$$

(30-2)

Equations 30-2 were ordinary Schrödinger equations for an electron moving in a given field. The first term was the kinetic energy plus the potential energy of the electron in the field of all the nuclei in the system. The second term was the coulomb energy of the electron in the field of all electrons in the system, including itself. The third term was what we now call an exchange-correlation term. In the first place, it subtracted from the remaining terms the quantity $\int u_i^*(2) u_i(2) g_{12} dv_2$, coming from the term $i = j$ in the coulomb energy. Thus it corresponded to Hartree's assumption that an electron did not act on itself, or removed the self-interaction, which does not belong in the equation. In the second place, it included all the exchange terms, which depend on spin through the restriction that it includes only terms from spin orbitals u_j whose spin is in the same direction as u_i.

I pointed out that this exchange-correlation term can be regarded as the potential, acting on electron 1 in the spin orbital $u_i(1)$, arising from an exchange-correlation charge density

exchange-correlation charge density

$$= - \sum_{j=1}^{n} (\text{spin } j = \text{spin } i) \frac{u_i^*(1) u_j^*(2) u_j(1) u_i(2)}{u_i^*(1) u_i(1)} \quad (30\text{-}3)$$

located in the volume element dv_2. This charge density integrates over all space to minus a single electronic charge, as we see by integrating over dv_2, noticing that on account of the orthogonality of the u's (which can be easily proved) the only term which does not integrate to zero is that for which $j = i$, and noticing that on account of normalization the term $j = i$

integrates to -1, showing that a single electron's charge is to be subtracted from the total electronic charge density to get the charge density producing the field acting on electron 1.

This exchange-correlation charge density is different for each value of i. I pointed out that for many purposes it is desirable to have a single potential appearing in the Hartree-Fock equations. Hence I suggested that in place of the different exchange-correlation charge densities for each i, one use a weighted mean, with a weighting factor

$$\frac{u_i^*(1)u_i(1)}{\sum(k)u_k^*(1)u_k(1)} \tag{30-4}$$

which represents the probability that an electron found at the position 1 should belong to the ith spin orbital. Thus I suggested replacing Eq. 30-2 by the alternative equation

$$\left[f_1(1) + \sum_{j=1}^{n} \int u_j^*(2)u_j(2)g_{12}dv_2 \right.$$

$$\left. - \frac{\sum_{k=1}^{n}\sum_{j=1}^{n}(\text{spin } j = \text{spin } k)\int u_k^*(1)u_j^*(2)u_j(1)u_k(2)g_{12}dv_2}{\sum(k)u_k^*(1)u_k(1)} \right] u_i(1) = \epsilon_i u_i(1)$$

$$\tag{30-5}$$

The form given by Eq. 30-5 is hardly more convenient to use than the original Hartree-Fock equations, but it gives an exchange-correlation term that represents the Fermi hole in the form suggested by the Hartree-Fock equations. To get a practical approximation, I then suggested replacing the Hartree-Fock exchange-correlation terms by the corresponding quantity for a homogeneous electron gas, as Wigner and Seitz had done in their cellular treatment, and as one does in the Thomas-Fermi-Dirac method. It seemed to me that by finding the kinetic energy properly, as well as the coulomb energy, one would do a good deal better than Wigner and Seitz and the Thomas-Fermi-Dirac method, which made the much more extreme approximation of using the homogeneous electron gas for the kinetic energy as well as for the exchange-correlation energy. Thus I went back to the work of Dirac, Bloch, and Wigner and Seitz to find an approximate form for the exchange-correlation potential.

In Fig. 22-6 we showed the form of the Fermi hole as calculated by the homogeneous gas model. It can be shown that this represents the averaged exchange-correlation charge density, which appears in Eq. 30-5, namely

averaged exchange-correlation charge density

$$= -\frac{\sum(k)\sum(j)(\text{spin } j = \text{spin } k)u_k^*(1)u_j^*(2)u_j(1)u_k(2)}{\sum(k)u_k^*(1)u_k(1)} \quad (30\text{-}6)$$

The exchange-correlation energy, which appears in Eq. 30-5, is very closely related to the exchange energy which Bloch had worked out in connection with his free-electron theory of ferromagnetism, which was described in Sec. 16. In the units we are using, the term for an electron of spin up is

$$-3\left(\frac{3}{4\pi}\right)^{1/3}(\rho\uparrow)^{1/3} \quad (30\text{-}7)$$

with a similar expression for spin down. Hence we can rewrite the proposed approximation to Eq. 30-5 in the form

$$\left[f_1(1) + \sum_{j=1}^{n}\int u_j^*(2)u_j(2)g_{12}dv_2 - 3\left(\frac{3}{4\pi}\right)^{1/3}(\rho\uparrow)^{1/3}\right]u_i(1) = \epsilon_i u_i(1) \quad (30\text{-}8)$$

for a spin orbital with spin up, and a similar equation for spin down. It was this statistical approximation which I proposed in the two papers in 1951, and which formed the basis of a great deal of calculation in subsequent years.

It was not long, however, before this derivation was challenged. In 1954, R. Gaspar, a member of Gombas' group in Budapest, wrote a paper in which he obtained the same type of dependence of the exchange-correlation energy on the charge density, but in place of the coefficient $3(3/4\pi)^{1/3}$ of Eq. 30-8, he got a coefficient only two-thirds times as great, or $2(3/4\pi)^{1/3}$. Very little attention was paid to Gaspar's paper at the time. Eleven years later, in 1965, Walter Kohn and L. J. Sham, unaware of Gaspar's work, rediscovered the same type of derivation that he had used, and again found the coefficient two-thirds times as large as mine. Only then was the discrepancy recognized by physicists in general, though I had noticed Gaspar's result when it came out, and had felt that his arguments were good.

The reason there was no notice taken of Gaspar's original work at the time was that in 1954, we had no way to make really practical use of Eq. 30-8, and until accurate calculations could be made, it was merely an academic matter which value of the exchange-correlation term should be used. By the time Kohn and Sham's paper appeared, there were very active workers making calculations of atomic and crystalline wave functions, using Eq. 30-8, and it was a matter of real importance to get the coefficient right. By then, the people making practical calculations usually tried out both values, sometimes writing the exchange-correlation term in the form

$$-3\alpha\left(\frac{3}{4\pi}\right)^{1/3}(\rho\uparrow)^{1/3} \qquad (30\text{-}9)$$

with $\alpha = 1$ for my original method, $\frac{2}{3}$ for that of Kohn and Sham. It was after that that the expression $X\alpha$ (X for eXchange, α for the parameter) came into use. But it was almost 1970 before the real significance of the discrepancy was properly understood. Consequently I shall pass over Gaspar's argument here, and shall bring it up in Sec. 33, in the discussion of the improved understanding of the method which we attained in the period around 1970.

There were, however, several additional developments concerning the exchange-correlation term which came up in the years between 1951 and 1955. First, there was the matter of the magnetic implication of the fact that the exchange-correlation potential for an electron of spin up involves $(\rho\uparrow)^{1/3}$, while we have $(\rho\downarrow)^{1/3}$ for an electron of spin down. It was of course this fact, known since Bloch's 1929 paper, that led to the discussion of ferromagnetism in the early days, as I described in Sec. 16. But by the time I made the suggestion of Eq. 30-8, in 1951, antiferromagnetism was beginning to be studied by neutron diffraction, and it was being discovered that many different substances showed this phenomenon. Consequently I made magnetic effects the topic of my principal contribution to the Shelter Island conference on quantum chemistry, in September 1951, as well as the topic of my second 1951 paper.

Clifford Shull and the neutron diffraction group at Oak Ridge were studying the antiferromagnetic behavior of the crystal of MnO and other similar oxides. At first sight this seems to have the sodium chloride structure. If one sets up planes through the crystal, perpendicular to a space diagonal or 111 direction, one plane will contain only Mn^{2+} ions, the next plane only O^{2-} ions, and so on. But closer investigation, which neutron diffraction allows, shows that alternate Mn^{2+} planes of ions behave differently: In the first such plane, all the spins point in one direction, which we may describe as spin up, while in the next plane (separated from the first by a plane of O^{2-}), all spins point down. An individual Mn^{2+} ion has five $3d$ electrons, and the experiments indicated that all five had their spins in the same direction, up in the first type of plane, down in the second type, and alternating through the crystal. Thus a single Mn^{2+} ion was in a ^6S atomic state.

The exchange-correlation potential of Eq. 30-9, then, will be numerically larger (that is, more negative) for spin-up electrons than for spin-down on those Mn^{2+} planes which are known to be occupied by spin-up electrons than on the other Mn^{2+} planes. Similarly a spin-down electron will find a lower potential energy on those planes known to be occupied by spin-down electrons. Such an alternating potential produces results which are well known from energy-band theory. In its presence, one must use a unit cell which is twice as large as one would otherwise use, since the Mn^{2+} ions with spin up act like different types of ions from those with spin down. The corresponding Brillouin zone is then only half as large as it would otherwise be, and gaps appear around the surface of this smaller Brillouin zone. The net result is that the $3d$ energy bands, which would be able to hold five electrons of each spin in the absence of the spin polarization, become split into two bands, rather widely separated from each other. Since each Mn^{2+} ion actually contains only five $3d$ electrons rather than ten, there are just enough so that only the lower of these two bands is occupied, the upper one being empty. In Sec. 35 we shall say something about recent calculations of the energy bands in cases such as this, and the method results in a quantitative description of these bands.

We can have, then, a self-consistent type of solution which shows the alternating potential and the corresponding antiferromagnetic behavior. It corresponds to having the lower bands occupied, then a gap, and upper bands empty, so that it would predict that MnO should be an insulator, as it is in fact known to be. On the contrary, if we had not taken account of the spin polarization, we should have found a $3d$ band capable of holding ten electrons per Mn^{2+} ion, which would have been only half-filled, and we would have found a conductor. Presumably, though it was not possible at the time to verify the fact, the total energy of the antiferromagnetic structure would be lower than that of a nonmagnetic structure, which would explain why one observes the antiferromagnetism in nature. It was this type of explanation which I proposed in the 1951 paper and Shelter Island report.

I mentioned in these papers that a similar idea had been used by Coulson and his student Miss Fischer in 1949, in discussing the problem of the hydrogen molecule. They pointed out that if one looks for a self-consistent solution of the H_2 molecule at large internuclear distance, in which the spin-up charge is largely concentrated on one atom, say atom a, and the spin-down charge is concentrated on atom b, one will find a lower potential for spin-up electrons on atom a than on atom b, and lower for spin-down electrons on atom b. Then the orbitals would no longer be the symmetric or antisymmetric combinations of atomic orbitals we discussed in Sec. 14, and which arise only when the potential is symmetric in the two atoms. Instead, the lowest orbital for spin up would be on the atom a, with perhaps a very slight contribution from atom b, while the lowest orbital for spin down would be largely on atom b. One can make a self-consistent solution of this type, which goes to the correct total energy at infinite internuclear distance, since each atom has just one electron, with no mixture of ionic states.

This striking example shows that by putting a restriction on the orbitals, such as the restriction in H_2 that the orbitals be either symmetric or antisymmetric, and that the spin-up and spin-down orbitals should have the same space dependence, one can in some cases be led to energies which are substantially higher than the lowest energy of which a single determinantal function is capable. It became common at this time to distinguish carefully between restricted and unrestricted Hartree-Fock calculations, and it began to be realized that an unrestricted or spin-polarized Hartree-Fock solution, such as that of Coulson and Fischer, in general will have a lower energy, and hence will be closer to the truth, than a restricted solution such as the conventional molecular orbital solution.

Naturally this intriguing type of theory was looked at with a great deal of care. Coulson and Fischer had investigated the behavior of this model, using the LCAO method, as a function of internuclear distance, and had shown that for H_2 it leads to something similar to the Heitler-London energy at large internuclear distances. At small distances, including the equilibrium separation, the ordinary molecular orbital solution with symmetric molecular orbitals gave the lower energy, so that this method led to no improvement over the standard methods of calculation. One of the early students in the SSMTG, George Pratt, worked out the details of the problem, and found the same type of results. In Fig. 30-1, we show Pratt's results for the total energy of H_2 by the unrestricted LCAO Hartree-Fock method, compared with the ordinary molecular orbital or restricted Hartree-Fock LCAO solution. The unrestricted curve merges with the restricted curve at an internuclear distance of about $2.30a_0$.

One topic that was much discussed at this 1951 Shelter Island conference was correlation energy, the effect by which, in the helium atom, the two electrons are kept apart, even though they have opposite spin, and are not affected by the exclusion principle. Lennard-Jones discussed this problem for molecules, as did Mulliken, Parr, and others. The distinctions between radial correlation, angular correlation, and various other types, were brought out carefully. Also it was clearly realized that the use of configuration interaction, a wave function made up as a linear combination of many determinantal functions, was capable of giving an exact description of correlation, without the need of including terms in r_{12} explicitly in the wave function, as Hylleraas had done.

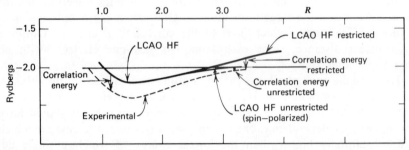

Fig. 30-1. Comparison of LCAO HF restricted and unrestricted (spin-polarized) ground-state energies for H_2. From G. W. Pratt, Jr. (unpublished).

I remember a spirited discussion, not reported in the proceedings of the conference, between myself and Löwdin as to the proper definition of correlation. I was upholding the view that the correlation energy was such an unclear concept, so poorly defined, that I felt that it would be better to abandon the term altogether. Löwdin, on the contrary, suggested what seemed at first sight to be an obvious definition. He pointed out that a Hartree-Fock wave function certainly contained no correlation effects. Consequently one could define the correlation energy as the additional lowering of the energy of the system when one starts with a Hartree-Fock solution, and then superposes enough configuration interaction to get an exact solution. In other words, it was simply the energy difference between the Hartree-Fock energy and the true energy. It is this point of view that has come into general use.

I did not argue further, but I felt that Löwdin had not really settled the matter, for a reason which I can easily indicate by reference to the Coulson-Fischer work on H_2. The point was that the Hartree-Fock method is not unique, but on the contrary, it depends greatly on whether or not restrictions are applied. We know that a restricted Hartree-Fock solution for H_2, at large internuclear distances, leads to having both electrons, one of each spin, in the symmetric molecular orbital. And we know that its energy is a mixture of the correct energy of the ground state, arising from two neutral hydrogen atoms, and of an ionic state, arising from a positive and a negative ion. Thus if we use this interpretation of the Hartree-Fock solution, the whole depression of energy required to push this energy down to the correct value, as shown in Fig. 30-1, has to be regarded as correlation energy. This is a correlation energy that has a numerically larger value at infinite separation than at finite distances. On the other hand, the unrestricted Hartree-Fock solution of Coulson and Fischer, with one electron of spin up on one atom, one with spin down on the other, goes to the correct energy at infinite separation, so that the correlation energy falls to zero at this limit. One must keep this situation constantly in mind in speaking of correlation energy, and unfortunately many of those who work in the field are not very clear about this point. Fortunately both Löwdin and Coulson, who have done much work in the field, understood it fully, but it is not always well brought out in discussions.

There is one thing concerning these solutions of the unrestricted Hartree-Fock method which was realized at once. In general, they do not have the correct symmetry properties to describe a single multiplet state of the molecule with which we are working. Thus a determinantal function formed from a spin-up spin orbital on atom a, and a spin-down spin orbital on atom b, represents a mixture of a singlet and triplet wave function. It is not the product of a function of spins and a function of orbitals, which we found in Sec. 14 was correct for a function of definite multiplicity. It is easy to set up a wave function that does represent a singlet, by making a linear combination of the two functions, one the original one, and the other one with a spin-down orbital on a, a spin up on b. But the interesting thing is that the average energy of the unrestricted Hartree-Fock method, or the single determinant with different orbitals for spins up and down, as shown in Fig. 30-1, is very nearly as low as the energy of the function of definite multiplicity.

Extensions of this idea of using different orbitals for different spins were soon made. Thus, in 1953, Hurley, Lennard-Jones, and Pople suggested a scheme which had widespread applications for molecules held together by covalent bonds. We can describe it in terms of the methane molecule, which was a subject of much discussion in the SSMTG. The central carbon atom in this molecule is surrounded by four hydrogens, forming a tetrahedron. The four electrons of the carbon outside its K shell and the four hydrogen electrons furnish enough electrons to form four electron-pair bonds, between the carbon and each hydrogen. From the work which Pauling and I had done on directed orbitals, back in the early 1930s, and described in Sec. 14, we knew that from a carbon $2s$ orbital and the three $2p$'s, one could make four tetrahedral hybrid orbitals, one pointing along each of the four tetrahedral bonds. Somehow an electron from the carbon in such a bond, and the electron from the hydrogen in the same direction, should be set up with opposite spins, to form a covalent bond similar to that in the H_2 molecule. But the number of ways of arranging the electrons is enormous. During the period around 1953, when we were studying this problem, I worked out the number of determinantal functions which would be required with methane, to get the same sort of accuracy in the wave function of the ground state which one got with the hydrogen molecule. Instead of two functions which one must combine in hydrogen, it proved to be the case that 104 determinants would have to be combined to get a ground-state wave function of the right symmetry for methane.

Great simplifications could be made, however, if one were willing to use different orbitals for different spins. One could put an electron of spin up on each of the four tetrahedral carbon orbitals, and an electron of spin down on each of the hydrogens. A single determinant formed from such orbitals would have the property of giving just two electrons, one of each spin, in each of the bonds. By allowing mixing between the tetrahedral orbitals and the hydrogen orbitals, similar to that found by Coulson and Fischer, and by Pratt in H_2, one could permit the possibility that, as the hydrogens get closer to the carbon, each of the orbitals might take on the properties of a bonding orbital between hydrogen and carbon, formed from a linear combination of a tetrahedral hybrid orbital on the carbon and a hydrogen orbital. A single such determinantal function would have an energy which would go to that of four neutral hydrogens, plus a neutral carbon in a quintet state with the configuration $2s\uparrow 2p\uparrow^3$, at large distances, and would presumably approach a non-spin-polarized molecular orbital state at small distances, as in the H_2 case. Hurley, Lennard-Jones, and Pople, proceeding along these lines, showed how one could set up a comparatively simple combination of determinantal functions which not only had these properties, but would correspond to a singlet state. I was already suggesting that the use of the statistical exchange might make it possible to make actual calculations for such an approximation. But when we came to look into the difficulties involved in the calculation, we saw that it was altogether too hard to work out with the methods we were then using. This is one of the types of problem we can now handle with the $X\alpha$ method, as we describe in Sec. 36.

Löwdin visited the SSMTG for extended periods several times during the 1950s, and made important contributions to the ideas of the simplified or averaged exchange. He was much interested in the ideas suggested by the antiferromagnetic work, and this led him to the idea of "alternant molecular orbitals," which he and Reuben Pauncz worked out in detail in later years. I had made some use of this general idea in my 1930 paper on the cohesive energy of the alkali metals (we return to these ideas later, in the detailed discussion of the Xα method). But he also worked out general formulations of the self-consistent field, which he published in the *Physical Review* for 1955.

For one thing, Löwdin proved that it is always possible to introduce what he called "natural spin orbitals" u_i, which had the property that the total charge density could be given rigorously by the equation

$$\rho(1) = \sum_i (i) n_i u_i^*(1) u_i(1) \qquad (30\text{-}10)$$

where the n_i's are occupation numbers which must have values between zero and unity, measuring the part of the total charge arising from the ith spin orbital. The proof started with Eq. 24-1 for the charge density, and from the assumption that the wave function ψ was built up by configuration interaction, as a linear combination of determinantal functions made up of spin orbitals. One does not in general find the simple form of Eq. 30-10 for the charge density when expressed in terms of arbitrary spin orbitals. Rather, in addition to the diagonal terms $u_i^* u_i$, one also has nondiagonal terms like $u_i^* u_j$, with coefficients c_{ij}, so that the charge density is given by a double sum. But Löwdin proved that one can make linear combinations of the original spin orbitals, such that only the terms $i=j$ are present, as in Eq. 30-10. These natural spin orbitals play an important role in thinking about the Xα method, as we shall see later.

He made another interesting suggestion regarding the approximate self-consistent field. I had written a short paper in 1953, suggesting that the potential field of Eq. 24-2 was a very plausible one to use for a general self-consistent field. This is not very far from the coulomb plus exchange potential used in Eq. 30-5, except that Eq. 24-2 is not limited to the Hartree-Fock or single-determinantal case, as Eq. 30-5 is. Löwdin showed that it is not in general possible to find a potential depending only on position which one can use in such a one-electron equation as Eq. 30-5 to give a completely exact derivation of the spin orbitals, but on the other hand he showed that if one looked for the most accurate possible potential of this type, one would be led to something like Eq. 24-2. There are some questions about the validity of this proof, but it at least suggests that the Xα method, which closely approximates the method of Eq. 24-2 has a fairly fundamental significance. We shall be able to understand the questions about the validity of this potential better at a later point, when we have gone into detail on the question of whether α should be unity, $\frac{2}{3}$, or an intermediate value.

There was one practical application of the Xα method in the SSMTG during the years 1951–1955. Pratt, whom I have mentioned already, solved the equations of the method, Eq. 30-8, using $\alpha = 1$ for the Cu^+ ion. This was the largest atom or ion for which Hartree up to that time had computed Hartree-Fock wave functions and energies, and we felt it was a good case to use for comparing the results of the HF and $X\alpha$ methods. The agreement was fairly good. One significant aspect of Pratt's calculations was that they were the first ones made in the SSMTG with the help of an electronic digital computer. Pratt investigated a very primitive IBM machine in the registrar's office at MIT, used for tabulating student grades, and found that it could be programmed for solving Eq. 30-8. We should mention at this point that almost a decade later, Frank Herman and S. Skillman, at the RCA Laboratory, programmed the same method for a much more sophisticated digital computer, and carried through and published calculations for the whole periodic system. The results both for eigenvalues, as tested by their agreement with ionization potentials, and for eigenfunctions, as tested by the charge density, were in quite remarkable agreement with experiment. Herman and Skillman found that if α departed from the value unity by an appreciable extent, the agreement with experiment was poorer.

31. Computers, Energy Bands, and Molecules, 1951–1964

There are two aspects to the problem of the self-consistent field: the determination of the potential in which the electrons move, given the u_i's, and the integration of the Schrödinger equation, Eq. 30-1 or 30-8, to get the eigenfunctions for a known potential. At each stage of the iteration of Hartree's method, one must carry through calculations of each sort. The main attention of the SSMTG during the period 1951–1964 was devoted to the second of these problems, for it was further from a satisfactory state than the determination of the potential. For an isolated atom, like Pratt's Cu^+, we can make a separation of variables in spherical coordinates, reduce the problem to that of an ordinary differential equation for the radial function, and solve that equation very easily by methods of numerical integration, which Pratt set up in satisfactory form for use with the digital computer. But for a molecule or crystal, the problem was very much more difficult, as we have seen in Book 1.

The first really successful postwar calculation of energy bands was not done at the SSMTG, but rather was the work of Herman and Callaway, using IBM computers at the Watson Laboratory, Columbia University. They used the OPW (orthogonalized plane wave) method, on diamond and germanium. We have already mentioned Frank Herman, who was for a number of years at the RCA Laboratory at Princeton, New Jersey, and who had not yet taken his Ph.D. in 1952, when his first calculations were made. He had done his undergraduate work at Columbia, and was keeping scientific contact with Columbia and the Watson Laboratory, with RCA, and with Conyers Herring at Bell Laboratories at Murray Hill, New Jersey, all close together geographically. His degree, for which the diamond work formed the thesis, was taken at Princeton. Joseph Callaway was a somewhat younger graduate student, also at Princeton, who collaborated on the germanium work. Both of them have gone on to distinguished careers in solid-state theory.

Diamond and germanium of course were very interesting for practical purposes in the technology of semiconductors, and the energy bands which Herman and Callaway found have formed the basis of the whole study of semiconductors since 1952. The OPW method proved to be eminently fitted to handle the problem of solving the one-electron equations for these materials. Herman from the first used the $X\alpha$ exchange-correlation of Eq. 30-8, with $\alpha = 1$. Later studies, not only on these crystals but on 3-5 compounds and others, convinced him that the value $\alpha = 1$ was the one needed to get best agreements between the energy bands and experiment. In the course of time, a great many different types of experiments were brought to bear on these energy bands, and the agreement between theory and experiment was really remarkable.

We did not have at MIT any such sophisticated digital computers at our disposal as those at the Watson Laboratory, which Herman and Callaway were using, and which had the support of IBM. Our first work at the SSMTG was done with desk computers; there was in existence a group of computer operators who could do computing for us, so that the members of the SSMTG did not have to do all their own computing. But the possibilities of the calculations we could undertake were really very limited. One of the first things I did when we started operations was to suggest to one of the students, Robert Parmenter, who was interested in energy bands, that he try to use what we now call the augmented plane wave method (APW), which Chodorow had tried out before the war. But Chodorow had also been using desk computers, and had had to extend himself about as far as was practical to get the few results he did on the copper crystal. Parmenter soon concluded that techniques had not advanced very far in the meantime, and he saw no prospect that the method could be made practical. He was interested in the lithium crystal, and he explored various methods. He first tried the LCAO, or linear combination of atomic orbitals, but soon found that it was more efficient to use this method for the $1s$ or K electrons of lithium, but to expand the conduction electrons, of $2s$ type, in plane waves. We soon realized that he was simply reinventing the OPW method, from a slightly different point of view, and he went on with that. For a simple crystal like Li, the method was quite workable.

I was not interested, however, in putting much effort into the OPW method. It is well adapted to a crystal in which the orbitals can be divided quite sharply into two classes: the core orbitals, like the $1s$ of Li or diamond, or the inner shells of germanium, which do not overlap the neighboring atoms, and the outer valence orbitals. But it is very difficult to adapt it to such a case as a $3d$ transition element, for which the $3d$ orbitals have a character midway between the inner, core electrons and valence electrons. They have their maxima well inside the atom, but the tails of the wave functions extend far out. It is the overlap of these tails from one atom to another which is responsible for the ferromagnetism, but at the same time this overlap causes trouble with the OPW method. I had never lost my interest in ferromagnetism, and by the 1950s I was equally interested in antiferromagnetism. I was not willing to settle for any scheme that did not seem to be well adapted to such problems. It was with this in view that I had first suggested the APW method in 1937, and it and Korringa's scattered-wave method were the only ones that really seemed suited to attack these problems. Since we already had some experience with the APW method, I continued to press for making this method practical. But it was several years later that this became possible, and in the meantime we had developed new capacity for calculation with digital computers, through some of the work on molecules.

One of the firm principles which I used in planning the research of the SSMTG was to carry work on molecules and solids along in parallel. I had always done this; the reader will realize that even in the 1920s and 1930s I had felt that the methods to be used for these two problems were fundamentally the same, and that advances in working one of the problems would lead to advances in the other. We were one of the very few research groups in the world which consistently followed this principle, and I believe that it was one of the most important things leading to our success. Since I was interested in ferromagnetism, I looked around to see if there might be some molecular problem that would be interesting in its own right, and yet might throw light on the ferromagnetism of solids. It was at once obvious that the simplest example was a well-known one, the oxygen molecule, O_2. It had been known for years that the ground state of this molecule had a magnetic moment: It was a $^3\Sigma_g$ state. Lennard-Jones, in his first paper on the molecular orbital method in 1929, had given a qualitative explanation of why it was a triplet rather than a singlet, each of the atoms having a single net spin, and those in the two atoms setting themselves parallel to each other. It is one of the few molecules whose ground state is not a singlet. Lennard-Jones' explanation was a straightforward molecular orbital argument, analogous to an energy-band method, so that it seemed to me that this might well throw light on how to bring the details of ferromagnetism into a band theory.

Accordingly I suggested to one of the graduate students, Alvin Meckler, that he try to do a sufficiently complete job on the oxygen molecule so as to understand it in detail, from both a molecular orbital and an atomic orbital point of view. In other words, I felt that it was not out of the question to carry through the same sort of analysis for it which I had done in 1930 for hydrogen, described in Sec. 14. I wished to have him set up molecular orbitals as linear combinations of atomic orbitals (for the outer electrons), set up determinantal functions from these, solve the resulting secular equations, and actually get the energy levels of the various multiplet states of the molecule as functions of the internuclear separation. In that way we would see what atomic states they went into at large distances, and just what went into the $^3\Sigma_g$ ground state. We realized from the beginning what an enormous and difficult task this would be. The molecular orbitals we were concerned with were the $1\sigma_g$, $1\sigma_u$, $2\sigma_g$, $2\sigma_u$, each of which we assumed would hold two electrons, one of each spin, and in addition the $1\pi_u$, $3\sigma_g$, $1\pi_g$, and $3\sigma_u$, which could hold 12 electrons if each one had an electron of each spin, but which actually had to hold only eight electrons in the neutral molecule. There were then $12!/8!4!=495$ determinantal functions, instead of the $4!/2!2!=6$ determinantal functions in the H_2 problem.

Of course, the first thing we did was to analyze these determinantal functions, to see how many of them had the same symmetry as the $^3\Sigma_g$ state, the ground state which principally interested us. It turned out that there were nine such determinants, so that it would be necessary to solve a secular equation of the ninth degree for the ground state, rather than a quadratic as in hydrogen. When we began to count the number of exchange and other integrals that would have to be computed, it became obvious that it would be enormously large. Clearly a digital computer more powerful than the IBM machine which Pratt was using was needed. Meckler found that the MIT electrical engineers were developing a very advanced research computer, the Whirlwind I, to be used in testing ideas of computer design. Meckler made the acquaintance of the people working with this machine, and obtained permission to use it during nighttime hours, when they were not using it for their own research. This computer was years ahead of its time, and it made it possible for us to get into many aspects of our work before commercial computers were available. Meckler started using it in the spring of 1952, and it was still used for some of our work as late as 1959, though the MIT Computation Center installed an IBM 704 computer in 1957, which was soon replaced by a succession of larger computers, and we shifted over to them as rapidly as possible.

In spite of these advanced computers, the difficulty of the large number of exchange integrals still remained. At about this time, S. F. Boys of Cambridge University, a bachelor of retiring habits who devoted his whole life to computing methods in chemistry, was developing more powerful methods for making molecular calculations. One of the important phases of his work was that instead of expanding the atomic orbitals in a series of exponential functions $\exp(-ar)$ times powers of r and spherical harmonics of the angle, he preferred to use Gaussian exponential functions, $\exp(-ar^2)$. These functions are not so practical for making the expansions; it takes two to three times as many such functions to get a good expansion. But the great advantage which these functions have over the ordinary exponentials is that the various exchange and other integrals are very much easier to compute with them. This advantage is so great that it often pays to use them instead of the ordinary exponentials. Meckler decided to use the Gaussians, but to use a minimum set of functions, not enough to get a good approximation to the atomic orbitals, but still enough to give a qualitative idea as to how the problem would work out. By making this concession in the matter of accuracy, and by using the Whirlwind computer, the calculations became possible. Meckler's thesis, written in 1952 and published in 1953, was really a landmark in the theory of complicated diatomic molecules.

In Fig. 31-1 I give one of the figures from Meckler's calculations. This gives the computed energy levels of the $^3\Sigma_g^-$ states of the O_2 molecule (the minus sign represents a complication of nomenclature which we shall not bother to go into) as a function of internuclear distance, according to his calculations. The figure shows eight out of the nine computed levels; the remaining one comes above the top of the figure. The lowest curve in the figure is the one representing the ground state. The particular point of this figure is that the calculations are carried far enough so that one can indicate, with each curve, what states of the oxygen atom each of the molecular states goes into at infinite internuclear separation.

The oxygen atom in its ground state has two $1s$ electrons, two $2s$, and four $2p$. A shell of four equivalent $2p$ electrons, according to Hund's study of the effect of the exclusion principle on equivalent electrons, which we considered in Sec. 4, would have three possible multiplets, ^3P, ^1D, and ^1S,

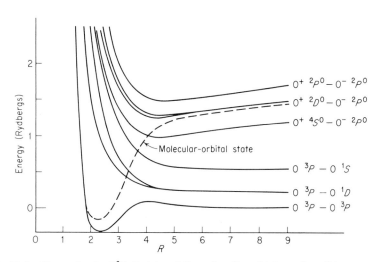

Fig. 31-1. Energy levels of $^3\Sigma_g^-$ states of O_2, as function of internuclear distance, according to Meckler. Highest level, which goes to $O^{2+}\,^3P$–$O^{2-}\,^1S$ at infinite distance, lies above the top of the figure. Dashed curve shows diagonal energy of molecular orbital state. From *Quantum Theory of Molecules and Solids*, Volume 1, by J. C. Slater. Copyright 1963 by McGraw-Hill Book Company, Inc. Used with permission of McGraw-Hill Book Company.

the ^3P being the lowest. In Fig. 31-1, we see molecular states dissociating into pairs of oxygen atoms with different combinations of these atomic multiplets: ^3P on one atom, combined with either ^3P, ^1D, or ^1S on the other. The reason why we do not see other combinations is that they do not lead to $^3\Sigma_g^-$ molecular multiplets, which are the only ones shown in the figure. But we also see molecular states which dissociate into ionic states, formed from O^{1+} and O^{1-} ions. The O^{1+} ion has multiplet states $^4S^0$, $^2D^0$, $^2P^0$, and the O^{1-} ion has only the state $^2P^0$. In addition to these states, there is one molecular state dissociating into O^{2+} and O^{2-}, of which the first has ^3P, ^1D, and ^1S, and the second has only a ^1S multiplet. This one state, which lies too high for the figure, dissociates into O^{2+} ^3P and O^{2-} ^1S. The single molecular orbital state representing the ground state, with use of the restricted Hartree-Fock method, is shown in the figure by a dashed curve, and goes to an energy at infinity, which is near the middle of the ionic levels. This is similar to the H_2 problem, where we know that the molecular orbital state dissociates into a partially ionic state.

In addition to these $^3\Sigma_g^-$ states, Meckler also computed the corresponding singlet states, $^1\Sigma_g^+$, to verify that those states really lay higher, and that the $^3\Sigma_g^-$ was in fact the ground state. This turned out all right, so that Lennard-Jones' qualitative prediction of many years earlier was verified. At the same time I made some use of this result in a talk on ferromagnetism which I gave in a conference during 1952. I noted that if Heisenberg's ideas were right, the two states would have energies which differed by the Heisenberg exchange integral, and since the triplet lay below the singlet, one would have to treat this as arising from a positive exchange integral, as described in Sec. 16. But I felt that there must be some analogy between this molecule and a ferromagnetic solid, and as I have explained earlier, I felt that as a crystal was squeezed, the exchange integral would necessarily become negative, and ferromagnetism would disappear. I suspected, in other words, that if the internuclear distance in the O_2 molecule were decreased, the energy separation between the singlet and triplet would decrease, and at some distance smaller than the equilibrium distance, the separation would vanish, and the singlet would lie below the triplet. It was possible to draw a curve for this energy difference as a function of internuclear separation, from Meckler's calculation. This curve is shown in Fig. 31-2, and it surely gives the impression that this represented the true situation.

These were very interesting results, but one of the most useful features of Meckler's work was that he had found how to get the use of the Whirlwind I for quantum-theoretical computations. It was not long before we were shifting over completely from the use of the desk computers and the more primitive IBM machine which Pratt had been using, to the Whirlwind.

This made a change in the situation regarding the use of the APW method for energy-band calculations. A student named Melvin Saffren joined the SSMTG in 1953, and indicated interest in working on this method. Since Parmenter had felt that the 1937 version of the method was too complicated to program, I devised, with Saffren's help, a modification which we thought might be more practical. Saffren programmed it for the Whirlwind, but found when he compared it with the original 1937 method, which he was also able to program, that the latter in fact was simpler. Saffren tested the method for the sodium crystal, finding results qualitatively similar to those of the cellular method. But he did not seem to be interested in carrying it further, and he took his degree and left for an industrial position, without really making use of the method.

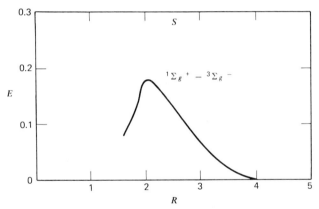

Fig. 31-2. Energy difference between magnetic and nonmagnetic state for oxygen molecule, from calculation of Meckler. From Slater, *Rev. Mod. Phys.*, **25**, 199 (1953).

In the meantime, however, another student, John Wood, joined the group, and was interested in the ferromagnetic problem. As a first step, he used the same method which Pratt had used for the Cu^+ ion, to get atomic orbitals for the isolated iron atom, and in doing so he became familiar with the use of the Whirlwind I for atomic calculations. Then he went on several years later to adapt the APW method, which Saffren had been using on the Whirlwind I, to the IBM 704, which was then becoming available in the MIT Computation Center. This program, which was not completed until about 1958, was one which he could apply to the iron crystal, and he made this APW calculation in both body-centered and face-centered iron the topic of his Ph.D. thesis in 1958, followed by a paper in the *Physical Review* in 1960. This was really the beginning of the practical calculation of energy bands in crystals involving transition elements.

Soon after this, Glenn Burdick, a graduate student who was with the SSMTG in 1960 and 1961, decided to use the APW method to get energy bands in copper. He used the potential which Chodorow had proposed before the war, and found good agreement with Chodorow for the few points which Chodorow had computed. But more important was an intercomparison with results of the Korringa-Kohn-Rostoker (KKR) method. Several workers at the General Electric laboratory at Schenectady had taken up that method and were applying it to energy bands of solids, and one of them, Benjamin Segall, was working on copper at the same time as Burdick. I was able to induce both of them to make calculations using identical potentials, so that we could get a real intercomparison of the accuracy of the APW and KKR methods. The results, which were published in 1962 and 1963, showed essentially perfect agreement between the two. This for the first time convinced workers that it was possible to solve the problem of Schrödinger's equation in a muffin-tin potential with complete accuracy.

The potentials that Wood, Burdick, and others were using in this early period were not the $X\alpha$ potentials, but rather were modifications of potentials derived from individual atoms. These were perfectly satisfactory to use in studying the adequacy of the APW and KKR methods, but we were not satisfied with them for further work. Consequently I urged that some work be done using the $X\alpha$ potential, and the first piece of work which adequately tested it was that done by Leonard Mattheiss on the atoms from argon through the $3d$ transition series.

Mattheiss joined the group in 1958, and took his doctorate in 1960, his thesis being a very interesting study of a hexagonal H_6 molecule. He then spent a year at the Cavendish Laboratory in Cambridge, and came back in 1962 for an additional postdoctoral year. He is now a member of the technical staff of the Bell Laboratories. During his postdoctoral year, he decided to follow up the work of Wood and Burdick, and to study the $3d$ transition series crystals. He set up a program for using the $X\alpha$ potential, which is still the basis for much of the use of the method, and obtained very reasonable energy bands for these elements. He used $\alpha = 1$, as Herman and his colleagues did with their OPW work on diamond, germanium, and many compounds of similar structure, and all of them got reasonable results. None of these early calculations pretended to be self-consistent. It was not until after 1964 or 1965 that really self-consistent calculations were attempted, and about that time the attention of physicists was centered on the uncertainty as to whether to use $\alpha = 1$ or $\frac{2}{3}$. Thus a new period in the energy-band calculations started about 1964, about the time I transferred my main activities to the University of Florida.

At about the same time that these improvements in the technique of energy-band calculations were being made, the experimenters were making great progress in devising experiments to check the details of energy-band structure. The study of the Fermi surface for metals, the surface in the Brillouin zone connecting all **k** values whose energy equaled the Fermi energy, was advancing by major steps. Cyclotron resonance and various other types of microwave low-temperature experiments, all outgrowths of the efforts of the late 1940s to use the new microwave and low-temperature techniques which had been developed during the war, gave large amounts of new experimental information, and the very gratifying outcome was that the energy-band calculations and these new experimental results seemed to be in good agreement with each other. This gave everyone much more confidence in the reality of the energy bands that we were computing, and it encouraged other workers to go into the calculation of energy bands.

About this time, 1964–1965, a brilliant young student at Pennsylvania State University, Terry Loucks, who had been working with the OPW method, decided that the APW method was more promising. He took his degree, moved to Iowa State University at Ames, and started a very impressive program of work with the APW method. It was only a few years later that he decided to leave the academic life and go into industry, where he has been conspicuously successful, but during the interim he and his students made striking contributions to energy-band calculations. At about the same time, J. M. Waber of Los Alamos became interested in APW calculations, and started a research program there. Waber has since moved to Northwestern University, John Wood has gone to Los Alamos, but still their energy-band programs are very active. Naturally their interest has been largely in crystals containing heavy elements, a field in which the APW and KKR methods are practically the only possible approaches. Another outside group which has gone into the field has been that in Tokyo, under Yamashita, Wakoh, Asano, and others. They also started around 1965, and have made important contributions, agreeing in major respects with the conclusions of the workers in this country.

This has given a picture of the energy-band work done in the SSMTG during the years 1951–1964. But at the same time, as I have emphasized before, we were not neglecting the atoms and particularly the molecules. I have already mentioned Meckler's work on O_2. Other similar work on diatomic molecules was done by Arthur J. Freeman on OH, by Leland C. Allen on HF, and further study of Allen, Arnold Karo, and others, on LiH and LiF. We wished to know what had to be done to get fairly good calculations of the molecular orbitals of diatomic molecules. We were not the only group working along these lines. Mulliken, at Chicago, had a young colleague Clemens J. Roothaan who had made important contributions to the field, and later work at Chicago on diatomic molecules by Enrico Clementi and others, advanced that field far enough so that we did not feel that we had to continue with it. Kotani and his group in Tokyo were also making valuable contributions to the study of diatomic and other molecules. I had a chance to visit that group, and those of other Japanese workers, during a trip to Japan in 1953.

The work that I was more interested in, however, was that on polyatomic molecules, a field in which at first almost no one was working. I wanted to bring out the connections between molecular and solid-state theory, and it was obvious that a molecule had to be a good deal more complicated than a diatomic one to show very close analogies to a solid. We analyzed the problem, as I had done with the methane molecule, and as Mattheiss had done with his hexagonal H_6 molecule, and it was clear that the problem was a prodigious one, explaining why nobody was seriously attacking it. Furthermore, I wanted to be able to deal with molecules at least one of whose atoms was a $3d$ transition or other heavy atom. This, with the LCAO methods we were using at the time, demanded analytic atomic orbitals, and Richard E. Watson, another of the students, undertook to set up such wave functions for atoms as heavy as the $3d$ transition series. This started a program of work which led him and Freeman to collaborate in studies of magnetic effects, particularly hyperfine structure, in crystals containing $3d$ atoms, a study which led to much useful work after they left the SSMTG.

We wished to analyze the group-theoretical aspects of the polyatomic molecules, so as to derive such help as we could from group theory. In this, George F. Koster, who had joined the SSMTG at the outset as a postdoctoral worker, and who soon was a faculty member, proved invaluable. He was a master of group theory, and he helped a great deal in teaching the methods to his colleagues. But the conclusion of our study of the symmetry properties of such ions as SO_4^{2-} was to show that a straightforward calculation, using Hartree-Fock methods and configuration interaction, was practically out of the question with the computational equipment we had available. Our general study of such problems was greatly helped by Robert K. Nesbet, who joined the group in 1954. He had taken his bachelor's degree at Harvard, then had gone to Cambridge University to take his Ph.D. under Boys, whom I have mentioned earlier. In his work since leaving the group, first at Boston University, now with the IBM laboratory at San Jose, he has shown himself an expert in the use of configuration interaction and other accurate methods for studying atoms and molecules.

All these efforts, and others by a number of other colleagues, had not carried us very far toward the goal of handling polyatomic molecules. In 1958, when we were faced with the prospect of reprogramming all of our Whirlwind programs for the IBM 704 and later the 709, we felt that we really needed additional help with the polyatomic molecule work. Fortunately, we found that Michael P. Barnett, a dynamic young Englishman who had taken his Ph.D. with Coulson, and whom I had met at the Shelter Island conference in 1951 and again at a similar conference in Austin, Texas, in 1955, and at other meetings, was available for a position. He had worked with Coulson on a very efficient method of calculating the molecular integrals that would be needed in polyatomic molecules. There are mathematical analogies between these methods, those which Löwdin had used in his calculations of the cohesive energy of the alkali halides, and those which arise in the KKR method. I felt it would be desirable to have someone like Barnett with this background in the group, particularly since he was an expert with computers, having spent several years with the United Kingdom branch of IBM. Accordingly we induced our sponsors, the Office of Naval Research, to bestow a special grant to allow us to get him, and to have him build up a group of several people who might be able to work out practical computer programs for polyatomic molecules. He came to the SSMTG in 1958, and was a member of it until 1964, part of the time as an associate professor.

Barnett soon had a number of people working with him, but it became evident that the MIT Computation Center was not able to provide the computing facilities which they needed. At about this time they were giving up their IBM 709, and getting a 7090 in its place, and the IBM company came to us with a proposal that they would sell us the old 709, as a used machine, at a price which was reasonable enough so that it seemed practical to establish our own computing group. Julius Stratton had become president of MIT in 1959, and he was sufficiently interested in our work so that he managed to make arrangements to do this. In 1961 this computing group, with Barnett in charge, came into operation, doing computing for the SSMTG and a number of other research groups around the Institute who could not get adequate computing time at the Computation Center. It operated for several years, and was an invaluable help in getting much computing done.

The thing which more than anything else was accomplished by this effort was the production of a set of computer programs, the so-called POLYATOM system, for carrying through the complete calculations of the structure of polyatomic molecules. This was the first such workable program that became generally available. The MIT group did not survive long enough to make much use of it, but some of those who had worked with it there, and others, have since done some excellent work with the method. Among these are Professor Jules Moskowitz, now in the chemistry department at New York University, and Professor I. G. Csizmadia, now in the chemistry department of the University of Toronto. They and their students have carried on much useful work, on organic molecules, of a type which has been carried further by Enrico Clementi, formerly with Mulliken, now at the IBM laboratory at San Jose, California, where Frank Herman and Nesbet are also now located.

We thus succeeded in our desire to get work started in the field of polyatomic molecules, though most of it was not actually done at MIT. But the experience was very illuminating. The work did not continue in the SSMTG simply because it was too expensive. We could not continue to afford our own computing center, and the members of the group received offers elsewhere which we could not meet. The costs of the calculations, in manpower and computing time, were of a sort which I could compare with the cost of corresponding calculations which we were making, in the same group and with the same computers, on crystals. It was clear that the combination of the APW method and the Xα exchange-correlation potential was making a much more efficient way of studying the properties of crystals than was the LCAO method, with analytic atomic orbitals, which was the standard method being used by the chemists, and which Barnett and his colleagues had worked out. I felt that there was no point in trying to continue the effort on polyatomic molecules, which was clearly uneconomical, but instead I resolved to try to apply the methods we were already using very successfully for solids, to molecular problems. This is the step which has only been taken in the last few years, in Florida.

The molecular work was not the only thing that inclined me in this direction. During the period around 1954, George Koster and I had looked into the problem of a localized impurity in a crystal. As I have indicated several times, and as will become very clear in later sections, this is a very important problem in solid-state theory, and yet a localized impurity is much more like a polyatomic molecule than it is like a perfectly periodic crystal. We were handling the localized impurities by Wannier functions, by methods which were essentially those of the LCAO method. It was easy to formulate the problem in general language, but when it came to computing the integrals which would be needed to make the theory quantitative, the problem was just as hard as a polyatomic molecule with the same number of atoms. Here too, I felt that we had to use the methods which had proved so successful with the energy-band problem, rather than sticking to the rather unrewarding ones which the chemists were using. But here too, it is only in the last few years that we have been able to get the new methods going.

During the years when all this scientific advance was going on, there were two other types of things which concerned me particularly. One was the wish to write down the information we were accumulating, in a form that would be useful to students and to workers in other institutions. We had started a system of Quarterly Progress Reports of the SSMTG, and this gave us our own method of making results available to the world. We did not regard it as a substitute for formal publication, but it was excellent practice for members of the group to prepare papers for the QPRs.

In addition we had a set of Technical Reports, of which I wrote a number during the period 1952–1954, trying to put the whole field into a written form which I felt would be useful. I regarded this as a first draft for later published books. It was obvious that the whole project would have to cover many volumes, and I did not want to hurry it. I talked with McGraw-Hill about the project of publishing it as early as 1952, and we decided to go ahead with it. But it was 1960 when the first publication came out of it: my two-volume *Quantum Theory of Atomic Structure*. The later volumes have followed at longer and longer intervals, made necessary by the practically breathtaking rate at which new material was being worked out. The next was the first volume of *Quantum Theory of Molecules and Solids*, in 1963, dealing entirely with the theory of molecules. Volume 2 of the same series, with the subtitle *Symmetry and Energy Bands in Crystals*, came out in 1965. The third volume, entitled *Insulators, Semiconductors, and Metals*, was published in 1967. The fourth and concluding volume of the series, entitled *The Self-Consistent Field for Molecules and Solids*, has been published in 1974. In between, I included much of the material in a more condensed form in a one-volume second edition of my *Quantum Theory of Matter*, published in 1968. These volumes between them contain essentially the whole of the work of the SSMTG, as well as much more besides, so that there is no loss to science from the fact that the QPRs and the Technical Reports of the SSMTG were never formally published.

The other type of activity during the years 1951–1964 was that leading to establishment of the Center for Materials Science and Engineering at MIT. I have mentioned earlier the steps which I had taken in helping to get the Research Laboratory of Electronics and the Laboratory for Nuclear Science established in the late 1940s. The story of the establishment of the Materials Center was a much longer one, and I shall go into it in the next section.

32. The MIT Materials Center, 1956–1964

Ever since the organization of the SSMTG in 1951, I had not lost sight of the goal of bringing together the widespread research going on at MIT in the field of the properties of materials. There was interest in Washington in such ventures. During the middle 1950s, Thomas H. Johnson was director of research for the Atomic Energy Commission. He had been head of the physics department at Brookhaven before going to the AEC, and I had known him many years before that, while he had been at the Bartol Laboratory. There are close relations between the nuclear problems of the AEC and the properties of materials, as I mentioned earlier in Sec. 28 where I was describing my work with the AEC in organizing a panel to investigate radiation damage to materials. Johnson realized the importance of the study of materials for the AEC, and during his period in Washington, on December 5, 1955, he called together a panel to look into the question of whether the AEC should support university laboratories in the field of properties of materials. This never came to anything—they decided they did not have any funds for the purpose. But it did serve to bring together a number of physicists who were deeply impressed with the need of more research on these problems. The members of this panel were Frederick Seitz, Harvey Brooks of Harvard, John Howe of North American Aviation, Morris Cohen of the metallurgy department of MIT, M. Gensamer of Columbia, A. Kaufmann, formerly of MIT, and by then running his own company called Nuclear Metals, as well as myself. Of these, Howe and Kaufmann had both been members of my 1949 study group on radiation damage.

The two MIT representatives at this meeting were myself and Morris Cohen. We began talking together after the meeting, and discovered that we had rather similar ideas about the organization of an interdepartmental laboratory dealing with materials research. Cohen and his colleagues had even made a survey of all the many research projects in the field of materials at MIT. With my experience in connection with getting the Research Laboratory of Electronics and the Nuclear Laboratory started after the war, I had fairly definite ideas as to how to proceed with such an organization. Cohen and I decided to work toward the organization of such an interdepartmental laboratory. But of course at that time funds were not nearly so plentiful as they had been in the mid-1940s, when the electronics and nuclear laboratories were started, and it was a matter of several years before we and our colleagues were able to put the materials center into operation.

The first step I took was to get a substantial annual grant from the National Science Foundation, starting in 1959, for supporting research in the properties of materials. This was the beginning of support from the NSF which I have been fortunate enough to receive ever since, up to the present. The grant was not nearly enough to allow the creation of a new laboratory, but was still enough so that I could go ahead and get the cooperation of a considerable number of members of several departments in the venture. I held organizational meetings of a selected group of staff members of the physics, chemistry, and metallurgy departments, not only to talk over the problems, but also to decide how the relatively small funds at our disposal could be used to help out various research projects in a small way. We called our organization the Laboratory of Chemical and Solid-State Physics, and issued annual reports in July of 1959, 1960, and 1961. The only obligation of members of the organization was to prepare an annual report. It had been clear that the Quarterly Progress Reports of the Research Laboratory of Electronics, of the Laboratory for Nuclear Science, and of the Solid-State and Molecular Theory Group were invaluable in bringing together the various members of those laboratories, and I felt that such reports, even if they came only once a year, might well have the same effect in our proposed laboratory.

It was not long before the possibility of further development became evident. The Advanced Research Projects Agency (ARPA) of the Department of Defense had become convinced of the value of interdisciplinary laboratories in the universities, and wished to use the field of materials as a first step in setting up such laboratories. Unlike the earlier effort of AEC, they had funds, though not as much as they had hoped. They sent a small group to talk to the universities, and I well remember a meeting I had with them, and with some of the members of the metallurgy department, early in 1959, where we discussed our ideas as to how such a laboratory should be organized. They were impressed, and on April 30, 1959, Mr. Kincaid of ARPA called me on the telephone, following this up by a letter, saying that the government was really interested in setting up interdisciplinary laboratories in the materials field, and asking a number of specific questions regarding our plans. Stratton, who was then president of MIT, appointed a small group of us to get together to formulate answers to Mr. Kincaid's questions, and on May 6 I sent him a letter, outlining our views. It was at the same time that we were preparing the proposals which resulted in the NSF grant which allowed us to set up the Laboratory of Chemical and Solid-State Physics. Also at the same time we had been doing a lot of thinking regarding a possible new laboratory building at MIT.

A good deal had been going on in the preceding years in the field of solid-state physics, indicating the need of new space. Clifford Shull had joined the MIT faculty in the fall of 1956, to make use of a reactor which we had decided to build, and which would be ready in 1957. Shull spent the year 1956–1957 at Brookhaven, on leave from the MIT appointment, while the reactor construction was going on. We built the reactor several blocks away from the main campus, in an industrial area, and Shull started his research program on neutron diffraction with the new reactor in the fall of 1957. Then there was a project for starting a magnet laboratory, inspired by Francis Bitter, with various members of the Lincoln Laboratory, including Benjamin Lax, involved in the planning. It had not been decided, in 1959 when we were planning a building, where this magnet laboratory would be built, and we worked on schemes for constructing a large building to combine the magnet laboratory, the proposed materials center, and some additional work, on Vassar Street, at the rear of the main campus. Stratton had had a committee looking into this during the spring of 1959. The final decision, resulting mostly from technical opinions of the engineers in charge of the construction, was to put the magnet laboratory adjacent to the reactor, rather than adjacent to the materials center. But in any case, much active thought about building plans was going on at this time, in the spring of 1959.

The ARPA, as a result of our letter of May 6, 1959, felt that our plans were sufficiently well formulated to suggest that we make a formal proposal for supporting a Materials Center at MIT. We did so in 1960, but were turned down. Several other universities were chosen instead. It was rather disheartening to have worked for so many years to bring something into being that presumably would have been quite easy to do in the more affluent days of the 1940s, but we could not afford to give up. Again in 1961 there was an opportunity to submit proposals to ARPA, and this time, finally, we had success.

We used the Laboratory of Chemical and Solid-State Physics as a nucleus around which to work, but a great deal of negotiation was necessary to bring some very valuable and necessary elements into the organization. In many ways the most important was the Laboratory for Insulation Research, which Professor von Hippel had established in 1936 in the electrical engineering department. A final disposition was reached in which the major part of the research in this laboratory was incorporated into the new Center for Materials Science and Engineering, which we were setting up. Much of the research in metallurgy and materials science, a good deal of electrical engineering work in the solid-state field, some chemistry, and some physics, including the SSMTG, were to be part of the newly formed center.

Most important was the size of the annual grants we received from ARPA. The annual rate was large enough so that we could use some of it for amortizing the cost of a new building. It was fortunate that we had had all the consideration of building plans which I have described, for this made it possible in the spring of 1961, when the ARPA contract was signed, to call in architects, and start very promptly with the planning of the building. The architects' suggestion was that instead of building on Vassar Street, the new building occupy vacant space in the middle of the campus, in a central location with respect to the various departments which would be involved in the research of the Materials Center, and this suggestion was universally approved. It was only in 1973 that the new electrical engineering complex of buildings, designed by the same architects, was completed, using the Vassar Street site which we had earlier suggested for the Materials Center, and it is obvious that the location of the various buildings which was finally decided on was ideal. The construction of the Materials Center building was started as soon as plans could be perfected; it was dedicated on October 1, 1965, and was named the Vannevar Bush building. This seemed like a very appropriate name; it was Bush's original appointment of a committee to coordinate materials research at MIT, of which I was appointed chariman back in 1934, which as far as I was concerned was the first step that eventually led to the new Materials Center.

The organization of the center started long before the building was completed. Its first Annual Report came out in 1962, as an expanded version of the form which we had been using for the Laboratory of Chemical and Solid-State Physics. At the time of this first report there was no director, only a committee, of which I was chairman, and of which N. J. Grant of the department of metallurgy was one of the members. A year later, R. A. Smith, from England, whom I have mentioned several times previously, came to the Institute as professor of physics, to be director of the center, a post which he held until 1969 when he returned to his native Scotland as a college president. At that time Professor Grant took over as director, a position he still holds. The soundness of our planning of this center is shown by the fact that its financing has held substantially without diminution all through the present period of financial stringency. The ARPA has now discontinued its support of these interdisciplinary materials centers, and has turned them over to the National Science Foundation. The NSF has reconsidered the laboratories in the various universities where they were located, and MIT is one of the very few in which the financing has continued even to the present at its full value.

By the middle 1960s, I felt that it was time to make plans for retirement. The year 1965–1966 would be my last full year, according to MIT rules, and it would be very difficult to contnue the SSMTG at the level it then had much beyond my retirement. My friend Per-Olov Löwdin had been so much impressed with the philosophy of the SSMTG that he had organized two similar groups himself: one at the University of Uppsala, where he is now Professor of Quantum Chemistry, and the other at the University of Florida in Gainesville. He divides his time between the two, and it is the latter group which I joined in 1964, and with which I am still affiliated. When Löwdin was considering going to Florida, in 1959, he telephoned me, and I half-jokingly told him that if he would set up a group there, I might well move down and join him. He kept reminding me, and consistently invited me to the Winter Institutes which he ran every January on Sanibel Island, off the west coast of Florida near Ft. Myers, a spot full of palm trees and pelicans. The first Winter Institute I was able to attend was in 1963. At once he and the Gainesville authorities asked me to move down. I said no, I was too busy.

But in 1964 I began to think retirement was getting close enough so that I might consider it. My wife and I both liked it, and we decided to go. The Institute worked out an arrangement so that I could be on part-time leave for the two years until my retirement, and then could continue to come for several months a year to MIT for five further years. The Florida retiring age was 70, not 65. So I managed to arrange things, and in September 1964 we moved to Florida. There in Gainesville I proceeded to set up something much like my Solid-State and Molecular Theory Group, as part of Löwdin's Quantum Theory Project. It is still going as strongly as ever. Though beyond the Florida retiring age, they have made a special dispensation, and I am still half-time on the active list. It is during the ten years that I have been in Florida that the $X\alpha$-SCF method has really come to fruition, and this is the topic of the next book.

BOOK V

*The Xα-SCF Method,
1964–1973*

33. Energy Bands, Magnetism, and the Transition State, 1964–1970

The Quantum Theory Project in Gainesville was a joint project of the departments of physics and chemistry. Löwdin had a Graduate Research Professorship (Florida's equivalent of the MIT Institute Professorship) in chemistry and physics, and I was offered a similar professorship in physics and chemistry (in each case the primary responsibility was to the first-named department). Since there was already an organization operating for the project, I did not have to build that up from the beginning. The physics department was under the directorship of Stanley S. Ballard, whom I had known for years. He had been one of the officials of the IUPAP while I was connected with it, and had later been head of the physics department at Tufts University, located in greater Boston, so I had known him as a local physicist in those days. He is an expert in optics, and had been associated with George Harrison in war work. It was not a situation of stepping into a group of unknown people.

The Florida physics department was a congenial one, with main emphasis on solid-state physics, statistical physics, and related fields. It reminded me of the MIT department in the days when I had been department head there. It was a far cry from the MIT physics department which I was leaving in the later 1960s. By then, the department had been literally captured by the nuclear theorists. Weisskopf became department head, following Buechner, in 1967, and during the 1960s so many faculty appointments were made in nuclear and high-energy theory that the department became completely unbalanced, in a way I would never have allowed while I was department head. Not only was the Florida department much better balanced than MIT's, but the administration of the university and of the state university system was much more friendly than at MIT. Stratton retired as president of MIT in 1966, to be succeeded by Howard Johnson, an expert in business administration. I felt no confidence that any research I was interested in could prosper under the atmosphere which the MIT physics department had at that time. On the other hand, Löwdin's powerful influence had already raised the Quantum Theory Project of the University of Florida to a position of prestige where it was regarded as the best example of scholarship in the state university system. Robert Mautz, who was vice-president for academic affairs when I went to Florida, and whom I came to know very well and admire very much, has now become chancellor of the Florida university system. The atmosphere in the state was friendly toward the sort of enterprise that Löwdin was trying to establish in Gainesville.

As one of the conditions of going down, it was understood that I could bring two younger colleagues as assistant professors, to help start the research project. During the first years, I arranged the appointment of James B. Conklin, Jr. He had just taken his degree at MIT on a problem in energy bands, working under George Pratt, who by then had become Professor of Electrical Engineering. He was not only familiar with APW calculations, but was a computer expert, and was invaluable in the project. We did not get a seond colleague the first year, but later Donald E. Ellis, one of our students from the SSMTG, joined us, but stayed only a year. Arthur Freeman, who had maintained close contact with the SSMTG while he had been associate director of the Magnet Laboratory, had worked closely with Ellis. He left to become head of the physics department at Northwestern University, and persuaded Ellis to join him there. It was our loss in Florida, but the result was to help build up a strong group at Northwestern, with interests similar to ours at Gainesville, so that probably it was a net gain for the country.

We fortunately had a very bright graduate student when I went down to Gainesville, a young Canadian, John W. D.. Connolly, who carried out an important piece of research during the first couple of years I was there. He then left to join the research laboratory of the Pratt and Whitney Aircraft Company at East Hartford, Connecticut, and acquired good practical experience there, but was induced to come back to our group as an associate professor when Ellis left us. In addition to these staff members, I managed to get a grant from the National Science Foundation to get our research going, and we started at once to have a number of graduate students and postdoctoral workers. Though the group never grew to quite the size the SSMTG had at its largest point, nevertheless it has been a very effective group. We have had close relations with the local computing center, which is now under the direction of Conklin, and our computing situation is in general very satisfactory.

My immediate suggestion regarding research projects was to start work on energy bands of ferromagnetic crystals, using the best $X\alpha$ techniques that were available. Connolly undertook to do the nickel crystal, which I had tried to work on back in 1936, as I described in Sec. 16. He very rapidly started the spin-polarized APW programs, with Conklin's help. This was the first time we had tried to see whether the spin-polarized potential suggested by the $X\alpha$ method was quantitatively able to explain real ferromagnetic energy bands. Connolly not only used these techniques, but from the first he iterated until he obtained self-consistency. This had hardly been done anywhere before that time.

His conclusions were very interesting. If he used $\alpha = 1$, as had been done with all $X\alpha$ calculations up to that time, the energy bands were not capable of explaining the ferromagnetism. But if he used a smaller value, in the neighborhood of the Gaspar-Kohn-Sham value $\frac{2}{3}$, he found good agreement with experiment. This was just at the time when physicists, following the Kohn-Sham paper of 1965, were becoming very conscious of the question as to the proper value of α. The Japanese group under Yamashita was also studying nickel at about the same time, and was coming to the same conclusion, that α definitely had to be much smaller than unity. This was the first direct indication that the smaller value was correct.

This result was definitely contrary to the indication found by Herman and Skillman in their atomic calculations and by a variety of workers in energy-band calculations, to the effect that α should be taken equal to unity. For several years this apparent contradiction puzzled us greatly. We tried to tie the difficulty up with the discrepancy between the derivation of α from a homogeneous electron gas which I had given in 1951, and the derivation which Gaspar, Kohn, and Sham had given later. My derivation definitely led to $\alpha = 1$, the other derivation to $\alpha = \frac{2}{3}$. I did not discuss the reason for this discrepancy earlier, in Sec. 30, but it is time now to do so.

The point was that the exchange-correlation energy appears in two different places in the derivation of the Hartree-Fock equation. It appears in the Hartree-Fock equation 30-1, or in its form giving an averaged exchange-correlation potential, Eq. 30-2. It was this expression which I had replaced by the statistical expression of Eqs. 30-7 and 30-8. But Gaspar and Kohn and Sham had started further back. They had started with the expression for the total energy of a determinantal function, Eq. 9-13, which also contains an expression for the exchange-correlation potential, and they had replaced it with a statistical expression at this point. Thus, in the notation we are using, they found for the total energy, computed statistically,

$$E_{\text{stat}} = \Sigma(i) n_i \int u_i^*(1) f_1 u_i(1) dv_1$$

$$+ \frac{1}{2} \int \rho(1)\rho(2) g_{12} dv_1 dv_2$$

$$- \frac{3}{2} \left(\frac{3}{4\pi}\right)^{1/3} \int \left\{ [\rho\uparrow(1)]^{4/3} + [\rho\downarrow(1)]^{4/3} \right\} dv_1 \qquad (33\text{-}1)$$

where $\rho(1) = \Sigma(i) n_i u_i^*(1) u_i(1)$, as in Eq. 30-10. We recognize that the exchange-correlation energy is the same one given in Eq. 16-6 for a homogeneous electron gas, according to Bloch's calculation of 1929.

Gaspar and Kohn and Sham then varied the spin orbitals in Eq. 33-1 to minimize the statistical energy, just as we varied spin orbitals in Eq. 10-2 to derive the Hartree equations. In the last term of Eq. 33-1, they had to vary the expression $[\rho\uparrow(1)]^{4/3}$, which contains $u_i^*(1)$, through Eq. 30-10. In doing this, one has $4/3$ times the $1/3$ power. The coefficient then comes out to be $(3/2)(4/3)=2$, instead of the coefficient 3 of Eq. 30-7. It was this discrepancy of coefficients which I had failed to notice in my derivation. In the derivation of the Hartree-Fock equations, there is no such discrepancy; the same coefficient appears both in the total energy and in the one-electron equations. One might consider that my method had given the α which would work best for the one-electron energies, while the variation method would give the best many-electron energy of the whole atomic system. But it was surely a very unsatisfactory situation if one had to use different α's for different purposes.

This difficulty was not cleared up until 1970, when I noticed a fact which had been previously overlooked. The one-electron energy, or eigenvalue, ϵ_i of the statistical one-electron equation,

$$\left\{ f_1(1) + \int \rho(2) g_{12} dv_2 - 3\alpha \left(\frac{3}{4\pi} \right)^{1/3} [\rho\uparrow(1)]^{1/3} \right\} u_i(1) = \epsilon_i u_i(1) \quad (33\text{-}2)$$

which is analogous to Eq. 30-8, is given by a different formula in terms of the total energy of Eq. 33-1 from that found with the Hartree-Fock equation. Thus the eigenvalue of the Xα method means something different from that met in the Hartree-Fock method.

Specifically, Koopmans' theorem, Eq. 10-5, shows that in the Hartree-Fock method, the eigenvalue $\epsilon_{i\text{HF}}$ equals the energy of the atom minus the energy of the ion found when the ith electron is removed, without allowing the orbitals of the other electrons to adjust themselves or relax to the new conditions found in the ion. That is, we may write

$$\epsilon_{i\text{HF}} = E_{\text{HF}}(n_i = 1) - E_{\text{HF}}(n_i = 0) \quad (33\text{-}3)$$

where n_i is the occupation number of the ith orbital, which is occupied in the atom and empty in the ion. On the other hand, one can prove straightforwardly that in the Xα method, the corresponding relation is

$$\epsilon_{i\text{X}\alpha} = \frac{\partial E_{\text{X}\alpha}}{\partial n_i} \quad (33\text{-}4)$$

The total energy, with either the Hartree-Fock or the Xα calculation, is not a linear function of the occupation numbers. Hence the finite difference of Eq. 33-3 is not equal to the partial derivative of Eq. 33-4. Thus even if the statistical method gave precisely the same value for total energy as a function of the occupation numbers as the Hartree-Fock method, which is actually true to quite a good approximation, one would find different values for the eigenvalues by the two methods. If one value represents the ionization energy correctly, the other value will not. The actual situation is that neither method is just right.

These results, which were first published in 1971, were entirely unexpected, but they cleared up the paradoxical situation which had existed for a number of years. The first thing to notice about Eq. 33-4 is that it is the same as Eq. 11-16, expressed in different notation. That is, the eigenvalues of the $X\alpha$ method satisfy the requirement which must be fulfilled if the Fermi statistics are to hold. The Hartree-Fock method does not. In other words, in the ground state of a system, as described by the $X\alpha$ method, all the orbitals with eigenvalues below the Fermi energy will be filled, all those with eigenvalues above will be empty. This does not exactly hold for the Hartree-Fock eigenvalues, though it is approximately correct. As we go through the atoms of the periodic system and examine the Hartree-Fock eigenvalues, we find the lower states to be filled, the higher ones empty, until we get to the $3d$ transition series. There the $3d$ eigenvalue lies below the $4s$, and yet the ground state has some empty $3d$ eigenfunctions, as well as one empty $4s$, one filled, in most cases. With the $X\alpha$ method, on the contrary, if we use a spin-polarized calculation, the eigenvalues are just different enough from the Hartree-Fock ones so that the occupation numbers fall into line perfectly with experiment.

It is significant that this is the first case in which the workers who were calculating energy bands ran into difficulty with the value $\alpha = 1$. This value gives energies agreeing rather closely with the Hartree-Fock values, as Herman and Skillman had discovered some years earlier. Therefore the application of the Fermi statistics to the Hartree-Fock eigenvalues, which is what everyone had been doing without justification, gave the wrong occupancy of spin-up and spin-down orbitals, and hence the wrong magnetic behavior. When we use $\alpha = \frac{2}{3}$, the order of the levels changes appreciably, the $3d$ rising with respect to the $4s$, and the occupancy leads to the correct explanation of the magnetism. But then the $X\alpha$ eigenvalues can no longer be used to calculate the energy differences involved in optical transitions. What do we do instead? The discrepancies prove to be quite large for the inner electrons, the $1s$ and $2s$, and are still very appreciable for the outer electrons of a $3d$ transition atom.

One way to proceed is to make separate self-consistent $X\alpha$ calculations for the energy of the atom and of the ion with one electron removed. The energy difference computed in this way, with $\alpha = \frac{2}{3}$ approximately, gives quite good values for the excitation or ionization energies, better than are found by the use of Koopmans' theorem and the Hartree-Fock method. The reason is that if we use Koopmans' theorem, we are varying the orbitals of the ground state to minimize the energy, but are not varying those found in the excited state. This results in the ionization or excitation energies being appreciably too large numerically. The total energy in both atom and ion, as computed by the Xα method, agrees well enough with the Hartree-Fock total energy so that this scheme of subtracting total energies is perfectly reliable. But it takes very accurate calculations, correct to many significant figures. The reason is that we are often subtracting total energies of tens of thousands of hartree units to get a difference of a few hundredths of a hartree or less. Obviously this requires many significant figures.

An alternative procedure soon occurred to me. It had not often been appreciated that the total energy of an atom or molecule could be written as a function of occupation numbers n_i, such as we have encountered in Eq. 33-1, where these numbers are capable of varying continuously between the limits zero and unity determined by the exclusion principle. There is nothing in the mathematics involved in setting up Eqs. 33-2 and 33-4 which demands that each n_i be an integer, either zero or unity, as is required in the Hartree-Fock method. The reason it is demanded there is simple. Each determinantal function in an N-electron system contains just N spin orbitals. These spin orbitals have occupation numbers of unity, while all others have occupation numbers of zero.

But a continuous dependence of the energy on the n_i's is obviously demanded if one is to give any meaning to Eq. 33-4, where one takes the derivative of the energy with respect to the n_i's. It suggested itself to me, therefore, to examine the dependence of the energy on the n_i's, regarded as continuous variables. From Eq. 33-1, we see that the first term, involving the operator f_1, is explicitly written as a linear function of the n_i's, while the second term, the coulomb energy, is a quadratic function, since the charge density depends linearly on the n_i's. But in addition to this explicit dependence, the various integrals involved in the calculation depend on the occupation numbers, since the potential energy in which the electrons move depends on the occupation numbers. Hence the real dependence of energy on n_i's is very complicated.

In trying to understand this dependence of energy on n_i's, I fortunately had the help of John Wood and the group at Los Alamos, where they had good access to a very large computer. They made calculations for a number of different nonintegral values of the n_i's, so that we could begin to plot out the nature of the functional dependence. We were in addition using a modification of the Hartree-Fock method, which we called the hyper-Hartree-Fock method (HHF), involving a linear combination of determinantal functions. This allows one to get the equivalent of nonintegral n_i's in a calculation of Hartree-Fock type. One of the workers at Los Alamos, J. B. Mann, had programmed this method, and had made calculations for all the atoms of the periodic system, using integral n_i's. His results were valuable in comparing the results of the $X\alpha$ method, which we had from Herman and Skillman's work, with the Hartree-Fock or HHF calculations. Fortunately Mann was willing to make HHF calculations of the energy of a particular atom as a continuous function of the n_i's, so that we had information about this functional dependence both for the $X\alpha$ and for HHF calculations. The results were similar by both methods, and Wood and I considered carefully the implications of the calculations.

The net result was that the energy was not far from a linear function of the n_i's, but with a quite appreciable quadratic term, and a much smaller but still finite cubic term. In the cases we studied, no further terms were required to get a good representation of the function. Even the quadratic approximation was good enough for most purposes. But in that case, one can understand the difference between the eigenvalues of Eqs. 33-3 and 33-4 very simply. The energy as a result of HHF calculations and that as a result of $X\alpha$ came out as such similar functions of the n_i's that the same formulation could be used in both cases. When we dealt with the HHF method, we defined a quantity E_i' by the equation

$$E'_{i\text{HHF}} = \frac{\partial E_{\text{HHF}}}{\partial n_i} \qquad (33\text{-}5)$$

This quantity has the same significance in the HHF method that $\epsilon_{iX\alpha}$ does in the $X\alpha$ method, and in fact it agrees rather well with the $X\alpha$ value. This is important, for it allows one to compute $E'_{i\text{HHF}}$ from Mann's tables, for atoms for which the $X\alpha$ method has not been used.

Let us now take this quadratic approximation, and find the difference between the quantities

$$E(n_i = 1) - E(n_i = 0) \qquad \text{and} \qquad \frac{\partial E}{\partial n_i}$$

in which the first gives the Hartree-Fock eigenvalue and the second gives the $X\alpha$ eigenvalue, if E represents E_{HHF} and $E_{X\alpha}$, respectively. Let us expand E as a quadratic, about the ground state $n_i = 1$. We have

$$E = E_1 + (n_i - 1)\frac{\partial E}{\partial n_i}\bigg|_1 + \frac{1}{2}(n_i - 1)^2 \frac{\partial^2 E}{\partial n_i^2}\bigg|_1 \qquad (33\text{-}6)$$

Then we find

$$E(n_i=1) - E(n_i=0) = \frac{\partial E}{\partial n_i}\bigg|_1 - \frac{1}{2}\frac{\partial^2 E}{\partial n_i^2}\bigg|_1 \qquad (33\text{-}7)$$

On the other hand, instead of expanding around the ground state, let us expand about the state with $n_i = \frac{1}{2}$, half-way between the initial and final states of the transition. We call this the transition state. Then we have

$$E = E_{1/2} + \left(n_i - \frac{1}{2}\right)\frac{\partial E}{\partial n_i}\bigg|_{1/2} + \frac{1}{2}\left(n_i - \frac{1}{2}\right)^2 \frac{\partial^2 E}{\partial n_i^2}\bigg|_{1/2} \qquad (33\text{-}8)$$

From this equation, we find

$$E(n_i=1) - E(n_i=0) = \frac{\partial E}{\partial n_i}\bigg|_{1/2} \qquad (33\text{-}9)$$

If, in other words, we use the energy as a function of n_i computed for the ground state, there will be a difference of $-\frac{1}{2}\partial^2 E/\partial n_i^2$ computed for the ground state, between the two types of eigenvalues, while if the calculation is made for the transition state the discrepancy disappears. We further verified that even when the third-order terms are considered, of the order of magnitude found in actual cases, the use of the transition state is still justified to a high accuracy. On the other hand, when the ground state is used for the basis of calculation, as in Koopmans' theorem, the error can sometimes be very considerable. The use of the transition state has proved in practice to be a great convenience in the use of the method.

To put the theorem in the form we actually use, where ordinarily two occupation numbers vary in a transition, one going from unity to zero, the other from zero to unity, we can state it as follows. If self-consistent calculations are made for a transition state in which each of these two occupation numbers is $\frac{1}{2}$, again half-way between initial and final state, then the energy difference between initial and final states equals the difference between the one-electron Xα eigenvalues of the electron that is having the transition between its initial and final quantum states.

The use of the transition state allows us to understand why such good eigenvalues were being found using $\alpha = 1$. Let us think of an ionization process, in which an inner, x-ray electron is being removed from the atom. The transition state is one in which the electron is half-removed. If we are thinking of the ionization of a K electron, we solve a self-consistent-field problem in which there are only one and a half K electrons, instead of two as in the ground state. Then these electrons will not shield or screen the effect of the nucleus as well as if we had the two K electrons present in the ground state. The electron will then be more strongly attracted to the nucleus than in the ground state. But when we use the transition state, we use α equal to something near 0.7 rather than $\alpha = 1$. This gives a numerically smaller exchange-correlation term than with $\alpha = 1$. This results in the electron being less strongly attracted than with $\alpha = 1$. The two effects, in other words, work in opposite directions, and the effect of going from ground state to transition state is approximately balanced by that of going from $\alpha = 1$ to $\alpha = 0.7$. The orbitals, however, are more accurate with $\alpha = 0.7$, and the $X\alpha$ eigenvalues are in the right relation to lead to the correct occupation numbers in the ground state. Hence we prefer the use of the transition state and $\alpha = 0.7$ to the earlier scheme of using the ground state with $\alpha = 1$.

34. The Multiple-Scattering Method, 1965–1973

In Sec. 31, I pointed out how desirable I felt it was to adapt the methods which we were using for crystals to the problem of polyatomic molecules. I suggested this in a short talk at the Sanibel Symposium in January 1965. Specifically, I suggested using the correlation-exchange correction proportional to the one-third power of the charge density, as described in Sec. 33, and in addition using a muffin-tin potential. No one took up this suggestion at once, but later in 1965, Dr. Keith H. Johnson joined our group, and he began work on it. He is a very able young worker, equally interested in physics and chemistry, who had done his undergraduate work at Princeton, and had gone on to graduate work at Temple University in Philadelphia. There he had used the KKR method for a study of the energy bands of brass, so he was familiar with the techniques of this method. He came to our group as a postdoctoral worker, spent over a year with us, and then went on to MIT, where he now holds an associate professorship in the department of metallurgy and materials science. As will appear in later sections, he is carrying on there, in the Materials Center, a program which in a very real sense is a continuation of the work I was doing earlier in the SSMTG. The latter group has practically gone out of existence, with no present members of the MIT physics department interested in continuing it.

Johnson showed how one could set up a computer program to solve the one-electron equation 33-2 in the approximate muffin-tin potential for a polyatomic molecule. Each atomic nucleus was to be surrounded by a sphere, inside which the potential was spherically symmetrical. These were called the atomic spheres (regions I). Outside these spheres was an interatomic region (region II), in which the potential was assumed to be constant. A larger sphere, called the outer sphere, was used to surround the entire molecule, and a spherical potential was assumed in the region outside this outer sphere, the extramolecular region (region III). The potentials in regions I and III were taken as the spherical averages of the actual potentials arising from the coulomb term and the exchange-correlation term in Eq. 33-2, and the potential in region II was taken as the average of the actual potential over that region.

Inside each of the atomic spheres, we can make an expansion of the wave function in spherical harmonics of the angles, and radial functions, as we did in Eq. 22-1, where we were discussing the cellular method of Wigner and Seitz. As in our discussion of Sec. 22, the radial Schrödinger equation here has two independent solutions for each l value, and for any energy E, of which one goes to zero or a finite value at the origin, while the other one goes infinite at the origin. Naturally we have to choose the solution which remains finite at the origin. Outside the sphere, in region II, we have a solution of the wave equation, which is well known. Again it can be expressed as an expansion in spherical harmonics of the angle and radial functions, where the latter are now the spherical Bessel and Neumann functions, which form the two independent solutions of the wave equation in spherical coordinates. The spherical Bessel function is regular at the origin, while the spherical Neumann function becomes infinite at the origin. But the behavior at the origin does not concern us, since we are using this solution only in region II, which does not include the origin.

The behavior of the spherical Bessel and Neumann functions at large r depends profoundly on whether they are functions of a real or an imaginary argument. We are dealing with the spherical Bessel or Neumann function of $[(E-V)r]^{1/2}$, where E is the energy and V is the constant potential energy in region II. Hence for positive $E-V$ we have a real argument, while for negative $E-V$ we have an imaginary argument. For a real argument the spherical Bessel and Neumann functions for large r are $1/r$ times the sine or cosine of $[(E-V)r]^{1/2}$ so that both of them stay finite, and any combination of the two is allowable. It is these functions which come into scattering theory. The value of u^*u approaches $1/r^2$, and this is the origin of the inverse square law for the intensity of a spherical wave. These solutions have to be considered in studying the excited states and the conduction band of a metallic conductor. However, for most finite molecules, the energy E is less than V for all bound states, so that we have to consider the behavior of the spherical Bessel and spherical Neumann functions for imaginary arguments.

For imaginary arguments, it is more convenient not to use the spherical Bessel and spherical Neumann functions directly, but instead to use multiples or linear combinations of them, which have simpler properties. Thus for the function which goes to zero at the origin, we take a function $i_l(x)$, proportional to the spherical Bessel function of ix, which has the properties

$$\lim(x=0)i_l(x) = \frac{x^l}{1 \cdot 3 \cdot 5 \cdots (2l+1)}$$
$$\lim(x=\infty)i_l(x) = \frac{\exp x}{2x}$$
(34-1)

For the function which goes to zero at infinity, we use $k_l^{(1)}(x)$, called the spherical Hankel function of the first kind, which has the properties

$$\lim(x=0)k_l^{(1)}(x) = (-1)^l \frac{1 \cdot 1 \cdot 3 \cdot 5 \cdots (2l-1)}{x^{l+1}}$$

$$\lim(x=\infty)k_l^{(1)}(x) = (-1)^l \frac{\exp(-x)}{2x}$$

(34-2)

This function is a linear combination of the spherical Bessel and Hankel functions of imaginary arguments. The quantity x is $[(V-E)r]^{1/2}$, in the case we are discussing, so that for energies E less than V, it is a real quantity.

If we were discussing a single atom, the region II outside the sphere 1 would extend to infinity, and we would have to use solutions of the type given in Eq. 34-1 or 34-2 in this region. Obviously we would have to use 34-2, since it becomes zero at infinity, whereas the function of Eq. 34-1 becomes infinite. Hence at the bounding sphere between regions I and II, we should have to have the functions inside I and II join smoothly over the surface of the atomic sphere. Both the function and its normal derivative would have to be continuous over the surface of the sphere, and it is convenient first to demand the continuity of the logarithmic derivative, $d\ln\psi/dr = (1/\psi)(d\psi/dr)$, where ψ is the wave function, and then later to demand that the function itself be continuous. This automatically brings about the continuity of the normal derivative as well. The practical way to secure this is to compute the logarithmic derivative both of the function inside the sphere I and of the function of the type of Eq. 34-2 outside the sphere I, and to plot them both as functions of the energy. The curves will intersect at certain values of E, which are the eigenvalues. For an isolated atom, it is only the radial function of one particular l value which will satisfy this condition at a given energy, so that the solution for an atom has a definite l value.

In a molecule, as Johnson was discussing it, this solution appropriate for a single atom would still be an exact solution of the Schrödinger equation inside the sphere representing the atom in question, and in the region II outside it. However, the tail of this wave function, decreasing exponentially according to Eq. 34-2, would extend out to neighboring atoms, and it would not join continuously onto solutions holding inside the spheres representing these other atoms. Instead, in the region II, the solution must be made up of exponentially decreasing tails of the form of Eq. 34-2, centered on each atom in the molecule. The sum of all these decreasing tails must represent, at the surface of each of the spheres, a function which joins continuously and with continuous normal derivative to the solution holding inside this sphere. And all of these solutions must be computed for the same E, which is an eigenvalue of the problem. One can see that this problem is very much more complicated than that of a single atom. A similar situation is found for energies E greater than the V in region II. These solutions, as we have seen, are like scattered waves, and the condition then becomes one of continuity between the wavelets scattered by all the atoms and the solution inside the atomic sphere of each atom. It is this scattering problem which is called a problem of multiple scattering, and the computational scheme of Johnson is often called a multiple-scattering method for this reason.

Fortunately there is a valuable mathematical theorem that allows this complicated problem to be worked out in an analytical way. Suppose we have two atomic centers, say a and b. Let us start with the wave function of the form of Eq. 34-2, centered on the atomic nucleus of atom b. This theorem allows us to expand this function as a linear combination of functions of the type of Eq. 34-1 centered on nucleus a, multiplied by spherical harmonics of the angles in a set of spherical harmonics centered on atom a. It is a theorem which is the basis of most of the computational methods that had been used in computing molecular problems in the past, including Löwdin's work on the cohesive energies of the alkali halide crystals and the Barnett-Coulson method of computing molecular integrals. It is also the basis of the KKR method for handling crystals, which is really a special case of the general program set up by Johnson.

By use of this theorem, one can express the complete solution of the Schrödinger equation in region II in terms of spherical harmonics of angles and radial functions of the form of Eq. 34-1, centered on a particular atom, as for example atom a. One can then easily apply the continuity condition that the logarithmic derivative of the radial function, for each l value, be continuous between this function we have been describing in the region II, and the radial function determined by numerical integration in region I for atom a. One applies this condition at the surface of each atomic sphere, and also one applies a similar condition at the surface of the outer sphere, separating region II and region III. There are as many separate equations as there are spheres, multiplied by the number of l values for which we assume nonvanishing contributions to the wave function. They are equations determining the coefficients of the radial functions of type 34-2, for each l value and each sphere, representing the exact solution of Schrödinger's equation in region II. These same coefficients also automatically determine the solutions inside the atomic spheres.

These equations for the coefficients are a set of simultaneous linear algebraic equations, which do not have a nonvanishing solution unless the determinant of the coefficients is zero. This gives the secular equation of the method. There will be a set of energies for which the determinant is zero, and we can get solutions. The only way to find these energies is to compute the value of the determinant as a function of the energy, and find where the curve crosses the axis. Obviously the programming of all these steps for the digital computer has been a major piece of work. Fortunately Johnson was able to get another of our former Gainesville postdoctoral workers, F. C. Smith, Jr., an expert programmer, to join him at MIT, and the programs now in use were the joint work of Johnson and Smith.

The programs include not only the one for solving the Schrödinger equation in the muffin-tin potential which we have been describing, but also the programs involved in attaining self-consistency: those of computing the charge density arising from the occupied spin orbitals, finding the spherical averages of charge density in regions I and III, and the constant average in region II (the so-called muffin-tin charge density approximation), and then the muffin-tin potential from this muffin-tin charge density. The final potential from one stage of iteration is fed into the starting conditions for the next stage, and the machine automatically repeats the process until the eigenvalues attain stable values, to some predetermined accuracy. Very similar programs can be used to compute the transition state self-consistent fields. There are also programs for displaying the charge density, in the form of contours, perspective drawings of density as a function of position, and so on. It is this set of computer programs which has shown itself capable of giving results more accurate than the standard procedure used by the chemists, and with very much less computer time required. Let us go on to describe some of the results which have already been attained by this method, before we look into its accuracy, and before we consider steps which are now underway to improve it.

Back in the early days of the SSMTG, when we had first thought about trying to understand polyatomic molecules, I mentioned in Sec. 31 that we looked into the problem of the sulfate ion, SO_4^{2-}, thinking that this would be a good place to start. In the early 1950s we soon found that this was quite impossible. But Johnson and Smith came back to it and to the perchlorate ion ClO_4^-, which has the same outer electrons, for their first test of the new methods. In 1970, they reported the eigenvalues of the occupied molecular orbitals for these ions, and compared them with other estimates and with experiment. These calculations came before we understood the correct value of α and the significance of the transition state, but Johnson and Smith made calculations for both $\alpha = 1$ and $\frac{2}{3}$, so that one could compare the two values.

The results they found for the occupancy of the various molecular orbitals were very interesting. In a formal way, one can set up an ionic picture of the structure of these ions. Thus, the sulfur atom has filled K

and L shells, two electrons in the $3s$ level, and four in the $3p$. The oxygen has two electrons in the K shell, two in the $2s$, and four in the $2p$. One can think of the sulfur as having a positive charge of six units, as if it had lost the two $3s$ and the four $3p$ electrons, and of each oxygen as having gained two electrons, so as to have a completed shell of six $2p$ orbitals. Thus the ions would have the same electronic structure as neon atoms, sulfur having lost and oxygen having gained enough electrons to have the same number of electrons as the neon atom. In the process, two extra electrons would have to be added, accounting for the charge on the ion. But no one would think that this extreme ionic picture should be taken very literally. It would require a tremendous amount of energy to remove six outer electrons from a sulfur atom, and two electrons cannot be held stably to form a doubly charged O^{2-} ion. However, a very large number of inorganic compounds can be described in the type of language we find from this ionic picture.

To understand what Johnson and Smith found, we must recall that the sulfate ion has a tetrahedral structure, and the types of symmetry of molecular orbitals which can be found in this symmetry can be understood from group theory. One finds that there are just five types of symmetry which are allowed. Basis functions for these five types, algebraic functions which have the same symmetry types are listed in Table 34-1. These polynomials, multiplied by functions of r, become atomic orbitals of atomic functions with the same sort of symmetry. Let us explain what this implies.

Table 34-1. Basis Functions for Irreducible Representations of the Tetrahedral Group

a_1	1 (alternatively xyz)
a_2	$x^4(y^2-z^2)+y^4(z^2-x^2)+z^4(x^2-y^2)$
e	$3x^2-r^2, \sqrt{3}(y^2-z^2)$
t_1	$x(y^2-z^2), y(z^2-x^2), z(x^2-y^2)$
t_2	x,y,z (alternatively yz, zx, xy, or $x(3r^2-5x^2)$
	$y(3r^2-5y^2), z(3r^2-5z^2)$

An atomic s function is independent of angles, and thus depends only on r (where $r^2 = x^2+y^2+z^2$). It thus has the symmetry a_1. An atomic p state has three orbitals, corresponding to the quantum numbers $m_l = 1, 0, -1$. Simple linear combinations of these three have the forms $xf(r), yf(r), zf(r)$, where $f(r)$ is a suitable function of r. Thus these three functions fall in the category t_2 in the table. A d state has five functions, corresponding to $m_l = 2, 1, 0, -1, -2$. Linear combinations of these five functions can be written in the form yz, zx, xy, $3x^2 - r^2$, and $\sqrt{3}\,(y^2 - z^2)$ times another function $g(r)$. In the spherical field of an atom, all five functions correspond to identical eigenvalues. However, in the tetrahedral field, three of them have the form t_2 and have one eigenvalue, whereas the other two have the form e with another eigenvalue. Thus we say that an atomic d state is split in a tetrahedral field into a threefold and a twofold degenerate state. This is the type of result which Bethe derived in 1929, in the study of the splitting of atomic energy levels in crystal fields which we mentioned in Sec. 20.

We can go on to higher l values from the results of Table 34-1. The polynomial for a multiplet arising from an atomic state of a given l value is of the lth degree in x, y, and z. This is why we had a function independent of x, y, and z for the s state, linear in x, y, z for a p state, and quadratic for a d state. For an f state we must have seven functions of the third power in x, y, and z, and these are an a_1, the three degenerate states of a t_1, and the three of a t_2. This is as far as we have carried Table 34-1, but we see that no atomic states with l less than 6 can lead to an a_2 level in the tetrahedral field. We now see that if we have the sulfur at the origin of coordinates, each s orbital will lead to an a_1 orbital in the molecule, each p will lead to a triply degenerate t_2, and there will be no functions of other symmetry arising from its occupied orbitals.

In addition to these levels arising from sulfur orbitals, however, we also have levels arising from the oxygen orbitals. Their analysis is much more complicated, since they are not located at the origin. For instance, if we have an oxygen s orbital, the molecular orbital which is the sum of the oxygen s's on all four atomic sites will be of a_1 form, but in addition we can make three linear combinations of these s's which lead to a t_2 symmetry. We shall not try to go further with the analysis, but obviously it must be worked out completely to give a proper description of the molecular orbitals which are occupied in the ion.

We can now state very simply the results of Johnson and Smith's study of the occupied molecular orbitals of the sulfate ion. They have the symmetries of those molecular orbitals which could be constructed by the LCAO method from the atomic orbitals of the S^{6+} neon-like ion and from the atomic orbitals of the four O^{2-} neon-like ions. In other words, the LCAO method based on an ionic picture gives a correct description of which orbitals lie lowest, and therefore are occupied in the ground state of the ion. But when we examine the actual construction of these molecular orbitals, we find that the occupied orbitals are such linear combinations of sulfur orbitals, and oxygen orbitals combined to have the same symmetry type, that they have the characteristics of bonding orbitals. In other words, instead of having the charge entirely concentrated on the sulfur, or on the oxygens, as if there were no mixing, we have an overlap between the sulfur and the oxygens, with a very considerable amount of overlap charge. This means that there is a good deal of charge on the sulfur, arising from the sulfur contribution to molecular orbitals which on the ionic picture would be completely concentrated on the oxygen. When one calculates the total amount of charge on the sulfur, instead of being six positive charges, it is a fraction of a unit, and the total charge on an oxygen, instead of being two negative charges, is a fraction of one unit.

These are the results which chemists had expected to find, and which earlier less complete calculations by other authors had found. They are characteristic of what is found in all cases of this sort, in which a reasonable ionic picture can be given for the structure of the ion or molecule. They give the explanation of the way in which an actual bond is not entirely ionic, but rather is partly ionic, partly covalent. Similar results were being found at about the same time for calculations on a number of crystals with the sodium chloride structure, which were being carried out by Conklin and his students, with the APW method, in Gainesville. These crystals, of which TiC is an example, again could be described on an extreme ionic basis, in this case as $Ti^{4+}C^{4-}$, and the occupied energy bands are such as would arise from this structure. The charge distribution, however, corresponds much more closely to a combination of neutral atoms than to the highly ionic description.

This first example of the use of the $X\alpha$ method for molecules, dating back to 1970, is enough to give a good idea of some of the results which the new method showed itself capable of giving. This was the beginning of a really explosive development. With each new molecule or ion considered, it became more obvious that the results were in better agreement with experiment than any of the previous calculations, and at the same time one could compare the computing time with other work which had been carried through in recent years, by the LCAO method, sometimes with the Hartree-Fock method, sometimes involving configuration interaction. The computing times by the $X\alpha$ method proved to be from 100 to 1000 times less than by the other methods. This verified my impression, dating back to the experience in the SSMTG in the early 1960s, that the combination of the $X\alpha$ exchange-correlation plus the muffin-tin approximation to the potential was a very much more efficient computing device than the methods the chemists were using.

Naturally the method attracted great interest. Workers in other laboratories around the world were anxious to get hold of the programs and use them. Everyone working with them was looking for applications that would be of practical value. By now the list of molecules and ions that have been worked on is very extensive, and it requires a list a page long even to say who is working on the method, and at what laboratory. We shall come back to these applications in Sec. 36. But before we do so, there are other directions in which the method has evolved, and we wish to discuss them. First we come to the use of the $X\alpha$ expression for total energy, to discuss cohesive energy and magnetic problems.

35. Total Energy and Magnetic Problems, 1967–1973

As soon as we realized that the method as we were using it was derived by applying a variation method to the total energy, it was obvious that it would be of very great interest to compute this total energy and verify its accuracy, rather than confining our attention solely to the one-electron equations and one-electron energies, as given in Eq. 33-2. The total energy obtained by introducing the parameter α into Eq. 33-1 is

$$E_{X\alpha} = \Sigma(i) n_i \int u_i^*(1) f_1 u_i(1) dv_1 + \frac{1}{2} \int \rho(1)\rho(2) g_{12} dv_1 dv_2$$

$$- \frac{9}{4}\alpha \left(\frac{3}{4\pi}\right)^{1/3} \int \left\{ [\rho\uparrow(1)]^{4/3} + [\rho\downarrow(1)]^{4/3} \right\} dv_1 \qquad (35\text{-}1)$$

In a way, this equation had been tested some years before by Peter DeCicco, in his Ph.D. thesis in the SSMTG in 1967. DeCicco had used Eq. 33-2, with $\alpha = 1$ and with the muffin-tin potential, to compute energy bands and wave functions for the KCl crystal. Then he had used Eq. 35-1, with $\alpha = \frac{2}{3}$, to find the total energy as a function of lattice spacing. This was before we properly understood the features underlying the choice of α, and his orbitals were certainly not as good as he would have found if he had used $\alpha = \frac{2}{3}$ in Eq. 33-2. But we must remember that the energy expression of Eq. 35-1 has a minimum property: The lowest value will come for just the right spin orbitals, but if slightly wrong spin orbitals and densities are used, it will make only a second-order error in the total energy. This is probably the explanation of the result of DeCicco's calculation, namely that the total energy as a function of lattice spacing was in very good agreement with the value which Löwdin had found 20 years earlier, in 1947. Löwdin's calculation, by the Hartree-Fock method, agreed well with experiment. This calculation of DeCicco's gave us grounds for hoping that we might get really good cohesive energies by use of Eq. 33-2 for the spin orbitals, and Eq. 35-1 for the total energy.

The first test we gave in Gainesville to this hypothesis was the calculation which Frank Averill, one of the graduate students, made for his thesis in 1971, on the compressibility of the cesium crystal. He, like DeCicco, was using the APW method for solving Eq. 33-2, since this program was available, and we had not made use of a KKR program. We were confident from the intercomparison between the APW and KKR calculations which Burdick and Segall had made on the copper crystal, mentioned in Sec. 31, that the two methods were essentially equivalent. Averill carried through the calculations using the value of α which gave the correct energy for the isolated atom. These α's, for the light atoms, had been computed shortly before by Karlheinz Schwarz, a young Viennese chemist who was with our group in Gainesville for some time as a postdoctoral worker, and Averill applied the same method to find α for an atom as heavy as cesium.

What he found when the calculations were completed, was that the energy of Eq. 35-1, as a function of volume, was in practically perfect agreement with experiment. We had had some hope that this might be the case. Löwdin in 1951 had made a Hartree-Fock calculation of the cohesive energy of metallic sodium, which agreed well with experiment, and led us to think that cesium might come out well. Later work by Averill, on the lighter alkali metals, gave results which were not quite as close to the experimental value as was found for cesium, but still indicated that our hypothesis was certainly correct, that the total energy of Eq. 35-1 gave results close enough to experiment to be very useful.

There was one very significant test which Averill applied to his results. This was a check of the accuracy of the virial theorem. In Eq. 24-4 I gave the relations following from the virial theorem for a diatomic molecule. For a solid under pressure p, with volume v, an analogous form of the theorem is

$$2KE + PE = 3pv \qquad (35\text{-}2)$$

where KE and PE are the kinetic and potential energies. Several workers had given proofs that the $X\alpha$ method satisfied the virial theorem exactly. Averill of course had the material for calculating the potential and kinetic energies separately. At the same time he had the total energy as a function of volume, from which he could get the pressure by differentiating the total energy with respect to the volume. When he substituted the kinetic and potential energies on the left side of Eq. 35-2, and the pressure as found in this way on the right, he found Eq. 35-2 to be satisfied to a very good accuracy. It would not be prefectly obvious that his calculations should have verified this theorem exactly, since he was using the muffin-tin approximation, both in satisfying the self-consistent field and in calculating the total energy, whereas the proofs of the virial theorem assume that Eqs. 33-2 and 35-1 were solved exactly. Averill in his thesis, however, gave a proof that the theorem should hold with the type of muffin-tin approximations he was using.

Still another check of the calculation of total energy has been recently given in work by Edward C. Snow on the copper crystal. Snow, who has been a member of the Los Alamos group for a number of years, and who had made various calculations of gradually increasing sophistication on copper, had never taken a doctorate. He decided to enter our group at Gainesville, to get the degree, and he chose for his thesis to make a calculation of the energy bands and cohesive energy of the copper crystal, using the $X\alpha$ method in its present most sophisticated form. In this work the calculated pressure as a function of volume is in practically perfect agreement with experiment. Copper is a material for which the pressure-volume relation is known experimentally up to extremely high pressures, from shock wave experiments, and the theory reproduces the observed results up to the highest pressures which have been reached.

A more surprising example of calculation of cohesive energy is that on solid argon, by Samuel B. Trickey, Averill, and F. R. Green, Jr., at Gainesville. Trickey, a faculty member of the Quantum Theory Project, has had much experience with the equations of state of the inert gas crystals. Solid argon forms a face-centered cubic crystal, the atoms according to usual views being held together by van der Waals attractions. It has been generally assumed that these forces cannot be explained in terms of the Hartree-Fock method. To get their behavior at large distances, it is necessary to carry out configuration interactions; no one has succeeded in deriving the forces from a single determinant. It was therefore naturally assumed that the $X\alpha$ method, which for a closed-shell atom like argon should correspond quite closely to the Hartree-Fock method, would lead to no bonding between the atoms. Trickey, a man of a rather skeptical nature, thought this would be a fine example on which to test the $X\alpha$ method. He was quite sure that he would get no cohesive energy.

He therefore sought the help of Averill, who had been making the cesium calculation, and of Green, a student, and they made a calculation on the argon crystal by the same methods used for cesium. The astonishing result was that the cohesive energy came out in about as good agreement with experiment as was found with the alkali metals. Similar results have more recently been found for the xenon crystal. These results have met with a great deal of skepticism on the part of the chemists, who thought they had been proving that the Hartree-Fock method could not lead to such results. It may be that the apparent discrepancy between the two points of view could come from the fact that the proofs generally given that the Hartree-Fock method cannot lead to an explanation of van der Waals attraction have been given for internuclear distances great enough so that the atomic clouds of the two atoms do not overlap. The van der Waals attraction is known to be proportional to R^{-6} at large internuclear distance R, in the nonoverlapping case. It is this behavior that must be explained by configuration interaction or by perturbation theory. But the bonding with the argon or xenon crystals comes at somewhat smaller R, where the overlap of the exponential tails of the charge density in the two atoms has started.

In a paper which I wrote in 1972, I gave a more rigorous proof of the virial theorem for the $X\alpha$ method than had been previously published, following closely the lines of my 1933 paper on the virial theorem. At the same time I showed that the Hellmann-Feynman theorem held equally rigorously for the $X\alpha$ method. This allowed us to use the sort of argument I outlined in Sec. 24. At the end of that section, I pointed out that if these two theorems are satisfied, we can start with the charge density as a function of nuclear positions. We can then find the forces on the nuclei from the Hellmann-Feynman theorem, integrate to get the total energy, and then use the virial theorem to get the kinetic and potential energies separately as functions of nuclear positions. All this follows from the charge density. If the $X\alpha$ method can lead to the same charge density as the exact solution of Schrödinger's equation, which it seems to be doing to a quite good approximation, then this means that the total, kinetic, and potential energies found from the $X\alpha$ method will all agree equally well with the exact values.

We have not yet had adequate opportunity to test the accuracy of the calculations of these quantities which follow from our present methods, but such a test is an important part of the future program with the method. In the 1972 paper, I used the case of the inert gas crystal as an example of these theorems, showing what sort of modification of the charge density had to occur to explain the known interatomic forces. These are the types of modification which we presumably will find from the $X\alpha$ method, when exact calculations are made.

The examples we have discussed so far are cases in which no magnetic effects would be expected. However, as the reader will realize, I have been interested for a great many years in understanding ferromagnetism, and have suspected that as a ferromagnetic crystal is compressed sufficiently, the ferromagnetism would disappear. With this in mind, I suggested to one of the graduate students, Thomas M. Hattox, that he study the magnetic behavior of the vanadium crystal as a function of lattice spacing. Vanadium, like chromium and iron, has a body-centered cubic crystal structure. However, iron is ferromagnetic, chromium slightly antiferromagnetic, and vanadium shows no magnetic behavior whatever. The neutron diffraction experiments are able to decide whether the individual atoms have a magnetic moment, but are oriented at random, as in a paramagnetic substance, or whether they have no magnetic moment at all. The decision is very definitely that they have no magnetic moment. We recall that the work which I did in 1936, described in Sec. 25, suggested that the energy bands of vanadium, chromium, and manganese were probably too wide to allow ferromagnetism, whereas iron, cobalt, and nickel, which come next, had narrower energy bands, allowing ferromagnetism. I suspected therefore that vanadium at a lattice spacing greater than the equilibrium distance, where its $3d$ bands would be narrower, might very possibly allow ferromagnetism. This hypothesis was made more plausible by the existence of various vanadium compounds in which the vanadium atoms are further apart on account of the intervening atoms, which are in fact ferromagnetic.

Hattox first carried out a spin-polarized $X\alpha$ calculation of the isolated vanadium atom, to see how it would behave at infinite lattice spacing. In Sec. 33 I mentioned that a spin-polarized calculation of the atoms by the $X\alpha$ method leads to a correct description of the occupancy of the various levels in the ground state of an atom. The vandium atom has five electrons outside the argon shell. Hattox found that in the spin-polarized calculation, the first energy level above those of the argon shell was that of the $4s\uparrow$, or the $4s$ orbital with spin up. Next came the $3d\uparrow$, then the $4s\downarrow$, and finally the $3d\downarrow$. With five electrons to be fitted into these levels, we would have one in the $4s\uparrow$, four in the $3d\uparrow$, and the other levels would be empty. This occupancy is self-consistent, in that it leads to the energy levels which Hattox found. It also agrees with experiment.

The four equivalent $3d$ electrons with spin up lead to a ^5D multiplet, with the maximum possible value of spin angular momentum along the axis, and when we combine this with the $4s\uparrow$, we are led to a ^6D state, which would then be assumed to form the lowest multiplet in the vanadium atom. It is not in fact the lowest multiplet. The configuration $3d^34s^2$ leads to two multiplets, ^4F and ^4P, of which the ^4F is in fact the lowest multiplet level in the atom. But I had shown in a paper in 1968 that the one-electron $X\alpha$ method cannot give predictions about individual multiplets, but only about the average energy of multiplets of a given multiplicity in the spectrum. The weighted mean of the ^4F and ^4P multiplet levels in vanadium lies very slightly higher than the $3d^44s^6D$ state, so that in this sense the $X\alpha$ method has led to the correct prediction of the lowest state of the atom.

Then, having found the behavior of the individual vanadium atoms, Hattox assumed a ferromagnetic orientation of the magnetic moments in the body-centered cubic structure, and he made calculations of the energy bands and total energy for a number of interatomic distances. What he found was exactly in accordance with our expectations. In Fig. 35-1 we show the curve he found for the magnetic moment as a function of internuclear distance. It falls to zero rather sharply at about 1.22 times the

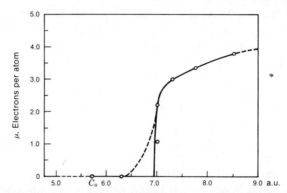

Fig. 35-1. The calculated magnetic moment of vanadium metal as a function of lattice parameter, as calculated by T. M. Hattox. From *Quantum Theory of Molecules and Solids*, Volume 4, by J. C. Slater. Copyright 1974 by McGraw-Hill, Inc. Used with permission of McGraw-Hill Book Company.

equilibrium distance in the crystal. Of course, there is no experimental way to check this fact, except that it definitely predicts that at the equilibrium distance there should be no magnetic moment. However, the value of 1.22 times the equilibrium distance is approximately the dividing interionic distance between those vanadium compounds that show no ferromagnetic behavior and those that do.

Next, in Fig. 35-2, we show what is happening to the energy bands when one goes through this critical internuclear distance. At larger distances, the $3d\downarrow$ band lies entirely above the Fermi energy. The Fermi energy falls within the $3d\uparrow$ band, which lies below the $3d\downarrow$. As the internuclear distance decreases, the bands broaden, and just about at the critical distance, the $3d\downarrow$ band has broadened so much that the bottom of it lies below the Fermi energy. Some electrons then are found in the $3d\downarrow$ band, so that the $3d\uparrow$ band does not hold as many electrons as before. The net magnetization decreases, and this brings the $3d\downarrow$ and $3d\uparrow$ bands closer together, thus enhancing the overlap. Very shortly this effect has gone far enough so that for distances appreciably smaller than the 1.22 times the equilibrium distance, all trace of spin polarization is lost. The total energy in this critical region shows some perturbations, as shown in Fig. 35-3. However, the energy is far above the equilibrium value, which comes at the minimum of the curve. This minimum agrees fairly well with the observed values. Both the value of the internuclear distance at the minimum, as calculated, and its energy differ from the experimental values by only a few percent, being close enough so that there is no doubt that we are expressing the general nature of the actual magnetic phenomena that would be found in a vanadium crystal, if there were any way to pull its atoms apart.

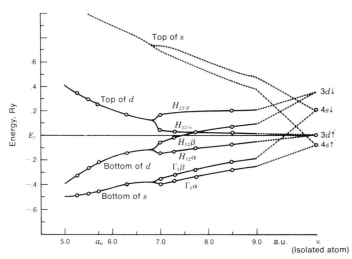

Fig. 35-2. Energy levels in vanadium as a function of lattice parameter. Atomic energy levels from a spin-polarized $X\alpha$ calculation for the $(3d\uparrow)^4 4s\uparrow$ configuration, as calculated by T. M. Hattox. From *Quantum Theory of Molecules and Solids*, Volume 4, by J. C. Slater. Copyright 1974 by McGraw-Hill, Inc. Used with permission of McGraw-Hill Book Company.

It is still a piece of unfinished business to make a corresponding calculation for the body-centered iron crystal. But there is no reason to doubt that a similar situation would be found, except that the internuclear distance where the ferromagnetism would disappear would be somewhat smaller than the observed equilibrium distance. In other words, this mechanism for the disappearance of magnetic behavior as the interatomic distance decreases, which was described in Sec. 25, appears to be thoroughly verified by these calculations by the $X\alpha$ method.

The case of chromium is somewhat different, on account of the fact that it is antiferromagnetic rather than ferromagnetic. We shall take up this case later. But the interesting feature of chromium, from the present point of view, is that neutron diffraction shows that the magnetic moment of a single chromium atom, alternating in orientation from one atom to the next, does not have a large value like the 5 Bohr magnetons which one would find for vanadium, but rather has a value of only about 0.6 Bohr magnetons. It is attractive to suppose that its internuclear separation comes almost at the same distance as the point of disappearance of

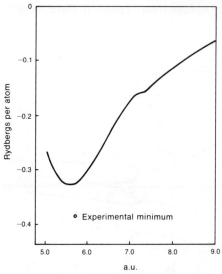

Fig. 35-3. The total energy of vanadium as a function of lattice parameter, relative to the spin-polarized Xα total energy of the $(3d\uparrow)^4 4s\uparrow$ configuration of the separated atoms, as calculated by T. M. Hattox. From *Quantum Theory of Molecules and Solids*, Volume 4, by J. C. Slater. Copyright 1974 by McGraw-Hill, Inc. Used with permission of McGraw-Hill Book Company.

magnetization, shown in Fig. 35-1, so that we are on the rapidly descending part of the curve, explaining the very small magnetic moment per atom. In this connection, we may mention that the magnetic moment per atom in iron is not quite as big as we should expect, from the study of the separated atoms. The indications are that there is a small amount of overlapping of the $3d\uparrow$ band above the Fermi energy here, so that it is not entirely filled, and at the same time the $3d\downarrow$ band is also not filled. This would be the case if we were near the top of the curve of Fig. 35-1, but not quite in the situation where the whole $3d\uparrow$ band was filled. In nickel and cobalt, on the other hand, the $3d\uparrow$ band seems to be entirely filled in the ground state of the crystal.

The calculation of Hattox on vanadium suggests a very important general situation in dealing with cohesive energy. We have noticed that the isolated vanadium atom, with its five electrons outside a closed shell, must be treated by the spin-polarized method. The lowest energy for the atom is found with the maximum possible number of electrons of spin up, which corresponds to a multiplet of maximum multiplicity. And yet the great majority of crystals do not have magnetic moments for the atoms, at the equilibrium separation. It must be, in other words, that a phenomenon such as Hattox has found, resulting in the disappearance of magnetization in the individual atoms when the energy bands grow too wide, will be found in the great majority of cases. The observed cohesive energy of a crystal would be the energy difference between the crystal at its equilibrium internuclear distance and the isolated atoms in their ground states. The spin-polarized calculation gives the best approximation to the ground state of the atom which can be found by a self-consistent-field method. Thus we see why it is not surprising for us to be getting results for the cohesive energy in good agreement with experiment by the $X\alpha$ method.

Even in such a simple case as an alkali metal, we run into problems of this sort. An alkali atom has only one outer electron. But if we think in terms of a spin-polarized calculation, we assume that this electron has its spin up. Then we find a lower energy for a spin-up orbital than for a spin down, and it is only the spin-up orbital corresponding to the valence electron which is occupied. What, then, do we find for the behavior of the crystal at very large internuclear separation? In Sec. 16 I stated that it had occurred to me in writing a paper on cohesion in monovalent metals, in 1930, that it might be that under these circumstances, the magnetic moments of the atoms might arrange themselves in an antiferromagnetic array at large internuclear separations. These metals have the body-centered cubic structure, and one could have spins up on the sites at the centers of the cubes, and spins down on the sites at the corners of the cubes.

In Sec. 30 there is some discussion of the effect of such an antiferromagnetic arrangement on the energy bands of the crystal. For spin-up orbitals, there will be one entirely filled energy band, in which the wave functions are large on the atomic sites known to have spins up, and above this will be another empty energy band, in which the wave functions are large on the intermediate atomic sites, which are known to have spin-down electrons. Similarly for spin-down orbitals, there will be a filled energy band in which the electrons are on the atomic sites known to have spin-down electrons, and a higher empty band on which the wave functions are large on the spin-up sites. The energy levels for the spin-up and spin-down bands will be identical, but the wave functions will be different.

The ground state of the metal, in this case, would correspond to an assignment of electrons to orbitals such as was described in the preceding paragraph. All electrons would be in filled energy bands, with gaps above. This assignment of electrons to orbitals could be described by a single determinantal function, and hence by a Hartree-Fock solution. But the wave function would correspond to the unrestricted Hartree-Fock case, similar to the Coulson-Fischer unrestricted wave function for the ground state of H_2, which we described in Sec. 30. We would then assume that as the internuclear distance was decreased, these energy bands would broaden, and at some critical internuclear separation, a catastrophic situation would occur, such as Hattox found for vanadium. The bands would broaden enough so that the spin-up and spin-down bands would start to overlap, electrons would transfer from the spin-up to the spin-down case, the energy separation between the bands would decrease, and at a slightly smaller internuclear distance this situation would go the whole way, and a non-spin-polarized case would be found.

J. L. Calais and collaborators have recently carried through a Hartree-Fock treatment of this case at Uppsala. They considered the lithium crystal, and found as we should suppose that there was a rather sudden change from the spin-polarized to the non-spin-polarized case at an interatomic distance that was well above the equilibrium distance. The phenomenon was similar to that illustrated in Fig. 30-1 for the hydrogen molecule. A treatment of the problem by the $X\alpha$ method was attempted by A. M. Boring in Gainesville in 1968. He made calculations on the sodium crystal, only at the observed internuclear separation, and seemed to find a very slight residual spin polarization of an antiferromagnetic sort at that distance. The general opinion at present is that this effect does not really occur at the observed internuclear distance, and that Boring's result probably came from computational difficulties. We expect, however, that if he had made the calculations at a sufficiently large internuclear separation, he would have found the spin polarization.

We must not forget, however, that the true situation must be very much more complicated than this simple model would suggest. In a sample of an alkali metal containing N atoms, and consequently N valence electrons each with a spin angular momentum of $\frac{1}{2}$ Bohr magneton, there are 2^N ways of assigning the N electrons to spin-up or spin-down states. One of these corresponds to the antiferromagnetic arrangement which we have just described, and one to a ferromagnetic arrangement. But in general the great number of states that exist correspond to the possibilities of different orientations of spins which result in paramagnetism. At infinite internuclear distance, it can obviously require no energy to rotate the spin of one atom, and consequently all 2^N assignments of electrons to states will be degenerate with each other. As the internuclear distance decreases, the energy levels of the whole atomic system will begin to spread out, and some particular one will lie lowest. This might be the antiferromagnetic arrangement, but it might not. At present it would be a practically hopeless task to carry out a thorough study of all of these levels, and to determine which actually is the lowest. But we know the energy at large internuclear distances—it is the energy of the isolated atoms. And we can be quite sure that no matter which combination corresponds to the lowest state, there will be a phenomenon such as Hattox has observed, if the internuclear distances get small enough. Hence at the observed internuclear distance, we are on quite safe ground in using a non-spin-polarized calculation, which is what Averill did.

The situation of antiferromagnetic chromium is not completely different from this hypothetical case of antiferromagnetism in an alkali metal. Here, however, it is the $3d$ orbitals which are responsible for the antiferromagnetism, and it is not unreasonable that the antiferromagnetic spin polarization might be just disappearing at the observed internuclear distance, as we postulated earlier. It is even possible, with chromium, to make some estimate of the way in which the energy gaps arising from the antiferromagnetic orientation come about. In 1967, Asano and Yamashita first carried out calculations of the non-spin-polarized energy bands of chromium, using the $X\alpha$ method, and then of the type of spin-polarized antiferromagnetic arrangement we have described. Certain gaps in the energy-band structure opened up in the antiferromagnetic case, just at the Fermi energy, so that the lower, occupied energies were pushed down by the transition to the antiferromagnetic arrangement. The reduction in energy arising in this way is presumably that which stabilizes the antiferromagnetic case.

Their calculation even throws light on a complicating feature of the antiferromagnetism of chromium, which was discovered by neutron diffraction in the 1950s, and was explained by Lomer in 1962. The true ground state does not seem to be exactly the arrangement we have suggested, in which there is an alternation of spin orientation when we go along from one atom to the nearest neighbor, resembling a sinusoidal modulation of magnetization with a half-wavelength equal to the interatomic distance. Rather, it is like a sinusoidal modulation of magnetization, with a slightly different wavelength. Similar noncommensurable oscillations of magnetization are now known to be the stable states in a great variety of crystals, including the magnetic structure of some of the crystals of rare earth elements. It would be too complicated to try to explain here the exact features of the energy bands which tend to stabilize these complicated magnetic structures, but the success of the explanation in the case of chromium suggests that in such problems we may hope in the future to make many applications of the spin-polarized $X\alpha$ calculations.

36. The Transition State and Localized Excitations

In Sec. 25 I described some of my thoughts dating from the 1930s on localized excitations and on their relation to magnetic problems. I carried these ideas somewhat further in the following years. In 1949, I wrote a paper called "Electrons in Perturbed Periodic Lattices," outlining the general nature of localized states in crystals, mostly aimed at the workers in the field of solid-state electronics. And in 1954, George Koster and I set up an analytic method of handling localized perturbations. But none of this led very directly to practical methods of calculation. I never lost interest in these problems, however, and when the $X\alpha$ method and the transition state began to be very successful, I wished to show how this method could be applied to quantitative study of both excitons and the Heisenberg exchange integral in magnetism.

The work of the 1930s showed that the optical transitions from a ground state of a crystal to localized excited states would have much greater oscillator strength than transitions to bandlike excited states. This had an immediate application to the idea of the transition state. We should be considering transition states in which an electron was, so to speak, half-excited from the ground state to a localized excited state. This transition state itself would then have a localized perturbation, and one would have to use the theory of such localized perturbations to study any optical absorption by a solid. The most striking illustrations came in the study of x-ray excitations.

Suppose, to be specific, that we are considering the absorption of x-rays with the excitation of a K electron in a fairly heavy atom in a crystal. The transition state will be one in which the atom in question has only $1\frac{1}{2}$ K electrons, as was mentioned at the end of Sec. 33. Thus the potential well at this atomic site will be considerably deeper than at other similar atoms in the crystal. From the work of the 1930s and of 1954, we know that in such a case the energy levels of the system will be divided into two quite different categories. First, there will be localized states, in which an electron moves in the neighborhood of the perturbed atom. There will be $1s$, $2s$,..., states with orbitals localized around this atom, and all with energy levels considerably lower than the ordinary band states. Second, there will be band states in which the wave functions are distributed through all the other atoms of the crystal, but in which they avoid the neighborhood of the perturbed atom.

It is the localized states which have the large oscillator strengths. Thus, for a K excitation, using the transition state functions, one would be interested in a one-electron transition from the $1s$ orbital up to a higher unoccupied orbital. There may be more than one of these higher localized orbitals into which the electron could go, and the one-electron energy of these orbitals would be lower than the bottom of the continuum of band states. These orbitals would be very much like those found for an isolated atom. For a great many years the experts in x-ray spectroscopy had felt that their experiments indicated just such localized excitations. They saw no evidence of the broadening of absorption (or emission) spectra such as would result if an electron were having a transition from one band to another. But the solid-state theorists, thinking only in terms of energy bands, were not able to understand these observations of the experimenters. For example, L. G. Parratt of Cornell and his students had examined the x-ray spectra of crystalline argon and of gaseous argon. The details of the spectra, which are fairly complicated, are practically identical whether one is dealing with the crystal or the atom.

For softer x-rays, and more particularly for far-ultraviolet absorptions, the depression of the potential well around the excited atom will not be nearly as great. This is because it will be an L or M electron which is half-removed in the transition state, and this has a much more extended orbital. The resulting exciton will then be more spread out, and the chemical surroundings of the excited atom will have more effect. This again is known experimentally. There is a pronounced chemical effect on these softer x-ray spectra. The excited states will have orbitals which are not entirely localized on the atom from which the x-ray electron is removed, but which may well correspond more to charge-transfer transitions, being localized more on a nearest neighbor atom. We have mentioned a case of this in Sec. 25, in connection with von Hippel's charge-transfer theory of the excitons in an alkali halide. But even in this case, the excited state is localized enough so that only the nearest neighboring atoms of the atom from which the electron is removed need be considered.

These facts suggested that for treating such localized excited states, it would be enough to treat a complex consisting of a central atom and its nearest neighbors, or if one must carry it further, a central atom, its nearest neighbors, and its next nearest neighbors. But such complexes are just the sort of problem that can be handled very successfully by the $X\alpha$ multiple-scattering technique of Johnson and Smith, which we described in Sec. 34. This opens up the field of x-ray spectroscopy, for which we now have a usable quantitative theory. Not many examples have yet been worked out to illustrate this application, but much work is underway to test out the method.

There is another technique which can be used experimentally to investigate the x-ray levels: photoemission. The general method has been known for many years, but recent improvements in experimental technique have focused attention on it. Described very simply, one shines monochromatic x-rays on a crystal or a molecule. The radiation ejects an electron from one of the inner levels of an atom. It is emitted from the surface of the crystal or from the molecule. One then uses electric or magnetic methods to conduct an analysis of the energy spectrum of the emitted electrons. One finds different emitted electrons, depending on which inner level they have been ejected from. Very accurate measurements can be made of the energy differences between these emitted electrons, and therefore between the x-ray levels. The measurements prove to be a good deal more accurate than those which can be made by direct x-ray experiments. The effect of chemical combination on these levels can be observed, and many polyatomic molecules have been investigated, as well as many crystals.

Many different experimenters have been working with this method, but the relations of our Gainesville theoretical group have been particularly close with the experimental group at Uppsala. There is an interesting story tying together this Uppsala work with both MIT and Gainesville. Professor Manne Siegbahn of Uppsala was one of the greatest pioneers in the field of x-ray spectroscopy. His tables of energy levels, dating back to the 1920s and earlier, were the basis of most of the early studies of the x-ray levels of the atoms. His work of course was well known at Harvard while I was there. His son, Kai Siegbahn, decided soon after the close of the second world war that he wished to go into the field of photoemission spectroscopy. He knew that at MIT one of the younger faculty members, Martin Deutsch, had worked out a very superior experimental technique for measuring energies of emitted electrons for use in the problem of beta-ray spectroscopy. This was used, of course, for studying nuclear energy levels rather than x-ray levels, and Kai Siegbahn visited MIT studying the techniques in the early postwar years, and I became acquainted with him at that time.

It still was a work of many years to get the photoemission technique well worked out at Uppsala. But in the late 1960s, we became aware of what was called ESCA, electron spectroscopy for chemical analysis, which was being developed in Kai Siegbahn's group. By then there was an impressive experimental group and a number of theorists working on the interpretation of the results. Two books, one on the application to crystals, the other to polyatomic molecules, were published, in 1968 and 1969. It was about at this time that Per-Olov Löwdin began to carry news of our $X\alpha$ multiple-scattering technique back to Uppsala, and the experimental group there became much interested in the method. By 1971, three members of the

Uppsala group were visiting Gainesville, namely U. Gelius, C. Nordling, and Hans Siegbahn, the son of Kai Siegbahn and the third generation of that illustrious family. They worked with John Connolly at Gainesville. Connolly has spent time in Uppsala, and the two groups have been working together very closely, with the computer programs available at both institutions.

The net result of this collaboration has been that the $X\alpha$ method has shown itself capable of explaining the energy levels more accurately than any of the other theoretical techniques which were being applied to the problem, and with much less computer time. Most of the work so far has been with polyatomic molecules. The first comparisons were with the sulfur hexafluoride molecule, SF_6, which Connolly and Johnson computed jointly, and which was done before the close collaboration with Uppsala started. In this calculation, as in other cases, it was found that the use of the transition state method was essential in getting the good results. A later paper, by Connolly, Gelius, Nordling, and Siegbahn, published in 1973, lists calculations on CF_4, NH_3, CO, NO, CO_2, N_2O, and C_3O_2, all by the same method, and all agreeing with experiment better than with any of the various current semi-empirical methods in use by the chemists, and about the same as ab initio MO-LCAO-SCF calculations. But, as usual, the computing effort is very much less than with the ab initio LCAO methods.

This may be as good a time as any to list some further molecules and clusters of atoms in crystals which have already been investigated by the $X\alpha$-SCF method, some in Gainesville, some at MIT, some by other groups. In addition to those listed above, and SF_6 and the $(SO_4)^{2-}$ and ClO_4^- ions we mentioned earlier, a partial list is the MnO_4^- ion, methane CH_4, ethane C_2H_6, H_2, H_2O, H_2O_2, Li_2, CO_2, benzene C_6H_6, C_4F_4S, P_4, and P_8. Among clusters of atoms from crystals which have been studied are $TiCl_4$, $(NiF_6)^{4-}$, $(CuCl_4)^{2-}$, $(PtCl_4)^{2-}$, $Ni(CO)_4$, $Fe(CO)_5$, and parts of the guanidinium and the phosphate ions, as they join to form guanidinium phosphate. We shall return to various features of some of these calculations and of others which have not yet been published.

These examples of which we have been speaking are cases in which the transition state is fairly well localized. However, it is known, from the theory of perturbed localized states, that if the perturbative potential on a given atomic site is too small, no localized states will split off from the continuum. This was brought out particularly clearly in the papers which Koster and I wrote in 1954. The reason is not at all abstruse, and can be explained simply in terms of wave mechanics, but it demands a little more mathematics than we wish to go into here. It suggests that there can be cases in which we do not have separate exciton formation, but in which instead one has direct excitation from a lower band to a higher band. It is this case that comes into the standard treatment of absorption which is met in the energy-band theory. There seem to be a good many cases where one finds this situation. Conklin and his students, as I have mentioned, had been studying the band structure of such crystals as TiC, using the $X\alpha$ method, and there seems to be no evidence that one encounters exciton formation in these cases. The ordinary energy-band theory gives a good account of the situation, and the $X\alpha$ energy bands are in good agreement with experiment.

In Sec. 25, we pointed out the close connection between exciton theory and the Heisenberg exchange integral. This connection becomes very much clarified if we think in terms of the transition state. Suppose we start with a ferromagnetic crystal, with all atomic moments having one spin orientation, which we may consider as spin up. Let us then consider a transition in which the magnetic moment of one atom is reversed, so that it points in the spin-down direction. We can find the increase in energy when we go to this state from the transition state concept, but we can also describe it in terms of the Heisenberg exchange integral. If we intercompare these two methods of calculation, we have a method of computing the Heisenberg exchange integral from the $X\alpha$ method.

Let us first consider the transition state picture. It is simpler to describe this from an example of a nonmetallic ferromagnetic crystal, in which we can describe the magnetic moment as coming from a definite ion in the crystal. An example which is simple to think of, though not to compute, is the EuS crystal, a ferromagnetic insulator which has a sodium chloride structure. Its energy-band structure corresponds to an ionic picture, $Eu^{2+}S^{2-}$, though as in earlier cases there is some covalent bonding. The Eu^{2+} ion has seven electrons outside the xenon shell, and the saturation magnetization shows that all seven electrons have spin up, as if they were filling the spin-up $4f$ levels, while the spin-down levels were empty. Thus there is a resemblance between this case and MnO, which we discussed in Sec. 30. The Mn^{2+} ion, as we mentioned there, has a half-filled shell of $3d$ electrons, with five $3d$ spin-up orbitals occupied. However, MnO is antiferromagnetic, whereas EuS is ferromagnetic. Its Curie temperature is very low, and the similar crystal EuSe, which is antiferromagnetic, also has a low Néel temperature, the temperature at which the antiferromagnetism disappears. This suggests that the energy difference between the parallel and the antiparallel orientations of neighboring Eu^{2+} ions must be very small, and it is this energy difference which we should like to find from a transition state calculation.

If we think of the Eu^{2+} ion in terms of an isolated atom, we should then have seven $4f\uparrow$ spin orbitals, with $m_l = 3, 2, 1, 0, -1, -2, -3$, and all with $m_s = \frac{1}{2}$, occupied in the initial ferromagnetic state of the crystal, whereas the seven $4f\downarrow$ spin orbitals are empty. If the magnetic moment of the ion is reversed, we should have seven empty $4f\uparrow$ spin orbitals, seven filled $4f\downarrow$ orbitals. In the transition state, we should have half an electron in each of the 14 states with $m_l = 3, 2, \ldots, -3$, $m_s = \frac{1}{2}, -\frac{1}{2}$. Thus in the transition state, this ion would have no spin polarization, having equal numbers of spin-up and spin-down electrons. We should then have to solve the problem of a crystal, or at least of a cluster of atoms consisting of the central atom, its neighboring S^{2-} ions, and the nearest neighboring Eu^{2+} ions. These Eu^{2+} neighbors would all have their spins pointing up, so that the central atom, with no spin polarization, would have a quite different potential well from the other europium ions.

If it were not for these neighboring Eu^{2+} ions, an electron on the central non-spin-polarized europium ion would see exactly the same potentials for spin up and spin down. Hence its spin-up and spin-down energy levels would not be split apart at all. However, there will be a little spreading of the Eu $4f$ orbitals onto the neighboring S^{2-} ions, on account of the mixing arising from the covalent binding, and these orbitals in turn would extend slightly onto the nearest Eu neighbors. Thus the spin-up and spin-down orbitals would see slightly different potentials, and there would be a small amount of spin polarization of these localized levels of the central Eu^{2+} ion. This is the effect called superexchange.

To find the difference between the original energy of the crystal and that with the one europium ion with reversed spin, we are then directed to find the difference between the one-electron energies of spin-up and spin-down $4f$ orbitals on this localized non-spin-polarized europium ion. We have just seen that there will be a small effect. Since there are seven electrons which would have a transition from $m_s = \frac{1}{2}$ to $m_s = -\frac{1}{2}$, we should have to add the energy differences for all seven states to get the energy difference between the original energy of the crystal and that with one europium ion with reversed spin. When we now equate this quantity to the energy difference as computed from the Heisenberg exchange integral, we can get a value of this integral. The Heisenberg theory allows one to derive from this the value of the Curie temperature.

As far as order of magnitude is concerned, it proves to be the case that Boltzmann's constant k times θ, the Curie temperature, is roughly equal to the one-electron energy difference which we find for one of the orbitals in the transition state, between spin up and spin down. This allows us, since we know the experimental Curie temperature, to find about how large this spin polarization of the transition state orbitals would be. Unfortunately, for EuS, which is a simple case to describe, the energy difference is too small to hope to compute with the present accuracy of the computer programs. However, a similar relation would hold for a ferromagnetic metal such as Fe, Co, or Ni, and there the energy difference should be quite large enough to compute. Unfortunately this has not been done so far, but this is another case where the transition state technique opens up a field for $X\alpha$ calculations, where previously no calculations from first principles have been possible.

The case of MnO would presumably be an easier one to compute than EuS, but experimentally it is antiferromagnetic. T. M. Wilson, one of the postdoctoral workers at Gainesville, made calculations of the energy bands of both ferromagnetic and antiferromagnetic MnO, in the late 1960s. This was before we understood the transition state, and Wilson's calculations were not set up to find Heisenberg exchange integrals. There is one point about the antiferromagnetic nature of this and similar crystals which has caused some question. We have used the antiferromagnetic nature of the crystal to justify a spin-polarized calculation. It is this calculation which separates the spin-up and spin-down bands, and shows that the lower bands, which are occupied in the crystal, are separated by a considerable gap from the higher bands, which are empty. This leads to the observed insulating nature of MnO, and the energy gap which Wilson found from his band calculations agreed fairly well with the observed value.

However, it has been pointed out by Mott and others that above the Néel temperature the antiferromagnetism disappears, and yet the MnO crystal continues to be insulating. How does this come about? It used to be assumed, since the crystal was not magnetic above the Néel temperature, that a non-spin-polarized calculation was the correct one to use. In that case, the single band arising from the atomic $3d$ state would be capable of holding ten electrons per atom, and yet only five electrons were present. Hence it was assumed that the crystal should become conducting above the Néel temperature, but it did not. Mott tried to use this and similar examples as proofs of the inadequacy of the energy-band theory.

The true situation, however, is different. The MnO crystal is paramagnetic above the Néel temperature. A single Mn^{2+} ion continues to be in its ground state with five parallel spins, however. Hence if we define the orientation of its spin as being spin up, we must use a spin-polarized calculation for this ion. As in all such cases, the spin-up $3d$ energy levels will lie far below the spin-down $3d$ levels. But now if we consider the neighboring Mn^{2+} ions, they will have their magnetic moments oriented in any arbitrary direction, in the paramagnetic state. Thus we should use a non-spin-polarized description for them. The central ion will, as in the other cases we have considered, have a very different potential from its neighbors, so that there will be localized $3d$ states attached to this central ion. Similarly, in this case, there will be localized states attached to each ion of the crystal. It is the existence of these localized states which leads to the nonconducting properties of the crystal above the Néel point. The optical absorption, however, is determined by the local cluster of atoms around the central ion, and it will have a very similar spectrum to that of an isolated ion, very similar above and below the Néel point. We have, then, in a way, the localized states which Mott wished, but they are produced by the magnetic fluctuations, and would not be found in the same way below the Néel temperature.

In this section we have seen a few examples of the illumination which the study of localized clusters of atoms and their localized excitations can throw on the properties of crystals. As examples of further calculations which are now underway, and which can be expected to give much further information in the future, we may mention two. First, colleagues of Johnson at MIT have studied the optical properties of a cluster of copper atoms consisting of a central atom and its 12 nearest neighbors, such as would be found in the face-centered cubic crystal. When a self-consistent calculation was made, and the spectrum was investigated by the transition state technique, there was a striking similarity between the resulting absorption spectrum and that of a copper crystal. In other words, many of the properties of an infinite crystal are really determined by an atom and its near neighbors, and are not really dependent on the crystalline nature of the material. This has been realized by some workers for a long time. Many of the Soviet workers, in particular, for many years have been studying amorphous materials, and have found that their properties are more like those of infinite crystals than anyone had imagined. But this means that the cluster method, using even such small clusters as the 13 atoms used in this copper calculation, can often give very useful information about infinite crystals.

The other example is an outgrowth of the one we have just mentioned. There is a very large amount of information which has been accumulated over the past few years about the nature of magnetic impurities in nonmagnetic crystals. In some cases the magnetization is located only on the magnetic impurity atom, but in other cases it spreads onto neighboring atoms. The cluster method is very well adapted to studying such problems. A nickel atom in a copper crystal, for instance, can be well simulated by a nickel atom surrounded by the 12 nearest neighboring copper atoms, and the problem is then no more difficult than the one we have been describing. Work along these lines is in progress, and we may hope that the whole field of magnetic impurities will be opened up for theoretical calculation by the $X\alpha$ method.

37. Improving the Muffin-Tin Potential

Most of the examples which we have so far given of the applications of the $X\alpha$ method have dealt with cases where the results were in very good agreement with experiment. But this has not been true every time, and the study of these relatively unsuccessful cases has given us a great deal of information. To put it briefly, it has shown us that the main errors in the method as we are using it come from the muffin-tin approximation to the potential, not from the $X\alpha$ method itself. And second, the greatest errors with the muffin-tin potential come with the simplest molecules, diatomics and others with only a few atoms. These are just the cases where the classical LCAO methods have been most successful, and consequently the comparative lack of success of the $X\alpha$ method has thrown an undeserved doubt on that method.

We have expected from the outset that these small molecules would be the most difficult ones. The reason is simple. The approximation made in the muffin-tin potential that the potential is spherically symmetric within the spheres surrounding the nuclei is very good. But the other approximation, that it is constant through the region outside the spheres, is rather poor. It is then to be expected that the method will be poorest for those molecules where this second region (II in the notation of Johnson and Smith) is proportionally largest.

Suppose we take a diatomic molecule, and for simplicity take the case where spheres of identical size are to be used for each of the two atoms. Each of these spheres will have a radius equal to 1/2 of the radius of the outer sphere, assuming that these spheres touch. Thus the volume of one of the atomic spheres will be 1/8 times the volume of the outer spheres, or the volume of the outer sphere will be four times the volume of the two atomic spheres taken together. The volume II will then be three times as large as the total volume of the atomic spheres. Or think of a linear triatomic molecule, like CO_2. If again we use three spheres of identical size, each one will have a radius 1/3 of that of the outer sphere, the volume of an atomic sphere will be 1/27 of the volume of the outer sphere, and only 1/9 of the volume of the outer sphere will be occupied by atomic spheres. Thus the volume II will be eight times as large as the total volume of the atomic spheres. Obviously things are getting rapidly worse. In contrast, if we had a molecule with octahedral symmetry, like SF_6, a central atom with six neighbors along $\pm x$, $\pm y$, $\pm z$, and if again we chose all seven spheres to be the same size, the volume of the seven atomic spheres would be 7/27 times the volume of the outer sphere, and the volume of the region II will be 20/7, or a little less than three, times the total volume of the atomic spheres. A molecule consisting of a central atom, and the 12 neighbors surrounding it which one can have with close packing, has even a smaller ratio of the volume II to the total volume of the spheres, namely 14/13. It is these latter cases where the muffin-tin method works best.

The relatively poor behavior of the muffin-tin method for small molecules is shown more strikingly in the total energy than in the one-electron energies. The latter values are not very bad, even for very small molecules. But in some cases the total energy of the molecule, at its minimum, comes out to be larger than the energies of the separated atoms, and in this case it is obvious that no information regarding the cohesive energy is obtained.

To estimate these errors, J. B. Danese, one of the Gainesville graduate students who is now doing postdoctoral work with Connolly, has written a program from which he can estimate the errors arising from the muffin-tin approximation. He is working with diatomic molecules, and starts with the spin orbitals found from the muffin-tin approximation. The standard computer programs of the muffin-tin method do not really calculate the total energy of Eq. 35-1 properly. One notes that in the coulomb energy term, $\frac{1}{2}\int \rho(1)\rho(2)g_{12}dv_1 dv_2$, one can rewrite this in the form of $\frac{1}{2}\int \rho(1)U(1) dv_1$, where $U(1)$ is the potential $\int \rho(2)g_{12}dv_2$. Neither the charge density $\rho(1)$ nor the potential $U(1)$ determined from it really has muffin-tin form, and the charge density found from our spin orbitals is probably much closer to the correct density than it is to the muffin-tin approximation. However, the computer program for the coulomb energy first finds muffin-tin charge density, then muffin-tin potential, and it uses these muffin-tin quantities in computing the coulomb energy.

Danese is attempting to compute the exact energy of Eq. 35-1, but he uses the approximate spin orbitals found from the muffin-tin approximation. He notes that the first term in the coulomb energy is the one already computed. Next there are cross terms from the product of the muffin-tin $\rho(1)$ and the non-muffin-tin correction to $U(1)$, or vice versa. He has computed these cross terms, which are of the next smaller order of magnitude. Finally he has estimated the second-order small terms from the product of the non-muffin-tin part of $\rho(1)$ times the non-muffin-tin part of $U(1)$. He has shown that this last term is small compared to any of the others. When he makes this calculation, disregarding the second-order small terms, and does this for the various terms of Eq. 35-1, he finds that the errors of total energy encountered in the muffin-tin approximation practically disappear. For instance, the muffin-tin calculation of the diatomic molecule C_2 gives no bonding, the energy of the molecule being higher than that of the separated atoms. But when he puts in his first-order corrections, he finds in this case a binding of roughly the correct magnitude. In several other cases which have been tried, the same situation is found to hold.

A different type of test is given by computations of D. E. Ellis of Northwestern, who used to be with the SSMTG and later with the Quantum Theory Project, and of E. J. Baerends and P. Ros of Amsterdam. Ros had spent a year with the SSMTG some years earlier, and had collaborated with Ellis while he was there at MIT. Ellis has worked out a modification of the LCAO technique, which allows him to make calculations of small molecules using the $X\alpha$ exchange-correlation term, so that he is rigorously computing the energy of Eq. 35-1. He is not, however, in any way using the muffin-tin approximation. The calculation takes the form of computing the total energy in terms of the coefficients of the various functions making up the approximate spin orbitals, and then varying these coefficients to minimize the energy of Eq. 35-1. This calculation, like all LCAO calculations, requires much more computer time than the muffin-tin calculation, but it is interesting to make the comparison. Baerends and Ros have studied the molecules CO, LiF, NO, N_2, O_2, and C_2H_4. They give transition state calculations of the various ionization energies of these molecules, but their calculations do not have sufficient significant figures to give reliable values of total energy. However, they find that for the ionization energies, the results are considerably closer to experiment than those obtained for the same molecules using the muffin-tin programs.

These calculations of Ellis and his collaborators involve a great deal more computing time than the original muffin-tin procedure. Thus if it were necessary to use such methods, we should be surrendering one of the very important advantages of the original method, namely its ease of calculation. On the contrary, Johnson and two collaborators, N. Rösch of the Technical University of Munich and W. G. Klemperer of Columbia, have proceeded along quite a different direction which seems to have great promise. They note that the difficulty with the muffin-tin approximation seems to come from the large amount of space in the region II, between the atomic spheres. Why not, then, make the atomic spheres larger, and allow them to overlap?

If one tries to understand the problem of overlapping spheres in a direct way, one runs into quite severe complications. What Johnson and his collaborators did, however, was to try first a very naive approach: simply to use the identical computer programs which were being used for the nonoverlapping case, but to use larger atomic radii, so that in fact the spheres would overlap. What they found, in a number of cases where they tried the method, was a rather dramatic improvement in the agreement between the ionization energies and experiment. They found, for instance, that in the two molecules ethylene and benzene, they improved agreement with experiment decidedly by using the radius 1.55 bohr units for the carbon atom, rather than the value 1.2652 which is half the carbon-carbon distance in ethylene. A similar assumption for ferrocene, $Fe(C_5H_5)_2$, again improved the situation. Baerends and Ros had also made calculations on ferrocene, and the improvement which Johnson et al. found in this case was comparable with what Baerends and Ros found by their technique. These results, which are rather new at the time this is being written, seem so promising that I have been trying to understand why the technique of overlapping spheres is working so well. Nothing has been published on this yet, but the conclusions I have come to seem straightforward and simple enough so that they can be well described here.

First we have to understand how the coulomb potential is found in spherical coordinates. This is done most simply from Poisson's equation

$$\nabla^2 U = -4\pi\rho \qquad (37\text{-}1)$$

Here $-U$ is the potential energy of an electron at a given point of space, as computed by electrostatics from a charge distribution ρ. The charge density ρ is supposed to include all charges, nuclear as well as electronic, and the equation as we have written it is correct in atomic units. The exchange-correlation term in the energy is not included in Eq. 37-1.

Then, since we are working in spherical polar coordinates, we make an expansion of the charge density in products of spherical harmonics of the angles and radial functions. Thus we write

$$\rho(r,\theta,\phi) = \sum (l,m)\rho_{lm}(r)Y_{lm}(\theta,\phi) \qquad (37\text{-}2)$$

which is the same sort of expansion which was used in Eq. 22-1 for the wave function. There are standard methods for finding this expansion from a known charge density, and the special case for $l=0$, $m=0$, or the spherical average, is included as a part of the ordinary muffin-tin computer program. Once one has this expansion, it is then a simple matter to find the coulomb potential U. It is given by the equation

$$U(r,\theta,\phi) = \sum (l,m) Y_{lm}(\theta,\phi) \frac{4\pi}{2l+1} \left[\frac{1}{r^{l+1}} \int_0^r (r')^{l+2} \rho_{lm}(r') dr' \right.$$

$$\left. + r^l \int_r^\infty (r')^{-l+1} \rho_{lm}(r') dr' \right] \qquad (37\text{-}3)$$

In case the reader is unfamiliar with Eq. 37-3, which is not found in all textbooks on electrostatics, he can easily verify it by direct substitution in Eq. 37-1. It is easy to set up computer programs to evaluate the integrals of Eq. 37-3. Once one has this potential, it is easy to set up expressions for the total energy. It is to be noted that the U which we are using here includes the potential energy of an electron both in the field of all electrons, as in the term $\int \rho(2)g_{12}dv_2$ in Eq. 33-2, and in the field of all nuclei included in the term $f_1(1)$ in the same equation.

Let us now look at Eq. 37-3 so as to get a feeling for what it involves. First we consider the term $l=0$, giving the spherical average potential, which is what appears in the muffin-tin approximation. Since Y_{lm} is constant for $l=0$, this term is

$$\frac{1}{r}\int_0^r \rho_0(r')4\pi r'^2 dr' + \int_r^\infty \frac{1}{r'}\rho_0(r')4\pi r'^2 dr' \qquad (37\text{-}4)$$

In the first term of Eq. 37-4, the integral from zero to r is simply the total charge inside the sphere of radius r; ρ_0 is the spherically averaged charge density. If the origin of the spherical coordinates is chosen to be the location of a nucleus of atomic number Z, the integration over an infinitesimal sphere surrounding the origin will include the integration of the infinite charge density of a point-charge nucleus times the infinitesimal volume of this nucleus, or in atomic units will be just Z. The remaining integral over the sphere will give the shielding from those electrons inside a sphere of radius r. Since the electronic charge density is negative, this contribution will be negative. Hence the first term of Eq. 37-4 will be $1/r$ times (Z − number of electrons inside sphere). The potential of a spherical charge distribution inside a sphere, in other words, is what would be found by concentrating all the charge at the origin and treating it like a point charge, leading to the potential proportional to $1/r$. This familiar result goes back to Newton. The contribution of these electrons inside the sphere to the shielding is commonly known as inner shielding.

The second term of Eq. 37-4 describes the outer shielding, or the contribution of the outer charge density to the shielding. Each shell of thickness dr' contains a spherically averaged charge of total amount $\rho_0(r')4\pi r'^2 dr'$. This shell, at all points outside it, would lead to a potential equal to its total charge divided by r, according to the result of the preceding paragraph. At the surface of the shell this would be the total charge divided by r'. The statement of the second term of Eq. 37-4 is then that this potential arising from outer shielding is constant everywhere inside the spherical shell. This result again goes back to Newton.

Now let us consider the terms for higher l and m values in Eq. 37-3. They are the terms disregarded in the muffin-tin approximation. The terms for $l=1$ are the dipole terms, those for $l=2$ the quadrupole terms, and so on, all multipoles being included in the summation over l. For the dipole terms, the first term of Eq. 37-3 tells us that the dipole potential at points outside a charge distribution contained in the sphere of radius r equals $1/r^2$ times a spherical harmonic of the angle. At points inside an outer spherical shell, the potential is proportional to r times a spherical harmonic of the angles. This can be easily shown to be a potential which is a linear function of x, y, and z, leading to a constant electric field inside the spherical cavity. It is this field which leads to the familiar $4\pi/3$ correction found in the Lorentz theory of dielectrics.

The terms for larger l, starting with the quadrupole terms $l=2$, fall off outside the sphere of radius r with higher inverse powers of the distance. It is only the terms for $l=0$ and $l=1$ which are ordinarily called long-range forces. They give outer-shielding contributions of finite size depending on the nature of the outer boundaries of the charge distribution. Thus it is well known that the potential inside a sample of material depends on the surface charge on the surface of the sample as well as on its shape, and similarly the electric field depends on the dipole distribution on the surface. The higher l values do not lead to any such dependence on surface behavior.

There is a well-known consequence of Eq. 37-3: The behavior of the potential outside a charge distribution confined to the interior of a spherical volume does not lead in any unique way to a knowledge of the charge distribution in the interior. Any distribution with the same total charge will give the same term in the $1/r$ dependence of the potential. Any distribution with the same dipole moment will give the same term in the $1/r^2$ distribution, and so on. These facts are well known to geophysicists, and have been known at least since the time of Gauss. No measurement of gravity outside the earth can tell us just where distributions of heavy material can be found, and no measurement of the magnetic field will tell us where iron mines must be dug, a disappointment to the practical men who had hoped Gauss could go further with the inverse square law than the equations allowed.

But coming closer to our present problem, a spherically symmetrical uncharged atom will exert no field whatever, or will have a zero potential, at any point outside the sphere containing all the electrons of the atom. Hence, in computing the potential in the regions II between atoms, we are allowed to replace the actual atoms, with their charge density varying enormously from point to point within the atomic spheres, by charge distributions which merely give the same potential outside, or which have the same total charge, total dipole moment, and so on. This means that the expansion of potential holding in the regions II between the atoms is not unique. If we use these smoothed-out atomic distributions, we shall have precisely the same potential throughout volume II, and yet we may have a very different potential inside the atomic spheres.

These facts can have a very profound effect on the applicability of the muffin-tin potential. To carry out an expansion of the actual charge density in a molecule or crystal, in the form of Eq. 37-2, would require terms of very high l values in order to lead to the detailed description of the charge density near the nuclei of distant atoms. But if we are interested only in an expansion which will work properly in the regions II between spherical atoms, we can replace these atoms by the smoothly varying charge densities which produce the same potential at external points. The resulting charge density will be a very much smoother function of angle, and the expansion of Eq. 37-2 can be much more rapidly convergent.

Let us, then, think of the expansion of the charge density around a particular nucleus in the molecule; we shall be working with similar expansions around each nucleus. For the atom centered on the origin, we use the actual charge density, but for all distant atoms we replace them by the type of smoothed distribution we have just described. Let us then ask what we should expect to find from Eq. 37-4, giving the spherically averaged potential. First we consider the inner shielding, the first term of Eq. 37-4. As the value of r increases, we include more and more electronic charge inside the sphere, and at some definite value of r, we have a sphere which includes precisely enough electrons to balance the nuclear charge. The value of $-U$, which is minus infinity at the origin where the atom of atomic number Z is located, will have risen to precisely zero at this sphere. But outside that sphere, we shall be dealing with the smoothed charge distributions. The total charge in the whole molecule outside this sphere is zero, if the molecule is uncharged; if it is a charged ion, there will be a surface charge, which is ordinarily balanced in Johnson's method by introducing an artificial charged sphere. Not only is the total charge zero, but the local charge density is approximately zero. It consists of small negative charges outside the other atomic spheres, as a result of the tails of the wave functions, and small positive charges inside these spheres, since a little of their charge is located outside the spheres. The spherical average of this volume charge, ordinarily cutting through both regions inside and those outside atomic spheres, will be very close to zero everywhere. In other words, the second term of Eq. 37-4 to a good approximation will be zero, aside from terms coming from the surface of the molecule. To a very good approximation, the potential will be a function of r within the sphere surrounding a given atom which just contains enough electrons to make it electrically neutral, and outside this sphere, the potential will be constant.

We have, then, a very good approximate justification of the muffin-tin type of potential which we have been using, with one exception: The sphere on which we should match the wave functions inside the sphere, where the potential depends on r, and those outside, where it is constant, should be a sphere just large enough to contain zero net charge, or to enclose a neutral atom. This sphere will be a good deal larger than the nonoverlapping spheres which have been used up to the present. What would happen if we were to use these larger spheres?

Here a small piece of current history is interesting. Our subject is advancing very rapidly at the moment. Ten days before this page was written I wrote Keith Johnson a letter outlining these arguments, which I had presented shortly before in a seminar at Gainesville. As I was ready to write this section, his answer arrived in the mail. It is a sufficiently interesting example of the way things are sometimes done that I shall quote verbatim from Johnson's letter:

Your theoretical argument for the use of larger overlapping spheres comes at just the right time. By coincidence, we and others have been discovering that the use of atomic spheres big enough to contain essentially all the electrons of the neutral atom leads to even better results for some systems than the use of smaller overlap. For example, just yesterday Joe Norman called me from Seattle to tell me that by using for PH_3 (phosphine) a phosphorus sphere large enough to contain an effective charge of 15 electrons and a H sphere large enough to contain one electron, he was able to calculate transition-state energies for PH_3 in almost *exact* agreement with experiment. The virial theorem is also almost exactly satisfied. To determine the radius at which each atomic sphere contains all its allotted electrons, he used the molecular potential generator program, which provides a tabulation of the effective charge as a function of radius for each nucleus. He also used the lower α-value of 0.78 for hydrogen. Using significantly less overlap or no overlap at all produces significantly poorer results. The outer sphere was kept at a radius corresponding to the usual nonoverlapping atomic spheres, so that it actually overlapped the larger spheres considerably. In principle, of course, the outer sphere can pass close to the nuclei of the outermost atoms or be closer to contain almost all the electrons of the molecule.

Frank Herman also called me recently to let me know that he found the use of much larger atomic spheres (large enough to contain the electrons of the neutral atom) for the organic molecule TCNQ (tetracyanoquinodimethane, Fig. 37-1) yields results in excellent agreement with photoemission data, and leads to the virial theorem being almost exactly satisfied. The use of significantly less overlap or no overlap at all gives results in poor agreement with experiment, and the virial theorem is not satisfied.

Thus the use of overlapping atomic spheres large enough to contain the electrons of the neutral atoms may be an excellent criterion for choosing sphere radii in polyatomic systems. It seems to yield results as good as what might be expected from a more exact non-muffin-tin treatment. Your physical argument for the use of such sphere radii in terms of electrostatics seems to justify the procedure.

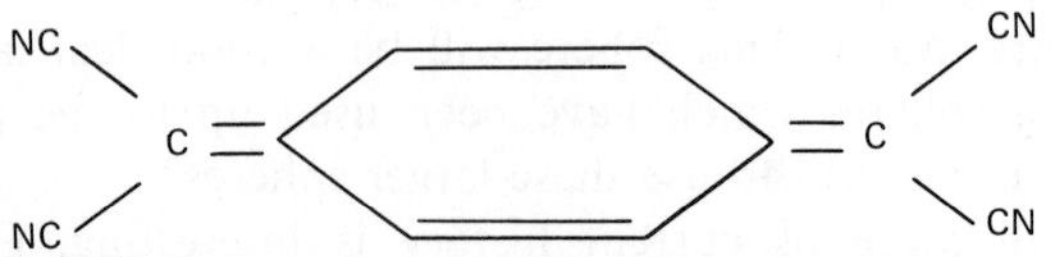

Fig. 37-1. The TCNQ (tetracyanoquinodimethane) molecule.

So that's that. However, there are still a number of points to be mentioned in connection with the overlapping spheres. The main computer program of the Johnson-Smith method simply joins the solutions inside the spheres smoothly onto the spherical Bessel and Hankel functions in the region of constant potential outside the spheres. This is still required with the overlapping spheres, and is still performed in the same way. But we must not overlook the fact that when we have two overlapping spheres, there will not be a region II between the spheres representing two neighboring atoms. Do we have any reason to think that the wave function will be continuous with continuous derivative between one of the overlapping spheres and the neighboring sphere? Nothing in the program directly ensures this.

Here I was in quite a quandary, until in the process of collecting material for this book, I remembered the von der Lage-Bethe paper of 1947, which was discussed in Sec. 28 and illustrated in Fig. 28-1. The reader will remember that von der Lage and Bethe were discussing the cellular method. They showed in a case like the sodium crystal that if the conditions of continuity between one Wigner-Seitz cell and the next were satisfied around the perimeter of a circle which was the intersection between the plane forming the boundary of the cell and the sphere of equal volume, the empty-lattice test would hold to a high approximation. With the overlapping spheres, we can well think of our problem in the language of the cellular method, using Wigner-Seitz cells. We should then use the muffin-tin wave function arising from a muffin-tin potential centered within one overlapping sphere out to the boundary of that Wigner-Seitz cell, and then in the next cell use a similar muffin-tin wave function centered in that cell. The spheres we are recommending would be about the same size as that used by von der Lage and Bethe.

Thus, if we think of Fig. 28-1, our method of satisfying boundary conditions would give exact continuity from the interior of a sphere to the region II outside it and back to the interior of the next sphere, everywhere over the surface of the cell of the figure lying outside the circles. We would, in other words, not only be doing as well as von der Lage and Bethe were doing, but a good deal better: We exactly satisfy the continuity condition all through the protruding corner regions of the cells. Hence we should certainly expect, following the experience of von der Lage and Bethe, that our wave functions would in fact satisfy the continuity conditions very satisfactorily through the overlapping-sphere region. Johnson and his colleagues in recent work have used the spheres of Rösch, Klemperer, and Johnson and the computer programs for synthesizing the wave functions, and have found that they get continuity of the wave function as accurately as their computer programs show. Hence it seems that we do not need to go to any additional effort to obtain continuity.

Let us think of a few further points regarding the overlapping spheres. It is not obvious that the program for computing the total energy, which was designed for nonoverlapping spheres and the muffin-tin potential, would hold without change for the overlapping spheres. Johnson and his colleagues have not carried this through, and it will certainly be a matter for future work to modify this program if necessary. The point is that the overlap region certainly must not be counted twice in performing volume integrations, and there is nothing in the existing program to prevent this. This is only an example of the further steps that must be taken before we are sure of the method.

Then in addition, it is certainly not justified to omit the dipole terms, $l=1$, in the potential. They are not included in the existing muffin-tin method, which works only with the terms in $l=0$. But as we have noted, long-range dipole forces occur. Many important physical phenomena depend on them. The most obvious is the existence of the dielectric constant. The whole theory of excitons, for instance, is dominated by the very high dielectric constants which some of the semiconductors have. Current theory puts in the dielectric constant as an empirical constant, and this would never satisfy us. The whole theory of dielectrics should be examined. There is probably a close relation between this and one of the considerable shortcomings of the muffin-tin theory for very small molecules. The muffin-tin calculation does not lead to the nonlinear nature of the water molecule or the nonplanar nature of the ammonia molecule. Debye, as far back as 1929, pointed out that by including dipole-dipole interactions between the oxygen or nitrogen atom and the hydrogens in these molecules, one could get a good explanation for the bent nature of these bonds. It is a piece of unfinished business to look into this.

Another application of the $l=1$ term is in the use of the Hellmann-Feynman theorem. We should note that the nuclei in a molecule are not acted on by the same forces which would act on an electron at the same point. The nuclei are acted on only by the ordinary electrostatic forces, while the electrons have also the forces arising from the exchange-correlation potentials. The potential of Eq. 37-3, in the limit as r goes to zero, is then that which acts on the nucleus at the origin. The term $l=1$, the dipole term, is the only one which leads to an electric field at the nucleus. It is just this field which gives the force on the nucleus, according to the Hellmann-Feyinman theorem.

For a molecule in which all atoms are at equilibrium, the forces on each atom are zero, and therefore the terms in $l=1$ at the various nuclei should all vanish. But for nuclei displaced from equilibrium, there will be forces given straightforwardly by the $l=1$ terms in the potential energy. This may well lead to the simplest and most direct method of computing the interatomic forces on displaced atoms, and hence of getting information about the energy as a function of internuclear position. The statement made in Johnson's letter quoted above, that the virial theorem was satisfied, meant merely that the calculation was made for the observed nuclear positions, and that $2KE + PE = 0$, a statement that the virial, arising from the sum of atomic forces times displacements, was zero. The fact that this is holding for the overlapping-sphere method gives us hope that the Hellmann-Feynman theorem may hold accurately enough to be useful. This could turn out to be a much simpler method of finding energy curves than the direct use of the total energy expression, just as the transition state technique has allowed us to get excitation energies without using the total energy.

We have now carried the $X\alpha$ method up to the present state of the art. It is a good place to stop and look ahead. I have not gone into the arguments which have led the various groups working on the method to choose the particular substances on which they have been computing. Enough has been said, I think, to demonstrate that the method can be used for problems of very considerable complexity. But we are not living in a vacuum; we are living in troubled times. We are hoping to apply our methods to problems of the present and the future, as discussed in the next book.

BOOK VI

Looking Ahead

38. Big Molecules, Energy, and the Computer

These pages are being written at a time when the energy crisis is sufficiently acute so that President Nixon is proposing not only short-range expedients for getting over the winter of 1973–1974, but also a long-range project as large as the Manhattan Project, to study the technology involved in the long-term problems associated with the crisis. By coincidence, Nixon's speech of November 7, 1973 came during a dinner which was being given by Harold P. Hanson, vice-president for academic affairs of the University of Florida, in honor of Edward Teller, who was scheduled to give an evening lecture about the laser. Teller was so concerned about the energy problem that, after the dinner guests had heard the president's speech on a loudspeaker, he gave a talk on energy instead of his announced topic.

I, as well as many others, have given a great deal of thought to the energy problem, and many of the applications of the $X\alpha$ method which have been made or are being proposed have a rather direct bearing on it. I felt, however, that those who would be setting up the proposed long-range research study probably would not be aware of what we were doing with the development of the method. Consequently during the week following the president's address I wrote a memorandum on the $X\alpha$ method and the energy crisis, and sent copies to a number of old friends who seemed likely to be consulted about the project. The memorandum was mostly a short description of what the method was, but there was one paragraph which I shall quote, since it suggested a few of the types of applications which I felt would be likely:

> The method is just as simple to apply to heavy atoms as to light ones. This means that it is uniquely fitted to attack such problems as catalysis produced by a heavy metal such as platinum, biological problems such as photosynthesis or the chemical processes involved in metabolism, the magnetic behavior of complexes of atoms surrounding a $3d$ transition atom or a rare earth or actinide atom, and so on. The use of such atomic complexes for high-intensity lasers is the sort of problem for which the method is perfectly adapted. There has been a greatly enlarged realization during the last several years of the importance of localized states such as impurity states and excitons in many aspects of solid-state physics, and the $X\alpha$ method is capable of treating such problems with clusters large enough to be significant. This indicates the reasons why those familiar with the method believe that it can be profitably applied to a great many of the technical problems which will be met in the energy study.

I talked this memorandum over with L. H. Nosanow, then the head of the physics department at Gainesville, an expert in the many-body theory, and with vice-president Hanson, a former x-ray spectroscopist, the successor of Robert Mautz, whom I mentioned in Sec. 33. They both were enthusiastic. But in addition, I sent copies to Keith Johnson, Julius Stratton, and Jerome Wiesner at MIT, as well as a few others; and to Seitz, J. R. Schrieffer of Pennsylvania (the recent Nobel Prize winner for his work in superconductivity, but also a former member of the SSMTG back in the early 1950's), Guy Stever, the present head of the National Science Foundation and science advisor to the president, whom I had known on my London visit in 1943; and a few others. The letter from Keith Johnson which I quoted earlier was in answer to the letter which I had written him when sending the memorandum. Let me quote additional paragraphs from his answer, to indicate the type of problem which he has in mind:

> Thank you for your letter and for the memorandum on the $X\alpha$ method and the energy crisis. I concur completely with the message of the memorandum, and I believe it is an excellent idea to send it to the selected group of individuals that you suggested. Hopefully it will have some impact, either directly or indirectly, on those involved in setting policy for the new energy research agency.
> Along the lines of energy-related research, I should tell you, if I have not already done so, about the interdisciplinary research proposal we have written on catalytic materials over the past six months. Following the workshop on catalytic materials held at Stanford last March, Nick Grant asked me to organize and coordinate an interdisciplinary research program on catalytic materials and to formulate a research proposal. I was able to find a number of faculty members in several departments (including metallurgy and materials science, chemical engineering, chemistry, electrical engineering, and mechanical engineering) who were interested in such an interdisciplinary effort. After working through much of the summer, we came up with a reasonably integrated package of proposed research focusing on a wide range of experimental techniques and the $X\alpha$ method on a few of the basic catalytic materials and reactions.... Nick Grant, who has taken the responsibility of contacting the various agencies, is very optimistic about its eventual sponsorship. We are excited about the prospects of research in this area, and it should be very important to the pollution and energy problems, for example in the development of catalysts for CO oxidation and for conversion of coal to gas.

This will give some idea of the types of problems we are considering. It has been known for a long time that nickel, for example, formed a very good catalyst. A small molecule adsorbed on a surface containing nickel can very easily be dissociated, and its constituent atoms can move along the surface, being regrouped into other molecules which then leave the surface. A great many reactions of great practical importance in chemical technology happen in this general way. The main emphasis on overcoming air pollution by automobiles is focused on the use of similar catalysts to facilitate the conversion of the harmful gases in the exhaust into harmless forms. The reason why the antipollution devices being planned at present are so expensive is that platinum seems to form the most effective catalyst known at present for this purpose.

In every case it is an atom with an unfilled inner shell, $3d$ or $4d$ or an f shell, which forms the catalyst. In a general way, it has been supposed that the reason was simple: These atoms have partly filled inner shells, and consequently many energy levels close together. It is very easy to have a process in which small amounts of energy are transferred back and forth between the adsorbed atoms and these energy levels in the catalyst. But until the $X\alpha$ method came along, no one of the theoretical methods available to the chemists was practical for studying the energy levels and really trying to understand what was going on. Now Johnson, and R. P. Messmer of the General Electric Company, among others, have been studying the problem with the $X\alpha$ method. One of the reasons why NiO is of such great interest is that clusters of nickel and oxygen atoms, which have already been studied by the $X\alpha$ method, give a good indication as to the chemical state of oxygen atoms adsorbed on nickel. Schrieffer is very much interested in these problems, and is carrying out a great deal of work bearing on it at the University of Pennsylvania. Surface chemistry and catalysis form one of the very active fields of practical interest at present.

But this is far from the only place where the heavy atoms and the large molecules and clusters come into the energy problem. How do we and other members of the animal kingdom take in oxygen from the air and use the food we eat to get the energy we need to operate with? Through metabolism, through breathing, through the hemoglobin in the blood going through the lungs and turning red, going through the blood vessels and turning blue, and then back to the lungs again. Why is the blood red, and why do we have to eat plenty of iron, to prevent anemia? Simply because the essential feature of hemoglobin is an iron atom surrounded by various neighbors, with energy levels close together, which permit a change in the state of the iron ion with exchange of energy. The structure of hemoglobin is a complicated one, but several scientists are working on applying the $X\alpha$ method to the problem, to get a better understanding of the energy levels and molecular orbitals of the essential atomic cluster involved in its action.

And how is it that the energy of the sun allows a plant to absorb sunlight and convert carbon dioxide and water into oxygen and living matter? Photosynthesis occurs on account of chlorophyll, and here again we have heavy atoms surrounded by clusters of atoms which allow energy to be absorbed and given up again. All sorts of mechanisms for converting radiant energy into forms which can be used by living matter involve similar kinds of chemical processes. And yet it is these processes that have been very poorly understood by the chemists. It is interesting to ask why. The reason is perfectly simple. The chemists have been sticking to the analytic atomic orbitals and the LCAO method. And those methods, as I have pointed out often in these pages, work well for small molecules and light atoms, but are practically useless for the heavy atoms and large molecules that are of real importance.

The chemists have done well with molecular orbitals for the familiar organic compounds, made of hydrogen, carbon, nitrogen, oxygen, and a few other light elements. But they can hardly do anything at all with an element heavy enough to have a few $3d$ electrons. I realized this shortcoming of the standard methods which they are using when I decided, way back in the 1930s, that it was absolutely necessary to get a different type of approach going, an approach which would handle heavy atoms as easily as light ones. This is where the $X\alpha$ method has come from, and it is at last, after many years of careful tending during its infancy, coming to a period of explosive growth. It is, of course, the computer that has made this possible. And yet the computer is something which many people view with awe, if not with positive hatred.

Fortunately I have lived close enough to computers for the past 50 years so that I can treat them in a completely matter-of-fact way. Back in the 1920s, I would use log tables or a slide rule for computations. A little later, when I was on the staff at Harvard, I had the use of a Monroe electric calculator, which was very useful; though once, when I inadvertently tried to make it divide something by zero, it clearly was going to take forever to get the answer, and when I tried to stop it I managed to blow the main fuse of the Jefferson Laboratory. But back in that same period, in 1925, I made the acquaintance of Vannevar Bush and Norbert Wiener, of MIT, two striking characters who played a great part in the development of computers.

My first experience with these two was in December 1925, when Max Born was lecturing at MIT. Before the Borns left, they gave a party at their apartment for some of their local friends. These included Kemble and myself, from Harvard, and several from MIT, including Wiener and Bush. It was quite a gay party. First we had dinner in a hotel around the corner from their apartment in the Back Bay, with a dinner dance in progress, the saxophone making so much noise that we could hardly hear ourselves talk. Then back to the apartment, to play with the Christmas toys that the Borns had bought to send back to Germany for their young relations. The main thing was an electric train, which was intended for their small boy. Professor Born had not been able to make it go, and we couldn't either, until finally Norbert, whom we expected to be the most impractical one of the group, suggested that the trouble might be that it worked with a transformer, and it might be that we were in the part of Boston that still had direct current in the lighting circuit. We telephoned the electric light company, and found that was the trouble.

That wasn't enough to stop MIT. Bush and Wiener knew that the electric train would work on direct current of the proper low voltage, which could be produced with a storage battery. Only of course no one had a battery. But being resourceful people, they telephoned around to neighboring garages until they located an automobile battery that could be borrowed for the occasion, and went out and got it. And sure enough the train went, and we had Born, his wife, physicists and mathematicians and engineers, all down on the floor playing with the train. After that all sorts of parlor tricks. Everybody enjoyed themselves, and decided that the Borns were very nice people.

Bush was a brilliant electrical engineer, 35 years old at the time, a native Yankee, and Wiener was a 31-year-old genius, a mathematician who had taken his Ph.D. before he was 20. After I went to MIT in 1930, I came to know Bush as a professor and then dean of engineering, and as the designer of the differential analyzer. This was a mechanical computer which filled a large room, a dizzying structure made up of shafts, gears, and integrating units built on the principle of a planimeter. Bush felt that the future belonged to the computer, but his machine had to be programmed with a screwdriver and a monkey wrench, and it took all day to set up to solve a problem. Even at that, he could compute the trajectory of a rifle shell so that he had the answer almost as rapidly as the shell would have followed through its orbit, and before I had been at MIT for many months, he was giving colloquia about how it could be used to solve the self-consistent-field problem. He and his young associate Samuel H. Caldwell managed to set up the helium atom so that it would be solved all in one step, without requiring the iteration that Hartree used. I have already mentioned how fascinated Hartree was with the machine, and how he often visited MIT to use it before he built his own.

Wiener was the typical absent-minded professor and universal genius. The classic story was of the time he stopped a friend at lunchtime on the walk connecting the main buildings and Walker Memorial, where the cafeteria was located. They talked a while, and when they parted, Norbert asked the friend, "Which way was I walking when I stopped?" "Toward the main building," said the friend. "Then I've had my lunch," said Norbert. He was fascinated with the differential analyzer and the implications of computers. He was the inventor of the word cybernetics, and probably more than anyone else he was responsible for the feeling that a computer is the next thing to an artificial brain. Bush was rather a more practical type, but I felt that both of them went rather too far in regarding computers in a mystical way. I have never regarded them as anything more than superior versions of desk calculators.

Toward the end of the '30s, the differential analyzer was getting considerably more sophisticated. Bush rigged up a tape input and an electrical switching scheme for putting a problem on the machine, obviating the monkey wrench. At about the same time, Howard Aiken at Harvard was working on computers along a different line. He started with the mechanical type of digital computers, namely Monroe or Marchant desk computers, but connected a large number of them together, with keys to cause them to add, subtract, multiply, divide, and so on, so that one could essentially program them. But here, though one had enough significant figures, which the integrating unit of the differential analyzer was incapable of, still the device was enormously clumsy. The ideas were there in these prewar computers, but essential features were lacking.

These features came during the war, and from groups with which I had no contact at the time. The essential point was to get electronic units rather than the mechanical ones which both Bush and Aiken were using. These electronic units were outgrowths of the electronic counters which the cosmic ray people had already been using before the war. By the time I became conscious of what was going on in postwar days, in the early days of the SSMTG, when Meckler learned about the Whirlwind computer, the devices were completely electronic and the main problems were with storage and switching. These were definitely in the field of electrical engineering, and von Hippel's Laboratory for Insulation Research was a leader in developing magnetic core storage. But the leader in developing the postwar computers at MIT was Jay W. Forrester, a young man who had come to the Institute after the war, and who was the one to build the Whirlwind. It was only a step from this to the commercial development of the same ideas by IBM and the other manufacturers of digital computers. Remarkably enough, I am not conscious that I had met Forrester, except very casually. But he is the one who has turned his interests in the direction of industrial management, and more recently into trying to predict the future course of civilization, and I shall come back to him in the next section.

As I mentioned earlier, I have never thought about computers in a mystical way, as so many people do, and consequently I have never considered them as anything but a tool. I have never personally used them, but this was only because the students were so good at it, and I felt my time was better used otherwise. I have had, however, a great many students who learned to use them, some of them extremely good programmers, and I have come to see both some of the good and some of the bad effects of computers on those who use them. As for the good features, I think that they are extremely good training in accuracy. Ordinarily a single error in programming will make a program fail to operate. There are few operations in these days in which a student learns that he simply must not make mistakes. Of course, this does not mean that mistakes are not made, but merely that endless checking is necessary. We seldom trust a program until the same problem has been set up independently on two different computers, by two different sets of people, at different places, and they get identical results on identical problems.

Even this test is not always enough, as we found with an early $X\alpha$ program for atoms. We had a program which computed self-consistent solutions of the light atoms up to silicon perfectly, but failed completely with germanium. What kind of error could this be? It turned out that as the first step in the self-consistent iterations, an input was fed into the machine which represented the best approximation to a self-consistent field which could easily be obtained. Then the machine converged rapidly to the answer. But through an error, what was being fed in was actually a purely random set of numbers instead of the desired approximation. With an atom as small as silicon, the program could take even these random numbers, and gradually converge toward the answer. But germanium was just too big. It never got started.

But the computers have bad features. The main one is that students get so fascinated with them, so immersed with what they can do, that they never get away from the computer far enough to think about what they are doing. This in a way is probably what has kept the chemists from thinking beyond the LCAO method. Ever since they started with computers, they have been using this method, enlarging it more and more, using bigger and bigger computers, without stoppng to ask whether there might not be some quite different approach to the problem. Probably if I had been personally using the computers, the same thing would have happened to me. But as it is, while the boys compute, I have to sit back and think.

I have mentioned the awe, and even horror, which some people have about computers. This arises from the fear that the computers, as mindless robots, will more and more take over the direction of civilization. This would frighten me as much as anyone, if I believed for a moment that it was true. But we must realize that it is not the computers, but rather those who decide how to use them, who might be frightening. I have mentioned Jay Forrester, and how he has been trying to use his computers to predict the future of civilization. If the computers can do this, we should all be gravely worried. Fortunately they cannot, and I am sure that Forrester realizes this as well as anyone. In the next section I present some views on these questions, which I think can be answered in a hopeful rather than a despairing way.

39. Computers and the Next Millennium

As the year 1000 A.D. approached, many of the inhabitants of Christendom looked to the forthcoming millennial year with awe, foreseeing the day of judgment predicted in the Book of Revelations. The year came and went, and nothing unusual happened. But as we look forward to the year 2000, we have much sounder reasons for thinking that this second millennial year will be the fateful one, when the human race may be facing extinction. Population explosion, ecological crisis, the exhaustion of natural resources, the moral collapse of humans, all seem to be approaching points where any one of them could lead to a breakdown of the civilization we have known. We see exponential curves growing without limit, and all approaching critical values within a few decades of the fateful year 2000. It seems as if great changes come over the world perhaps about every 500 years. Around 1000 B.C. civilization was getting well under way. About 500 B.C. Greece and Rome were flourishing, and then came the birth of Christ. The fall of Rome occurred not long before 500 A.D. and by 1000 A.D., though the day of judgment did not arrive, the dark ages were roughly over, and the awakening was ready to start. Around 1500 A.D., the discovery of the New World and the flourishing of the Renaissance brought new hope to the world. And here are are approaching 2000 A.D. Nothing looks good this time.

Is it really all this bad? I was born in 1900, and hence have been here to see what has happened in the century preceding the fateful year 2000. I have written down in this volume some of the things I have seen, both as a scientist with close relation to technology, and as a person who has tried to explore the underlying basis of things. I do not personally find things hopeless. In my own field I have seen real progress during this century, and I see no reason why if I were living a hundred years later I should not find progress continuing. I do not agree with the gloomy prophets. And yet much of what they say is true. Is it true that the world faces disaster as the year 2000 A.D. approaches? And if it is, is it in any way the fault of science and technology, as so many superficial thinkers are saying? None of us can escape realizing the effects of population explosion, pollution and ecological crisis, exhaustion of natural resources, and moral collapse. The newspapers are full of them, scientists are studying them, and everyone can evaluate for himself the effects of another 30 years of development like what we have seen in the preceding 30.

We have the extrapolations of such computer-oriented economists as Jay Forrester and Dennis Meadows at MIT. They have started with a few simple statistical laws which represent the best approximations they can make to the actually existing relations, put these together into a set of simultaneous equations, and solved them on the computer. Each of the effects they postulate, such as population increase or increase of industrial production, would lead to an exponential curve, but their interrelations are too complicated to understand without the full computer treatment. It is these extrapolations which tell us that the century following the year 2000 A.D. is approximately the catastrophic period. They are intelligent people, who realize fully that the resulting picture is oversimplified. But at least they are making a good try, and no matter what combination of assumptions they make, the final result is that a rather disastrous situation could arise within a few decades after 2000 A.D. This could take the form of a sudden large population decrease, which could result from famine, depletion of natural resources, or other disastrous effects.

There are a great many problems in mathematical physics and engineering in which one studies the building up of some effect to a final steady state. In the early stages of the process, an exponential buildup is very common. But as it goes further, other effects come in which cause the rate of increase to level off and start down, until finally a steady state is reached in which the buildup ceases. Other cases are also possible: The buildup can overshoot, then have to decrease, starting a transient oscillation that can diminish in amplitude as time goes on, eventually leading to a steady state, or can increase in amplitude, leading to a catastrophe. These cases are shown in Fig. 39-1. All of them can arise from quite simple mathematical systems. But they can give us some idea as to the sort of thing workers like Forrester and Meadows have run into from their study of much more complicated systems. What they are afraid of is the phenomenon of overshooting, leading to a catastrophic fall. What they recommend as the best hope for the world is to manage to arrive at the steady state in a smooth fashion, without the great transients that are all too easy to set up. This is the so-called S-curve (named from the shape of the curve), shown in Fig. 39-1b.

The difficulty is that it is very hard to set up this particular case, without regulating the various features of the system very carefully. Will it be necessary to have the dictatorship of Aldous Huxley's *Brave New World* or George Orwell's *1984* to achieve it, and if so, would life under these circumstances be worth living? Forrester and Meadows have very little to say about this; they realize that it is the most difficult problem that is faced, if catastrophe is to be avoided. People must realize without coercion that many unfamiliar decisions must be taken, and the present period of moral collapse is a poor time to be making these intelligent decisions, particularly when the time available for them is very short. To my way of thinking, it is the lack of intelligent thinking about philosophical questions which seems to be the most serious danger facing us.

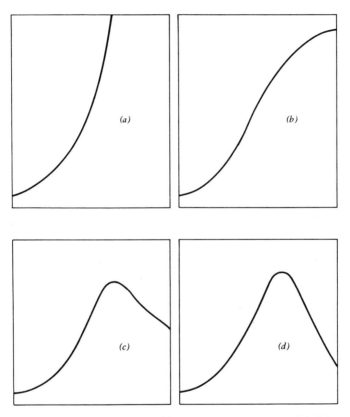

Fig. 39-1. Figure showing different possible results of an exponential buildup. (*a*) The simple exponential curve building up without limit. (*b*) The so-called S-curve, building up smoothly to a steady state. (*c*) Buildup with a transient which decreases with time. (*d*) Buildup with a transient of catastrophically increasing amplitude. In each case the horizontal axis represents time. The vertical axis could be population, or almost any other measure of activity.

A natural reaction to the proposal of a steady or equilibrium state, which many have expressed, is to feel that without growth, healthy existence is impossible. This is largely a relic of the thinking of past centuries, in which growth was always present, and in which its consequences were believed to be necessary. With an overall steady state, much change can occur, leading to healthy variety. The difficulty is that very few thinkers have sat down to explore steady-state solutions. Such situations do not in the least imply that techniques will permanently remain as they are at present. I fully expect scientific progress to continue, and with it improved technology, which should make it possible to solve the problems of pollution, ecological crisis, exhaustion of natural resources, and the other things which have frightened people. The danger in the coming years would come, not from too much technology, but from too little. In my way of thinking, however, the real danger to our civilization comes from people like John Kenneth Galbraith, who proclaim that we have an affluent society, seeming to make it unnecessary for anyone to work. Or like Lewis Mumford, who has developed such an irrational prejudice against science and technology that he would dearly love to see everyone desert them.

In my view, the social and political scientists are the ones responsible for our troubles. They are the inheritors of centuries of thought which seems to me to be based on fallacious premises. The fallacy is the assumption that social studies can be founded on the same sort of postulates that one uses for mechanics and similar branches of physics. I do not believe that social and political "sciences" ever can, or ever should try to be based on what we ordinarily think of as science.

The Greeks started the fallacy, but we can hardly blame them, for they were working in the very early days of human thought. Hobbes, in his *Leviathan* in 1651, was a contemporary of Newton and Galileo, and he tried very hard to set up his political and social thinking along the lines of Newtonian mechanics. This line of thought formed the basis of the philosophy of the age of enlightenment. Nineteenth century thinking perpetuated the same trends. Socialism can be traced back to Rousseau and the enlightenment, communism goes back to Hegel. All of these share a dogmatic approach, based on assumptions which bear very little resemblance to the behavior of real people. Classical economics is founded on the economic man, a ridiculous travesty of the way people act. Even a great deal of religion has the same built-in errors. Predestination, the supposition that one has no control over one's destiny, comes from the same assumption that if we start with the initial state of a person, everything that follows must be predictable from the laws of mechanics. This is as ridiculous a view as the old belief in the stars as determining human actions.

I have said that the fallacy involved in the social and political sciences is the assumption that they are sciences in the Newtonian sense. This does not imply at all, however, that I do not believe that living beings obey the same fundamental laws which hold for inanimate objects. The difference comes in the enormous complexity of all living things, and most of all, humans. The work we are doing in applying quantum theory to atoms, molecules, solids, biological systems, has carried us much closer to a real understanding of the nature of living matter than was ever possible before. Our greatest efforts with our best computers can only handle very simple molecular systems. To one with the experience that I and my colleagues have, it would seem preposterous to believe that even with many years of further work we could predict by means of a computer what an individual person would do. The more one works with these things, the more one becomes impressed with the infinite variety of nature. The social sciences of course recognize this variety, but they attempt to work by statistics, as Forrester and Meadows have to do. They assume that an average behavior is good enough to predict what will happen to a population.

To some extent this is true. But the really unexpected things that happen in a society are those which depart from the average, and which arise on account of the exceptional individual. The exceptional man produces ideas, the multitude follows them. The ideas may be right, in which case the man is a prophet, or they may be wrong, in which case a foolish fad will result. In either case far-reaching consequences follow, which could not be predicted from the average behavior of the human race. No computer will ever be made large enough to tell what an exceptional man will do, or where he will lead us. The only way we can study people is to study people, and this means essentially that we must study biography and history, not science, to understand human nature. And yet the fallacy has spread so far that most of the young people of the present college generation seem to think that the only way to truth is to study the social sciences.

These thoughts may be enough to indicate why I do not think one can take too seriously the details of the extrapolations of the computer-oriented economists. But at the same time they lead toward what I believe is the faith that can keep us working through a difficult time. The most striking thing that the space program has done for us is to allow men to stand on the moon, look back at the earth, and realize what a fine thing it is, and how it is all alone in space. I very strongly doubt if we shall ever find life like ours anywhere else that we can reach. The earth is all we have.

But we must realize that in the few billions of years that the earth has existed, nature started with something as lifeless as the moon, and built up from this more and more complex structures, culminating in man. The time during which man has existed, a few million years at most, is almost infinitesimally small compared to the age of the earth, and the ten thousand or so years in which anything approaching civilization has been going on is a small fraction of the age of man. We are living in a unique moment of the history not only of what we are familiar with, but of all the universe, or at any rate of all the history of the solar system. Maybe a higher intelligence has been standing by, helping along the development from a dead to a living world, or maybe pure chance has been able to produce the evolution. Modern quantum theory would not be outraged at the thought of a higher intelligence putting in its touch at the right moment, but does not demand it. But however that may be, we are at a point where our individual actions may affect the outcome of this most remarkable of all experiments, in a decisive way. Let us hope that our philosophers and political and social scientists can learn to take a more sensible view than they have done in the past, and that they can lead us to take the steps which may be necessary to carry us safely through the next, fateful millennial years.

SUGGESTED READING

No attempt has been made in this book to include references or indications for supplementary reading. The subject is a very broad one; to give references to all the papers mentioned would be an unnecessary burden. However, I have written a series of books covering in much more detail all the scientific topics we discussed here. These books have voluminous bibliographies. The following is a list of these books, with a short discussion of their contents, and a suggestion as to which ones are more appropriate to particular topics covered in this book:

Introduction to Theoretical Physics (with N. H. Frank), McGraw-Hill, New York, 1933. A general discussion of mechanics, heat, electricity and magnetism, optics, and wave mechanics, on a level suitable for an advanced undergraduate or beginning graduate student. A good introductory text for a reader who has not followed mathematical physics, but wishes to get a sufficient background to read the present volume.

Introduction to Chemical Physics, McGraw-Hill, New York, 1939; Dover, New York, 1970. Thermodynamics, statistical mechanics, kinetic theory, including Fermi statistics. Properties of gases, liquids, and solids. Elementary discussion of atoms, molecules, and the structure of matter, including elementary methods of wave mechanics. Again on a level suitable for an advanced undergraduate or beginning graduate student. A good introductory text for a reader who has not followed chemical physics, but wishes to get a sufficient background to read the present volume.

Microwave Transmission, McGraw-Hill, New York, 1942; Dover, New York, 1959. Discussion of transmission lines and impedance, electromagnetic waves in free space and wave guides, relation of wave guides to transmission lines, radiation from antennas, coupling of wave guides and coaxial lines. Fairly specialized, but the treatment of radiation could be useful in connection with optical applications.

Mechanics (with N. H. Frank), McGraw-Hill, New York, 1947. Revision of the material on mechanics from *Introduction to Theoretical Physics*.

Electromagnetism (with N. H. Frank), McGraw-Hill, New York, 1947; Dover, New York, 1969. Revision of the material on electromagnetism from *Introduction to Theoretical Physics*, incorporating some microwave treatment.

Microwave Electronics, Van Nostrand-Reinhold, Princeton, New Jersey, 1950; Dover, New York, 1969. Theory of wave guides and resonant cavities, coupling of guides to cavities, periodically loaded wave guides, general principles of klystron, linear accelerator, traveling wave amplifier, magnetron, cyclotron, synchrotron. Treatment is from point of view similar to that used in quantum electrodynamics.

Modern Physics, McGraw-Hill, New York, 1955; reprinted with title *Concepts and Development of Quantum Physics*, Dover, New York, 1969. Historical account of development of modern physics, starting with nineteenth-century physics, carrying through to wave mechanics, structure of atoms, molecules, and solids, and the atomic nucleus. Intended for advanced undergraduates or graduate students; a good introductory text for a reader who has not followed the development of modern physics, but wishes to understand the present volume.

Quantum Theory of Atomic Structure. Volumes 1 and 2, McGraw-Hill, New York, 1960. General principles of quantum theory and wave mechanics, with application to atomic theory. Volume 2 treats Hartree-Fock method, accurate methods of atomic calculation, group theory, angular momentum operators, Racah's method, Dirac's theory of electron spin, radiation transition probabilities, hyperfine structure. There is an extensive bibliography in Volume 2, of papers on wave mechanics and atomic theory, from the early days of quantum theory to the 1950's. Would carry the reader from the present volume up through more advanced treatment of wave mechanics. Intended for graduate students and advanced readers.

Quantum Theory of Molecules and Solids, Volume 1, *Electronic Structure of Molecules,* McGraw-Hill, New York, 1963. H_2^+, hydrogen molecule, both Heitler-London and molecular orbital methods, diatomic molecules, group theory and symmetry of wave functions, application to ammonia, methane, water, ethylene, benzene. Appendices treat Born-Oppenheimer method, Hellmann-Feynman and virial theorems, Hartree-Fock method, two-center integrals, methods for nonorthogonal orbitals, repulsion of helium atoms, oxygen molecule, group theory, ring of six hydrogen atoms, method of Hurley, Lennard-Jones, and Pople, and various other topics. Extensive bibliography of papers on molecular theory, from early days to 1960. This book would give the reader of the present volume a good introduction to the standard LCAO methods in use by the chemists. Intended for graduate students and advanced readers.

Quantum Theory of Molecules and Solids, Volume 2, *Symmetry and Energy Bands in Crystals,* McGraw-Hill, New York, 1965; Dover, New York, 1972. Crystals and their symmetry properties and space groups, atomic radii and the chemical bond, symmetry of electronic wave functions, discussion of plane-wave expansions, LCAO and OPW methods, cellular and APW methods, with examples of calculations of energy bands in crystals. Appendix covers detailed description of many space groups, treatment of momentum eigenfunctions, spin orbit and relativistic effects in energy bands, and other topics. Extensive bibliography of papers on solid-state theory, from early days to about 1963. Would give the reader of the present volume a good introduction to standard methods of calculating energy bands. Intended for graduate students and advanced readers.

Insulators, Semiconductors, and Metals, Volume 3 of *Quantum Theory of Molecules and Solids,* McGraw-Hill, New York, 1967. Theory of electrical conductivity, cyclotron resonance and related phenomena, Drude-Lorentz and Ewald theories of insulating crystals and of the optical properties of metals, theory of x-ray diffraction, including thermal vibrations, lattice vibrations of metals and insulators, lattice energy of ionic crystals, cohesive energy of metals by Thomas-Fermi and Wigner-Seitz methods, the crystal as a many-body problem. Appendix covers motion of wave packets in periodic potential, wave functions of impurity atoms, free electrons in a magnetic field in cylindrical coordinates, Ewald's method for crystalline fields, polarization and van der Waals attraction, and the virial theorem for solids. Very extensive bibliography covers not only theory of these various topics mentioned in the text, but many experimental papers as well, dealing with modern methods of studying energy bands.

Quantum Theory of Matter, 2nd edition (entirely revised from 1st edition, which was published in 1951), McGraw-Hill, New York, 1968. A condensed version of a great deal of the material in *Quantum Theory of Atomic Structure*, Volume 1, and *Quantum Theory of Molecules and Solids*, Volumes 1 to 4. Intended as a graduate text on the quantum theory and its application

Suggested Reading

to atoms, molecules, and solids. A very usable supplementary text for readers of the present volume who wish to go further into the theoretical background of the work, without going into the detail covered in *Quantum Theory of Atomic Structure* and *Quantum Theory of Molecules and Solids*. The bibliography covers only some more recent papers dealing with the topics of those books, up to about 1966.

The Self-Consistent Field for Molecules and Solids, Volume 4 of *Quantum Theory of Molecules and Solids,* McGraw-Hill, New York, 1974. A much more detailed treatment of the $X\alpha$-SCF method than given in the present volume. Statistical exchange approximation and the self-consistent field, dependence of total energy on occupation numbers, the spin-polarized method, self-consistent potential for molecules and crystals, the multiple-scattering cluster method for impurities and complex molecules, alkali halide crystals, energy levels of covalent and partially covalent crystals, ferromagnetism of iron, cobalt, and nickel, temperature dependence of ferromagnetism, spin waves and the Heisenberg exchange, antiferromagnetism of MnO and similar oxides, and of transition metals. Extensive appendix on theory of the exciton. Very extensive bibliography, including not only theory of topics treated in the text but also excitons and magnetism, from early days up to about 1970. Valuable supplementary reading for the reader of the present volume who wishes more detail. Intended for graduate students and advanced readers.

INDEX

Abraham, Max, 142
Absorption, of radiation, 18, 21, 156, 308
Action variables, 15–16
ARPA (Advanced Research Projects Agency), Dept. of Defense, 269–271
Aiken, Howard, 330
Albertson, Walter, 168
Algebra, matrix, 20
Alkali halide crystals, 5, 156, 200, 230
Alkali metals, 126–129, 301
Allen, Leland C., 263
Allis, William P., 164, 187
 and Morse, 187
Allison, Samuel K., 7
Alloys, 190
AMO (Alternant Molecular Orbital) method, 252
Alternating potential, 248
Altmann, S. L., and Bradley, 187
Amaldi, E., 233, 240
American Physical Society, 158, 193, 239, 242
Ammonia molecule NH_3, 321
Amorphous solids, 166, 311
Ampère, A. M., 23
Amsterdam, 228, 314
Analytic atomic orbitals, 128
Anderson, Carl, 145, 149
Anemia, 327
Angular correlation, 250
Angular momentum, 20–29, 38
Ann Arbor, Michigan, University of Michigan, 233
Annihilation operators, 141
Anomalous dispersion, 17
Antibonding orbitals, 104, 108, 110, 114, 127
Antiferromagnetism, 129, 203, 242, 247, 256, 296, 301–303, 310
Antisymmetric wave function, 46–51, 57, 58–61, 151, 249
Argon crystal, 295
Artificial brain, 330
Asano, S., 263, 303
Aston, Frederick W., 6, 193

Astrophysics, 22
Atom, two-electron, 43–51
Atom bomb, 3
Atombau und Spektrallinien, A. Sommerfeld, 22, 26
Atomic core ("Atomrumpf"), 23, 24, 53
AEC (Atomic Energy Commission), 170, 232, 268
Atomic number, 46
Atomic orbitals, 66, 90, 114, 290
Atomic spheres, 312
Atomic units, 47
"Atomrumpf," atomic core, 23, 24, 53
Attraction, electrostatic, 6
APW (Augmented Plane Wave) method, 189, 191, 230, 256, 257, 261, 263, 266, 276, 291, 293
Austin, Texas, University of Texas, 265
Austrian Tyrol, 159
Average energy, for Bloch sum, 114
Average value, 40
Averill, Frank W., 293–295, 303
Azimuthal quantum number, 16, 34

Baerends, E. J., 314
Ballard, Stanley S., 275
Banking crisis, 1933, 158
Bardeen, John, 172, 226, 227
 Cooper, and Schrieffer, 226
Barnett, Michael P., 241, 265, 266, 287
Barrow, W. L., 212
Bartol Research Laboratory, 168, 268
Bates, D. R., Ledsham, and Stewart, 103
BTL (Bell Telephone Laboratories), 165, 169, 170, 190, 212, 213, 215, 227, 231, 238, 241, 255, 262
Benzene molecule C_6H_6, 63, 107, 109, 110, 113, 114, 307
Berkeley, Calif., University of California, 7, 63, 104, 163, 171
Berlin, 62, 65, 158, 159
Beryllium crystal, 190
Bessel functions, 188, 230, 285, 320
Bethe, Hans A., 115, 157, 227, 232, 240, 241, 290, 320

343

Index

Biochemistry, 3
Biography, 337
Biophysics, 3
Bitter, Francis, 167, 226, 270
Bloch, Felix, 7, 62, 112, 115, 123, 126, 129, 181, 195, 243, 246, 247
   homogeneous electron gas, 126, 176, 195, 277
   sums, 112, 113
   theorem, 113, 118, 174, 177
Body-centered cubic structure, 119, 173
Bohr, Niels, 5, 6, 10, 11, 15, 18, 19, 23, 159, 193, 240
   atomic units, 70
   correspondence principle, 10, 16
   frequency condition, 16, 39, 40
   hydrogen atom, 5, 10, 24, 34
   -Kramers-Slater theory, 11, 15, 18
   magneton, 24, 299
   orbitals, 53
   periodic system of the elements, 23, 25, 52
Boltzmann, L., 83
   constant, 83, 309
Bomb, atomic, 3
Bonding orbital, 104, 110, 114, 127, 291
Borelius, G., 229, 240
Boring, A. M., 302
Born, Max, 6, 21, 34, 95, 140, 153, 159, 166, 328
   -Heisenberg-Jordan matrix method, 21, 24, 38
   -Oppenheimer approximation, 95, 96, 197
Bose-Einstein statistics, 138, 140
Boston Museum of Science, 170
Boston University, 264
Bothe, W., 12
Boyce, Joseph C., 166
Boys, S. Francis, 258, 264
Bragg, Sir Lawrence (W. L.), 56, 163, 166, 172, 229, 240
   James, and Bosanquet, 56
Bragg, Sir William (W. H.), 56
Brass, energy bands, 284
Brattain, Walter H., 227
Breit, Gregory, 150, 228
Bridgman, Percy W., 5, 230
Brillouin, Leon, 117, 240, 241
   zones, 118, 173, 190, 248, 263
Bristol, University of, 95, 190

Broadening, of spectral lines, 18
Brookhaven National Laboratory, AEC, 224, 233, 237, 241, 268, 270
Brooks, Harvey, 241, 268
Budapest, University of, 158, 195, 246
Buechner, William W., 170, 275
Burdick, Glenn W., 262, 294
Burrau, O., 102
Bush, Vannevar, 172, 216, 234, 271, 328, 329

Calais, Jean-Louis, 302
Calcium, atomic spectrum, 22–27
Caldwell, Samuel H., 329
California Institute of Technology (Cal. Tech.), 7, 171, 214
California, University of, Berkeley, 7, 63, 104, 163, 171
Callaway, Joseph, 255
Cambridge, Massachusetts, 6, 21, 163
Cambridge, University of, 6, 166, 190, 228, 241, 258, 264
Carbon dioxide molecule CO_2, 307, 312
Carbon molecule C_2, 313
Carbon monoxide molecule CO, 314, 326
Carlsberg brewery, Copenhagen, 240
Carnegie Institute of Technology, Pittsburgh, 173
Case Institute, Cleveland, 165, 171
Catalysis, 325, 326
Cavendish Laboratory, Cambridge University, 6, 9, 11
Cavendish Professor, Cambridge University, 166, 190, 229
Cavity, for electromagnetic problem, 136
Cell, Wigner-Seitz, 173
Cellular method, 173–184, 185–188, 243–246, 320
Cesium crystal, 293
Chalk River laboratory, 241
Charge density, 43, 55, 56, 87, 113
Charge transfer, 200
Charles W. Morgan, whaling ship, 170
Chemical effect, 305
Chemical physics, 3, 4
Cherwell, Lord, 229
Chicago, University of, 7, 163, 193, 233, 241, 264
Chlorophyll, 327
Chodorow, Marvin, 189, 256, 262

Index

Chromium, 25, 203, 296, 299, 303
Chu, L. J., 212
Churchill, Sir Winston, 229
Classical mechanics, 15
Clementi, Enrico, 264, 266
Closed shells, 27, 55
Cloud chamber, 145
Clusters of atoms, 325
Coal-gas conversion, 326
Cobalt, 25, 120, 123, 128, 185, 203, 297, 310
Cockcroft, Sir John D., 193
Cohen, Morris, 268
Coherent scattering of x-rays, 56
Cohesion, of metals, 127, 129, 301
Cohesive energy, 293
Collins, Samuel C., 226
Columbia University, 185, 209, 226, 255, 268, 314
Commutation rule, 20, 36, 140
Complete set of orbitals, 69
Complex conjugate, 37
Complex spectra, 22, 62, 66–80
Compressibility, sodium crystal, 176
Compton, Arthur H., 7, 8, 12
Compton effect, 8, 12, 15, 56
Compton, Karl T., 6, 8, 63, 64, 163–165, 170, 209, 215, 233
Computers, electronic digital, 4, 72, 172, 254, 255, 276, 288, 328–333
Conant, James B., 64
Condon, Edward U., 7, 63, 65, 165, 172, 212, 241
  and R. W. Gurney, tunnel effect, 132
  and G. H. Shortley, *Theory of Atomic Spectra*, 65, 71
Conduction band, 115, 156, 186
Configuration interaction, 69, 141, 154, 183, 250, 253
Conklin, James B., Jr., 276, 291, 308
Connolly, John W. D., 276, 307, 313
Conservation of energy and momentum, 8, 11, 117
Coolidge, Albert S., 155
Copenhagen, University of, 6, 11, 13, 19, 159, 233, 240
Copper, alloys, 191
  atom, 255
  atomic cluster, 311
  chloride, $(CuCl_4)^{2-}$, 307

  crystal, 185, 189, 202, 262, 295
  ion, Cu^+, 253
Core orbitals, 256
Corliss, Lester M., 242
Cornell University, 224, 228, 231, 305
Corpuscular theory, 8
Correlation energy, 181, 183, 249–251
Correspondence principle, Bohr's, 10, 16
Cosmic rays, 211, 330
Coulomb energy, 46, 125
Coulomb integral, 49, 92, 98
Coulomb potential, 33
Coulomb repulsion, 46, 70
Coulson, Charles A., 241, 248, 251, 265, 287
Coulson-Fischer method, 248, 250, 252, 301
Counters, 330
Coupling, of p electrons, 28
Covalent bonds, 93, 104, 166, 251, 309
Creation operators, 141
Crystal momentum, 116
Crystallography, 117, 166
Crystals, 3, 255
Csizmadia, I. G., 266
Cunningham, E., 131
Curie, Pierre, 120
Curie temperature, 121, 308, 309
Cusp, of helium wave function, 154, 155
Cybernetics, 330
Cyclotron, 169, 171
Cyclotron resonance, 263
3d electrons and ferromagnetism, 127, 128
3d energy bands, 248
3d transition elements, 185, 203, 256, 262, 279, 303, 308, 325, 328
  and optical absorption, 156

Danese, J. Bryan, 313
Darwin, Sir Charles G., 240
Daudel, Raymond, 241
de Boer, J. H., 240
de Broglie, Louis, 12, 13, 21, 31, 35, 40
de Broglie, Maurice, 12
de Broglie waves, 12, 13, 31–35, 187
Debye, Peter, 7, 13, 63, 152, 159, 166, 172, 321
DeCicco, Peter, 293
Degeneracy, 27, 48
Delft, Technical University, 228, 230
Demos, Peter, 226
Density matrix, 194–195

Density, of states, 82
Desk computers, 330
Determinant, definition and properties, 58
Determinantal method, 57, 141
Determinantal wave function, 57, 58, 60, 66, 73, 125
Deuterium, 6
Deutsch, Martin, 306
Diagonal matrix, 38, 41
Diagonal matrix component of Hamiltonian, 70, 125
Diagram techniques, 141
Diamagnetic susceptibility, 152
Diamond crystal, energy bands, 185–186, 255
Diatomic molecules, 95, 197–198, 263, 312
Dielectric constant, 321
Dielectrics, 317
Difference, finite, 18
Differential analyzer, 329
Differential operators, 35–39
Diffraction, 9, 45
Dipole, 317, 321
Dirac, P. A. M., density matrix, 194–195
  determinantal wave function, 58–61
  quantum theory, 21, 42, 46, 240, 246
  radiation theory, 19, 130–144
  statistics, 81–85, 89
  theory of the electron, 145–150, 164
Directed orbitals, 105
Dispersion formula, Kramers–Heisenberg, 15–20
Dissociation energy, 93
Domains, magnetic, 120
Doppler effect, 9
Double bonds, 106–109
Double quantization, 140–141
Drude, P., 17
Duane, William, 5, 56
  and Havighurst, R. J., 56
DuBridge, Lee A., 209, 220
Duke University, Durham, N. C., 239

East Germany, 158
Eastman, George, 167
Ecology, 333, 336
Economic man, 336
Effective mass, 115
Ehrenfest, Paul, 45

ETH (Eidgenossische technische Hochschule), Zurich, 13, 166
Eigenfunction, 33, 55
Eigenvalue, 33, 40, 55
"Eigenwert," 33
Einstein, Albert, 7, 8, 145
  derivation of Planck's law, A and B, 11, 17, 130, 140
  energy = mc^2, 142
Eisenschitz, R., and London, F., 152
Electric field, 137
Electrical conductivity, 115
Electrodynamics, quantum, 130–144
Electromagnetic energy, 137
Electromagnetic theory, 5, 8–11
Electron diffraction, 117
Electron gas, 124–126
Electron pair bonds, 104
Electron shells, 23, 25
ESCA (Electron Spectroscopy for Chemical Analysis), 306–307
Electron spin, 14, 22–30, 60–62, 66–77, 124–127, 181–182, 243–252, 279, 297–303, 308–310
  and Dirac's theory, 149
Electronic digital computer, 4, 72, 172, 254, 255, 276, 288, 328–333
Electronics, 168, 217
Electronics Corporation of America, 227
Electronics, solid-state, 4, 86
Electrostatics, 316
Ellis, Donald E., 276, 314
Empty lattice test, 187
Energy, 35, 41. *See also* Hamiltonian
Energy, of atom, 57
Energy bands, 156, 165, 173–184, 192, 200, 255, 263, 275, 279, 310
Energy density, 9, 11, 17
Energy gap, 116, 156, 310
Energy levels, quantized, 20
Energy, of molecules, 325
Ensemble, 44–45
Entropy, 82–83
Equivalent circuits, 212
Equivalent electrons, 29, 259
Ethane molecule C_2H_6, 109, 307
Ethylene molecule C_2H_4, 63, 106, 314
Europium sulfide EuS, 308–309
Evans, Robley D., 171, 209
Ewald, P. P., 230, 240

Index

Exceptional individual, 337
Exchange-correlation charge, 244
Exchange-correlation energy, 244–247, 277
Exchange-correlation hole, 182–193
Exchange energy, 125, 195
Exchange integral, 49–51, 91, 98, 122, 129, 204, 258, 260, 308, 309
Excitation energy, magnetic, 204
Excited states, 10
Excitons, 156–157, 200, 204–205, 308, 321, 325
Exclusion principle, Pauli, 14, 22, 25, 29, 50, 148
  and determinantal wave functions, 59
  and symmetry, 49
Exponential increase, 334
Eyring, Henry, 193, 241

Face-centered cubic structure, 119
Famine, 334
Faraday, Michael, 13
Fermi, Enrico, 81, 84, 86, 135, 143, 193, 241
Fermi-Dirac statistics, 81–86, 112, 115, 125, 138, 140, 168, 279
Fermi distribution function, 84–86
Fermi energy, 84, 87, 98, 115, 124, 298
Fermi hole, 182, 194, 243–245
Fermi surface, 168, 190–191, 263
Ferrocene, $Fe(C_5H_5)_2$, 315
Ferromagnetism, 62, 120–129, 181, 185, 202–204, 243, 247, 256, 260, 261, 279, 296–300, 308–311
Feshbach, Herman, 165
Feynman, Richard P., 196, 228
Field theory, 141
Finite electron, 142
Fischer-Hjalmars, Inge, 248
Fisk, James B., 170, 212, 213
Fitzgerald contraction, 143
Florence, Italy, 233
Florida, University of, 153, 263, 266, 272, 276, 288, 291, 293, 295, 302, 306, 313, 326
  computing center, 276
  Quantum Theory Project, 272, 275, 314
Fock, V., 79
Force, Lorentz, 139
Ford foundation, 165
Forrester, Jay W., 330, 331, 333, 337

Four-vector, 39, 145
Fourier series, 115
Fourier transform, 56
Fowler, Ralph H., 6, 11
Franck-Condon principle, 63
Franck, James, 159, 200
Frank, Nathaniel H., 164, 192, 219, 233, 238
Free electrons, 112
Free energy, Helmholtz, 82
Freeman, Arthur J., 263, 276
Frenkel, J., 157, 200
Frequency, 8, 16
Frequency breadth, 10
Fröhlich, H., 240–241

Gainesville, Florida, University of Florida, 153, 263, 266, 272, 276, 288, 291, 293, 295, 302, 306, 313, 326
Galbraith, John K., 336
Galena crystal, 227
Galileo, 233, 336
Gamow, George, 132
Gaspar, R., 196, 246, 247, 277
Gaunt coefficients, 71
Gauss, K. F., 317
Gaussian function, 32, 258
Geiger, W., 12
Gelius, Ulrik, 307
General Electric Company, 173, 262, 327
Gensamer, M., 268
German Physical Society (Deutsche physikalische Gesellschaft), 158
Germanium, 227, 256, 331
Germany, 3
g-factor, Lande, 24
Gibbs, J. Willard, 44
Gingrich, Newell S., 239
Glass, 166
Gombas, P., 195, 246
"Good" quantum number, 27
Gorter, C. J., 229, 240
Göttingen, University of, 21, 63, 153, 156, 159, 200
Goudsmit, Samuel A., 22, 24, 26, 149
Grant, Nicholas J., 271, 326
Greeks, 336
Green, Col. E. H. R., 169
Green, F. R., Jr., 295
Green, Hetty, 169

Group theory, 55, 60, 106
Growth, 336
"Gruppenpest," 58, 60, 62, 106, 141
Guanidinium phosphate, 307
Guggenheim fellowship, 6, 62
Gyromagnetic ratio, 24

Hagstrum, Homer D., 213
Half quantum numbers, 33, 138
Hall, Edwin H., 5
Hall effect, 5
Halpern, Julius, 226
Hamilton, Sir William Rowan, 31
Hamiltonian function, 15, 47, 210
  for electromagnetic field, 135–139, 212
  for electron, 146–149
  for N-electron problem, 70
Hamiltonian operator, 37, 67
Hankel functions, 286, 320
Hansen, William W., 228
Hanson, Harold P., 325, 326
Hardy, Arthur C., 166
Harmonic frequency, 16
Harrison, George R., 167, 237, 275
Hartman, Paul L., 213
Hartree, Douglas R., 7, 53–56, 62, 78–80, 86, 95, 143, 152, 153, 172, 174, 211, 214, 254, 329
Hartree, W., 54, 55
Hartree atomic units, 70
Hartree-Fock self-consistent-field method, 53–57, 78–79, 86–87, 124, 128, 143, 154, 181, 243, 249–250, 254, 260, 278–280, 291, 293, 301
Harvard University, 5, 6, 13, 19, 54, 56, 62–64, 95, 127, 150, 155, 163, 167, 171, 172, 192, 226, 237, 241, 264, 268, 328, 330
  Jefferson Physical Laboratory, 328
  Lyman Laboratory of Physics, 64, 172
  Society of Fellows, 170
  Theoretical seminar, 22, 30
Harvey, George G., 219
Hastings, Julius M., 242
Hattox, Thomas M., 296, 300–303
Haworth, Leland J., 228
Hegel, G. W. F., 336
Heisenberg, Werner, 6, 14, 19–21, 31, 35, 36, 38, 42, 45–48, 56–63, 90, 93, 112, 120, 122–123, 126, 128, 134, 151, 153, 159, 203, 228
  exchange integral, 50, 51, 204, 260, 308, 309
Heitler, Walter, 58, 60, 90, 95, 240
  The Quantum Theory of Radiation, 135
HL (Heitler-London) method, 90, 92, 95–101, 104, 108, 122, 126, 152, 204
HLSP (Heitler-London-Slater-Pauling) method, 105
Helium atom, 26, 43, 45–51, 57, 62, 69, 90, 128, 151–155, 183, 329
  energy, 154
  Hylleraas's wave function, 154
Hellmann-Feynman theorem, 195–197, 296, 321
Helmholtz free energy, 82
Hemoglobin, 327
Herman, Frank, 254, 255, 262, 266, 277, 319
Herman-Skillman atomic calculations, 254, 277, 279
Hermitean operator, 41, 130
Herring, Conyers, 172, 190, 255
High-energy physics, 4, 170–171, 217–220, 223–225
Hill, Albert G., 190, 220, 228, 233
Hilsch, R., and Pohl, 156, 159, 200
Hiroshima, 224
Hirschfelder, Joseph O., 241
History, 337
Hitler, Adolf, 3, 159, 200
Hobbes, Thomas, 336
Holes, in negative-energy states, 148
Holes, in valence band, 156
Homogeneous electron gas, 124–126
Hoover, Herbert, 158
Houston, William V., 163
Howe, John, 268
Hückel, Erich, 7, 63, 106, 108–111, 112
Hull, Albert W., 163
Hund, Friedrich, 7, 27, 50, 58, 60, 62, 95, 96, 112, 159, 190, 259
  and Mrowka, 185
Hund's rule, 50, 60, 61, 123
Hunter College, New York City, 220
Hurley, Andrew C., 251
Hurley, Lennard-Jones, and Pople method, 251–252
Huxley, Aldous, 334

Index

Huygens, Christian, 9
Hydrogen atom, 5, 10, 33, 90
  molecular ion H_2^+, 102–104
  molecule H_2, 43, 46, 90–94, 97–101, 122, 126, 127, 155, 183, 249, 252, 262, 307
  peroxide molecule H_2O_2, 307
Hydrogenic problem, 33, 48
Hylleraas, Egil, 153, 154, 250
HHF (Hyper-Hartree-Fock) method, 281
Hysteresis, 120

Illinois, University of, 173, 241
Imperial College of Science and Technology, London, 157
Impurities, localized, 266
Independent electrons, 57
Inert gas atoms, 151
Information theory, 240
Inner field, 121
Inner quantum number, 22
Inner shielding, 316–318
Institute for Advanced Study, Princeton, N. J., 172, 187, 203
Insulators, 115, 185, 310
Interference, 9
Internal energy, 82
IBM (International Business Machines Corp.), 240, 241, 330
  computers, 251, 258, 261, 265
  San Jose Laboratory, 264, 266
  United Kingdom Branch, 265
  Watson Laboratory, 255
IUCr (International Union of Crystallography), 233, 239
IUPAP (International Union of Pure and Applied Physics), 228, 233, 238, 239, 242, 275
Inversion symmetry, 97, 99, 106
Inyokern, 239
Ionic states, hydrogen molecule, 99, 127
Ionic structure, 289
Ionization energy, 56, 57, 279
Iowa State University, Ames, 263
Iron, 25, 120, 123, 128, 185, 203, 261, 296, 299, 327
  carbonyl Fe $(CO)_5$, 307
  group, 25, 120, 123, 128, 185, 203, 261, 310
Iterations, 55

James, Hubert M., 155, 232
  and Coolidge, hydrogen molecule, 155, 183
Jensen, H., 195
Jews, Nazi Persecution, 3
Johnson, Howard W., 275
Johnson, Keith H., 284, 286, 288, 307, 311, 314, 318–326
  and Smith, multiple-scattering method, 288, 290, 305, 312, 320
Johnson, Thomas H., 268
Jones, Harry, 190, 229
Jordan, Pascual, 21, 140, 153
Journal of Chemical Physics, 195

K shell, 25, 46, 47
Kaufman, A., 268
Kellner, G., 93, 128
Kemble, Edwin C., 5, 11, 22, 30, 96, 155, 192, 328
Kerst, Donald W., 228
Killian, James R., 233
Kimball, George E., 185
Kinetic energy, 35, 195
  in perfect gas, 124
Kip, Arthur F., 226
Kirkwood, John G., 153
Kittel, Charles, 226
Klein, Oskar, 140
Klemperer, W. G., 314, 321
Kohn, Walter, 191, 231, 246, 277
Koopmans, T. C., 80
Koopmans' theorem, 80, 278, 280, 282
Korean war, 234
Korringa, J., 191, 230, 257
KKR (Korringa-Kohn-Rostoker) method, 191, 231, 257, 262, 263, 265, 284, 287, 293
Koster, George F., 205, 264, 266, 304, 307
Kotani, Masao, 240, 241, 264
Kramers, Hans A., 6, 11, 15–19, 233, 240, 242
Kramers-Heisenberg dispersion formula, 15–20, 130
Kronig, Ralph de L., 23, 228, 230
Kronig-Penney band theory, 229
Krutter, Harry M., 185, 196
Kuhn, W., 19, 202
Kuhn-Thomas sum rule, 19, 20, 202

L shell, 26
Lagrange multipliers, 68, 78
Lamar, Edward S., 168
Landé, A., 6, 21, 22–24, 52
Langevin, P., 120
Lark-Horovitz, Karl, 227
Laser, 325
Lawrence, Ernest O., 169, 171, 193
Lax, Benjamin, 226, 270
Lead sulfide PbS, 227
Leiden, University of, 230
Leipzig, University of, 6, 62, 65, 78, 100, 105, 112, 126, 128, 158, 159, 185
Lennard-Jones, Sir John E., 7, 95, 96, 98, 241, 249, 251, 257, 260
Lenz, W., and H. Jensen, 195
Leviathan, by Thomas Robbes, 336
Lewis, Gilbert N., 104, 106
Lifetime, of stationary states, 10, 18
Lindsay, Robert B., 53
LCAO (Linear Combination of Atomic Orbitals) method, 98, 173, 191, 249, 256, 264, 266, 290, 312, 314, 328, 331
Linear combination of determinants, 67
Linear oscillator, 32
Liouville's theorem, 44
Lithium crystal, 256, 302
Lithium fluoride LiF, 314
Lithium molecule Li_2, 307
Liverpool, University of, 240
Livingston, M. Stanley, 169, 171, 224
Localized states, in crystals, 156, 200, 202, 304–305, 325
Lomer, W. M., 303
London, Fritz, 63, 90, 95, 228
Long-range forces, 317
Loomis, Alfred L., 209
Lorentz, H. A., 17, 139, 142, 317
Lorentz-Drude theory, 85, 115
Lorentz-Fitzgerald contraction, 143
Los Alamos Scientific Laboratory, AEC, 222, 224, 263, 280, 295
Loucks, Terry L., 263
Low temperatures, 228
Löwdin, Per-Olov, 129, 153, 229, 239, 241, 250–253, 265, 271, 275, 287, 293, 306
Lowell, A. Lawrence, 64
Luttinger, J. M., 226

Lyman, Theodore, 5, 6, 64, 167
Lyman series, 5

M shell, 26
McGraw-Hill Book Company, 192
MacInnes, Duncan A., 228
Madison, Wisconsin, University of Wisconsin, 233
Magnet, Bitter, 167
Magnet, permanent, 120
Magnetic field, 137
  impurities, 311
  moment, 23, 24, 120
  quantum number, 16, 34
  resonance, 4
Magnetism, 202–204, 247, 276, 296–301, 308–311
Magneton, Bohr, 24, 299
Magnetoresistance, 164
Magnetron, 209–214
Malvern, England, 214, 227, 228
Manchester, University of, 166, 172, 214
Manganese, 25, 203, 297, 310
Manganese oxide MnO, 247, 308
Manhattan Project, 219, 325
Mann, J. B., 281
Many-electron problem, 42, 43, 45, 141
Marchant desk calculator, 330
MIT (Massachusetts Institute of Technology), 8, 54, 63, 159, 328, 333
  Acoustics Laboratory, 171
  Aeronautical engineering department, 170, 214
  Center for materials science and engineering, 215, 225, 237, 267–271, 284
  Computation center, 258, 261, 265
  Cyclotron, 171
  Electrical engineering department, 170, 200, 276
  Francis Bitter National Magnet Laboratory, 167, 172
  George Eastman Research Laboratory, 167, 172
  Hayden Library, 238
  Institute professorships, 237
  Laboratory of Chemical and Solid-State Physics, 269–271
  Laboratory for Insulation Research, 200, 270, 330

Index

Laboratory for Nuclear Science and Engineering, 220, 226, 267, 268
Lincoln Laboratory, 234, 238, 270
Metallurgy and materials science department, 268, 284, 288, 311, 314, 318–322, 325–326
National Magnet Laboratory, 270, 276
Nuclear reactor, 270
Physics department, 6, 21, 53, 64–65, 127, 163–172, 185, 190, 226, 275, 306
Postwar planning, 217–225
Radiation laboratory, 209–213, 216–217, 221, 223
Radioactivity center, 171
RLE (Research Laboratory of Electronics), 216–223, 226, 234, 237, 267, 268
Research program on catalysis, 326
SSMTG (Solid-State and Molecular Theory Group), 225, 237, 239, 242, 249, 251–255, 261, 264, 265, 268, 270, 271, 276, 284, 288, 291, 293, 314, 326, 330
Quarterly Progress Reports, 237, 267
Vannevar Bush Building, 271
Walker Memorial, 329
Whirlwind computer, 258, 260, 265
Matrices for Dirac's theory (spinors), 147
Matrix components, 38, 67
of Hamiltonian, determinantal functions, 70
Matrix, diagonal, 38, 41
Matrix elements, 36
Matrix equation, 37
Matrix mechanics, 15, 20, 31, 35–37
Matrix multiplication, 20
Matrix, non-diagonal, 38, 41
Mattheiss, Leonard F., 262, 264
Mautz, Robert B., 275, 326
Maxwell, Emmanuel, 226
Maxwell, James Clerk, 13
Maxwell's equations, 3
Meadows, Dennis, 333, 337
Mechanics, classical, 31, 44
Mechanics, matrix, 15, 20, 31, 35–37
Mechanics and optics, 31
Mechanics, statistical, 44
Mechanics, wave, 3, 4, 12–13, 31–45
Meckler, Alvin, 257–260, 263, 330
Meissner effect, 210

Messmer, R. P., 327
Metabolism, 327
Metals, Fermi gas treatment, 85
Metallic conductor, 115
Metallurgy, 3
mks (meter-kilogram-second) units, 137
Methane molecule CH_4, 251, 307
Michelson, Albert A., 10
Michigan, University of, 23, 163
Microwaves, 209–214, 228, 263
Millennium, 333
Millikan, Robert A., 171, 193
Minnesota, University of, 63, 157, 158, 163
Missouri, University of, 233, 239
MO (Molecular Orbital) method, 95–111, 257, 290
MO-LCAO (Molecular Orbital-Linear Combination of Atomic Orbitals) method, 101–112, 115, 132, 307
Momentum, 8, 13, 32, 35, 39, 116
Monochromatic waves, 10
Monroe desk calculator, 328, 330
Morse, Philip M., 165, 171, 187, 224, 239
curve, 165
Morse-Allis theory, electron scattering, 187, 230
Moskowitz, Jules, 266
Mott, Sir Nevill F., 7, 190, 233, 242, 310
Mueller, Hans, 166
Muffin-tin approximation, 188, 230, 262, 284, 288, 312–322
Mulliken, Robert S., 7, 95–97, 106, 193, 233, 239, 241, 249, 264, 266
MS-SCF (Multiple-Scattering Self-Consistent-Field) method, 284–287
Multiplets, atomic, 14, 22, 62, 66–80, 259, 297
Multiply periodic motion, 15–17
Mumford, Lewis, 336
Munich, University of, 63, 159, 164, 187, 314
Museum of the History of Science, Florence, 233
Mystic, Conn., 170

Nagasaki, 224
Nancy, University of, 164
National Academy of Sciences, 173, 219
NDRC (National Research Committee), 209

NRC (National Research Council) fellowships, 172
NSF (National Science Foundation), 269, 271, 276
Natural resources, 333, 336
Natural spin orbitals, 253
Nature, 11, 15, 19
Nazi regime, 159
Néel temperature, 308, 310
Neon atom, 23, 26, 188, 310
Nesbet, Robert K., 264, 266
Neumann functions, 188, 285
Neutron diffraction, 4, 241, 247, 270, 299
New York University, 266
Newton, Sir Isaac, 9, 316, 336
Newton's rings, 9
Newtonian equations of motion, 138
Newtonian mechanics, 3
Nickel, 25, 120, 123, 128, 185, 202–203, 276, 326
Nickel carbonyl Ni(CO)$_4$, 307
Nickel hexafluoride (NiF$_6$)$^{4-}$, 307
Nickel oxide NiO, 327
Nitrogen molecule N$_2$, 314
Nitrous oxide NO, 314
Nixon, Richard M., 325
Nobel prize, 5, 6, 169, 172, 196, 326
Nordheim, Lothar, 115
Nordling, Carl, 307
Nordsieck, Arnold T., 213
Normalization, 34, 36
Norman, Joseph G., 319
North American Aviation Company, 268
Northwestern University, Evanston, Ill., 233, 263, 276
Nosanow, Lewis H., 326
Nottingham, Wayne B., 168–169, 227, 232
Nuclear and high-energy physics, 3, 4, 170–171, 217–220, 223–225
Numerical integration, 55
Nuremberg, 159

Oak Ridge National Laboratory, AEC, 233, 241, 247
O'Bryan, Henry M., 168
Occupation numbers, 279, 280
Octahedral symmetry, 312
ONR (Office of Naval Research), 237, 265
  Solid-state panel, 239
OSRD (Office of Scientific Research and Development), 222
Ohio State University, 230
One-electron energy, 79–80, 246, 278
One-electron integrals, 70
Operator, differential, 35–39
Oppenheimer, Robert, 95
Optical excitation, 52
Optical Society of America, 239
Optical term values, 56
Optics, 166
Orowan, Egon, 239
Orthogonality, 37, 54
OPW (Orthogonalized Plane Wave) method, 190, 255, 256
Orthonormal functions, 67, 71
Orwell, George, 334
Oscillator, linear, 20, 32, 137
  strength, 18, 19, 202, 304
  virtual, 11, 14, 18, 38, 39
Outer shielding, 316–318
Overlap, 91, 105, 110, 114, 256
Overtone, 16
Oxford University, 228, 241
Oxygen molecule O$_2$, 257, 263, 314

p Electrons, coupling, 28
Panofsky, Wolfgang K. H., 228
Paramagnetism, 120, 296, 302, 310
Paris, 242
Parmenter, Robert H., 256, 261
Parr, Robert G., 241, 249
Parratt, Lyman G., 305
Pasadena, California Institute of Technology, 163
Pauli, Wolfgang, Jr., 6, 21, 26, 61, 63, 140, 149, 228, 240
  exclusion principle, 14, 22, 25, 29, 50, 148
Pauling, Linus, 7, 61, 64, 105–106, 109, 165, 172, 193, 229
Pauncz, Reuben, 129, 252
Pearl Harbor, 213
Peierls, Rudolph, 63, 115, 240
Penney, Lord W. G., and Schlapp, 157
Pennsylvania, University of, 173, 226, 326, 327
Pennsylvania State University, 242, 263
Pepinsky, Raymond, 242
Perchlorate ion (ClO$_4$)$^-$, 288, 307
Perfect gas, 81
Periodic boundary conditions, 112

Index

Permanganate ion (MnO$_4$)$^-$, 307
Permutation group, 60
Philosophy, 335–338
Phosphine molecule PH$_3$, 319
Photoconductivity, 156, 201
Photoelectric effect, 8
Photoemission, 168, 306–307
Photons, 8, 9, 11
Photoswitch Corp., 227, 231
Photosynthesis, 325, 327
Physical Review, 65, 192
π (pi) bonds, 111
π (pi) orbitals, 108–111
Pierce, John R., 213
Pippard, A. B., 228
Pittsburgh, University of, 239
Planck's constant, 5, 8
Planck's law of radiation, 11
Plane wave expansion, 118, 141, 173
Planimeter, 329
Platinum, 325, 327
Platinum chloride (PtCl$_4$)$^{2-}$, 307
Pohl, R., 156, 159
Poisson's equation, 315
Political scientists, 336
Pollution, 333, 336
POLYATOM program, 265
Polyatomic molecules, 264–266, 284, 307
Pople, John A., 251
Population explosion, 333
Positron, 145, 148
Potassium chloride KCl crystal, 293
Potential energy, 35
Potential wells, 132
Potentials, scalar and vector, 136
Poynting's vector, 11
Prague, 158
Pratt, George W., Jr., 249–255, 258, 260, 261, 276
Pratt and Whitney Aircraft Co., 276
Precession, 16
Predestination, 336
Princeton, N. J., Institute for Advanced Study, 172, 187, 203
Princeton University, 7, 22, 63, 163, 165, 172, 173, 185, 204, 255
Principal quantum number, 26
Probability, 34, 41, 43
Prokofjew, W., potential, 175, 189
Pulsation, 134

Purdue University, 227

Quadrupole, 317
Quantum conditions, Sommerfeld, 10, 13, 15, 26, 33
Quantum defect, 52
Quantum electrodynamics, 130–144
Quantum number, 16, 34
Quantum numbers, diatomic molecule, 96–97
Quantum theory, 4, 6, 15–21, 31–42, 43–51
Quantum transition, 134
SS Queen Elizabeth, 228
SS Queen Mary, 228

Rabi, I. I., 209, 220
Radar, 209–214
Radial correlation, 250
Radial integrals, 72
Radial motion, 16
Radial quantum number, 16
Radiation, 10, 41, 130–144
Radiation, black body, 134
Radiation damage in solids, 232
Radiation field, Hamiltonian treatment, 130–144
Radiation resistance, 131
RCA (Radio Corporation of America), 254, 255
Ramsauer effect, 187–188
Rare earth absorption, 156
Raytheon Manufacturing Co., 227
Reactors, nuclear, 232, 270
Reciprocal lattice, 117
Recoil electrons, 8, 12
Reflection symmetry, 46, 106
Relativity, 8, 13, 39, 44, 146
Repulsive interactions, 6, 151, 152
Resonance, 134
Reviews of Modern Physics, 192, 195, 196
Rochester, University of, 173, 220, 224, 239
Rockefeller Institute, New York City, 228
Rockefeller University, New York City, 173
Rome, University of, 135
Roosevelt, Franklin D., 158
Roothaan, Clemens G., 241, 264
Ros, P., 314
Rösch, N., 314, 321

Rosen, Nathan, 94
Rossi, Bruno B., 224, 240
Rostoker, J., 191, 231
Rothenburg, 159
Round Hill, 169
Rousseau, Jean J., 336
Rubidium atom, 56
Rudberg, Erik, 168, 240
Russell, Henry Norris, 22, 24–25, 27, 168
  -Saunders coupling, 5, 22, 24–30, 50, 168
Russian revolution, 158
Rutherford, Sir Ernest, 6
Rydberg states, 52
Rydberg units, 52, 70

S States, 29
Saffren, Melvin, 261
St. Johns College, Cambridge, 131
Samuel, Arthur L., 213
Sanibel Island, Florida, symposium, 153, 272, 284
Saturation magnetization, 120–121
Saunders, Frederick A., 5, 22, 24–25, 27, 168
Scalar potential, 136
Scattering, of electrons, 12, 115
Scattering, Thomson, 19
Scattering, x-ray, 8, 19, 56, 230
Scherrer, Paul, 7, 172, 230
Schiff, Leonard, 188
Schrieffer, J. Robert, 226, 326, 327
Schrödinger, Erwin, 13, 21, 31–43, 45, 53, 55, 130, 146, 175
Schrödinger's equation, 3, 4, 30–45, 81, 95
  involving time, 39, 40, 130
  symmetry properties, 46
Schwarz, Karlheinz, 294
Second quantization, 140–141
Secular equation, 68, 113, 287
Segall, Benjamin, 262, 294
Seitz, Frederick A., 173, 185, 232, 241, 268, 326
Selection rule, 17
Self-consistent field, 53–57, 210, 253, 255, 288, 329
  for crystals, 173, 262, 276
Self-interaction of electrons, 142–144, 146
Semiconductor, 115, 185, 255
Separation of variables, 47
Sham, L. J., 246, 277

Shells of electrons, 25–26
Shelter Island conferences, 228, 233, 241, 247, 248, 265
Shielding of electrons, 316–318
Shockley, William, 169, 186, 187, 201, 203, 227
Shortley, George, 65, 172
Shull, Clifford G., 240, 241, 247, 269–270
Shull, Harrison, 241
Siegbahn, Hans, 307
Siegbahn, Kai, 306–307
Siegbahn, Manne, 306
σ (sigma), orbitals, 106, 108–109
Silicates, 166
Silicon atom, 331
Silicon rectifier, 227
Simon, A. W., 12
Simon, Sir Francis E., 229
Single bonds, 108, 109
Singlet wave function, H_2, 99
Singlets, 25, 29, 50, 60, 74–77, 92, 122, 259, 260
Skillman, S., 254, 277
Skinner, H. W. B., 168, 214–215
Slater, John C., Bohr and Kramers, joint paper (1924), 11–12
  Born, Max, cooperation with (1925), 21, 328–329
  Brookhaven National Laboratory (1951–1952), 224, 237–242
  Cambridge University, traveling fellow (1923), 6
  Copenhagen University, traveling fellow (1923–1924), 6, 11–12
  England in wartime (1943), 214–215
  Florida, University of, Quantum Theory Project (1964–), 272, 275–322
  Hartree, friendship with (1928–1958), 53–56, 62, 172, 211, 214, 329
  Harvard, graduate student (1920–1923), 5–6
  Harvard, staff member (1924–1930), 6–7, 22–30, 63–65, 192
  Institute for Advanced Study, Princeton (1937), 203–205
  IUPAP, vice president (1948–1954), 228–229, 233, 238–239, 242
  Leipzig, University of, Guggenheim fellow (1929), 6, 62–63, 78–79
  MIT Materials Center (1956–1964),

Index   355

225, 237, 267–271
MIT physics chairman (1930–1951), 63–65, 163–172, 237
MIT postwar reorganization (1945–1951), 215–228
MIT Solid-State and Molecular Theory Group (1951–1966), 225, 237–242, 243–254, 256–267
Published books (1933–1974), 267, 339–341
Radar research, BTL (1941–1944), 212–216
  MIT Radiation Lab., (1940–1941, 1944–1945), 209–212, 219–223
  Radiation damage survey (1949), 219–223
Smith, F. C., Jr., 288
Smith, R. A., 215, 227, 228, 271
Snow, Edward C., 295
Social scientists, 336
Socialism, 336
Sodium chloride NaCl, 5, 308
  energy bands, 186, 201
Sodium crystal, 173–184, 294, 302
Solid-state electronics, 4, 86
Solid-state physics, 4
Solids, 257
Sommerfeld, Arnold, 10, 13, 15, 22, 26, 85, 115, 123, 159, 164, 172, 187, 228
  quantum conditions, 10, 13, 15, 26, 33
Soviet scientists, 311
Soviet union, 158
Space program, 337
Specific heat, electrons in metal, 85
Spectroscopy, 166
Spectrum lines, 10, 18
Spherical harmonics, 33–34, 55, 66, 71, 176, 284, 315
Spherically symmetrical field, 27, 54–55, 173, 188, 316
Spin waves, 129, 204
Spinning electron, 14, 22–30, 60–62, 66–77, 124–127, 149, 181–182, 243–252, 279, 297–303, 308–310
Splitting of atomic levels in crystals, 157
Spontaneous radiation, 17, 135
Squire, Charles F., 226
Standing waves, 13
Stanford University, 7, 63, 167,

173, 188, 189
Stationary states, 10, 11, 18, 20, 38
Statistical exchange, 243–252
Statistical mechanics, 44
Statistical nature of quantum theory, 11
Statistics, 337
Statistics, Fermi-Dirac, 81–86, 112, 115, 125, 138, 140, 168, 279
Steady state, 334
Stever, H. Guyford, 214, 326
Stock market crash, 158
Stockbarger, Donald C., 166
Stockholm, 233, 239
Stoner, Edmund C., 26
Stratton, Julius A., 164, 212, 219, 223, 233, 265, 269, 275, 326
String, vibrating, 31, 32
Sugiura, Y., 92, 98
Sulfate ion $(SO_4)^{2-}$ 264, 288, 307
Sulfur hexafluoride SF_6, 307, 312
Sum rule, Kuhn-Thomas, 19, 20, 202
Superconductivity, 210
Superexchange, 309
Surface chemistry, 327
Swann, W. F. G., 9
Symmetric group, 60
Symmetry, of wave function, 46, 48, 74, 106, 151, 231, 249–251, 258

Tate, John T., 192
Technology, 170
Teller, Edward, 7, 63, 325
Temple University, Philadelphia, 284
TCNQ (Tetracyanoquinodemethane), 319–320
Tetrahedral symmetry group, 289
Tetrahedron, 251
Texas A. and M. University, 226
Texas, University of, 265
Thermal equilibrium, 84
Thermal vibration of crystals, 242
Thermodynamic probability, 83, 240
Thermodynamics, 5
Thomas, L. H., 86, 241
TF (Thomas-Fermi) method, 86–88, 195
TFD (Thomas-Fermi-Dirac) method, 89, 195, 196, 245
Thomas, W., sum rule, 19, 20, 202
Thomson, Sir George P., 163
Thomson, Sir J. J., 6, 19

Thomson scattering, 19
Time-dependent wave equation, 39, 131
Tinkham, Michael, 226
Titanium atom, 25
Titanium carbide crystal TiC, 291, 308
Titanium chloride $TiCl_4$, 307
Tokyo, University of, 263
Toronto, University of, 266
Torques between electrons, 27
Torricelli, 233
Transistor, 4, 227
Transition probability, 17
Transition state, 281–283, 304–311, 314, 319
Transitions, 11, 20, 38
Trickey, Samuel B., 295
Triplet energy, two-electron problem, 50, 74
Triplets, 25, 29, 50, 60, 74–77, 92, 122, 259, 260
Truman, Harry S., 222
Trump, John G., 170
Tufts University, Medford, Mass., 275
Tunnel effect, 132
Two-electron atom, 43–51, 70–77

Uhlenbeck, George, and Goudsmit, 22, 24, 26, 149
Uncertainty, principle of, 19, 45
Unsöld, Albrecht, 54–55
Uppsala, University of, 229, 239, 272, 302, 306
Urey, Harold C., 6, 193

Valence band, 115, 156, 186
Vallarta, Manuel S., 164, 211
Vanadium atom, 25
Vanadian crystal, 296–300
Van de Graaff, Robert J., 169
Van der Waals attraction, 152, 153, 295
Van Vleck, John H., 6, 7, 19, 63, 106, 123, 157, 228, 229, 241
Variation principle, 41, 67, 93
Vector coupling, 51
Vector potential, 136–138
Vibration of string, 31, 32
Vienna, 158
Virial theorem, 195–197, 294, 296, 319, 322
Virtual oscillators, 11, 14, 18, 38, 39

VJ Day, 224
von der Lage, F., and Bethe, 187, 231–232, 320, 321
von Hippel, Arthur, 200, 233, 239, 270, 305, 330
von Kármán, Theodore, 140
von Neumann, John, 140, 172

Waber, James T., 263
Wakoh, S., 263
Walker, Laurence R., 213
Waller, Ivar, 56, 62, 229, 230
Wang, S. C., 94, 153
Wannier, Gregory, 204–205, 266
Warren, Bertram E., 166, 209
Washington University, St. Louis, 8
Water molecule H_2O, 307, 321
Watson, Richard E., 264
Wave, de Broglie, 12, 13, 31, 32
Wave equation, 31, 32, 45, 136
Wave mechanics, 3, 4, 12–13, 31–45
Wave packet, 45
Wave theory, 9
Wave train, finite, 10, 19
Wavelength, 31, 32
Waves, standing, 13
Weiss, Pierre, 121, 122
Weisskopf, Victor F., 19, 224, 240, 275
Weyl, Hermann, 58–60
Whirlwind I computer, 258, 260, 265, 330
Wiener, Norbert, 328, 329
Wiesner, Jerome B., 220, 233, 326
Wigner, Eugene, 7, 19, 58, 60, 62–63, 65, 140, 172, 173, 231, 232
  and Seitz, cellular method, 173–184, 185–188, 243–246, 320
  and Seitz, homogeneous electron gas, 176, 181
Williams, John H., 228
Wilson, A. H., 115
Wilson, C. T. R., 6
Wilson, Timothy M., 310
Wisconsin, University of, 63, 157, 163
Wood, John H., 261, 280
WPA (Work Projects Administration), 168
World War II, 4

Xα-SCF (Xα-Self-Consistent-Field) method, 54, 86, 125, 151, 184, 196, 198, 204, 242–247, 252, 255, 262, 266, 272,

Index

275–322, 325
Xenon crystal, 295
X-ray electrons, 52, 114
X-ray excitation, 304–306
X-ray scattering, 8, 19, 56, 230
X-ray spectroscopy, soft, 168, 233
X-ray term values, 56

Yamashita, U., 263, 277, 303

Zacharias, Jerrold R., 220, 223, 224
Zeeman effect, 21, 24, 26, 168
Zener, Clarence, 127
Zinc alloys, 191
Zinc atom, 87
Zurich, 13, 62, 123, 164, 166, 228